Superstrings
and Other Things
A Guide to Physics

Second Edition

Superstrings
and Other Things
A Guide to Physics
Second Edition

Carlos I. Calle

CRC Press
Taylor & Francis Group
Boca Raton   London   New York

CRC Press is an imprint of the
Taylor & Francis Group, an **informa** business

A TAYLOR & FRANCIS BOOK

Taylor & Francis
6000 Broken Sound Parkway NW, Suite 300
Boca Raton, FL 33487-2742

© 2010 by Taylor and Francis Group, LLC
Taylor & Francis is an Informa business

No claim to original U.S. Government works

Printed in the United States of America on acid-free paper
10 9 8 7 6 5 4 3 2 1

International Standard Book Number: 978-1-4398-1073-6 (Paperback)

Library of Congress Cataloging-in-Publication Data

Calle, Carlos I.
  Superstrings and other things : a guide to physics / Carlos I. Calle. -- 2nd ed.
    p. cm.
  Includes bibliographical references and index.
  ISBN 978-1-4398-1073-6 (alk. paper)
  1. Physics. 2. Physical scientists. 3. Discoveries in science. I. Title.

QC21.3.C35 2009
530--dc22   2009022170

Visit the Taylor & Francis Web site at
http://www.taylorandfrancis.com

and the CRC Press Web site at
http://www.crcpress.com

To Dr. Luz Marina Calle,
fellow NASA scientist and wife,
and to our son Daniel.

Contents

PART I Introductory Concepts

PART II The Laws of Mechanics

PART III   The Structure of Matter

PART IV Thermodynamics

PART V Electricity and Magnetism

PART VI Waves

PART VII Modern Physics

Preface

As a research scientist at NASA Kennedy Space Center working on planetary exploration, I am very fortunate to be able to experience first hand the excitement of discovery. As a physicist, it is not surprising that I find science in general and physics in particular captivating. I have written this book to try to convey my excitement and fascination with physics to those who are curious about nature and who would like to get a feeling for the thrills that scientists experience at the moment of discovery.

The advances in physics that have taken place during the last 100 years have been astounding. In the early 1900s, Max Planck and Albert Einstein introduced the concept of the quantum of energy that made possible the development of quantum mechanics. This revolutionary theory opened the doors for the breathtaking pace of innovation and discovery that we have witnessed during the last 50 years.

At the beginning of the new century, physics continues its inexorable pace toward new discoveries. An exciting new theory might give us the "theory of everything," the unification of all the forces of nature into one single force which would reveal to us how the universe began, and perhaps how it will end.

Although these exciting new theories are highly mathematical, their conceptual foundations are not difficult to understand. As a college professor for many years during the early part of my career, I had the occasion to teach physics to nonscience students and to give public lectures on physics topics. In those lectures and presentations, I kept the mathematics to a minimum and concentrated on the concepts. The idea for this book grew out of those experiences.

This book is intended for the informed reader who is interested in learning about physics. It is also useful to scientists in other disciplines and to professionals in nonscientific fields. The book takes the reader from the basic introductory concepts to discussions about the current theories about the structure of matter, the nature of time, and the beginning of the universe. Since the book is conceptual, I have kept simple mathematical formulas to a minimum. I have used short, simple algebraic derivations in places where they would serve to illustrate the discovery process (for example, in describing Newton's incredible beautiful discovery of the universal law of gravitation). These short forays into elementary algebra can be skipped without loss of continuity. The reader who completes the book will be rewarded with a basic understanding of the fundamental concepts of physics and will have a very good idea of where the frontiers of physics lie at the present time.

I have divided the book into seven parts. Part I starts with some introductory concepts and sets the stage for our study of physics. Part II presents the science of mechanics and the study of energy. Part III follows with an introduction to the structure of matter, where we learn the story of the atom and its nucleus. The book continues with thermodynamics in Part IV, the conceptual development of electricity and magnetism in Part V, waves and light in Part VI, and finally, in Part VII, the rest of the story of modern physics, from the development of quantum theory and relativity to the present theories of the structure of matter.

NEW IN THIS EDITION

In the eight years since the publication of the first edition of this book, M-theory—the more encompassing version of superstring theory—has continued to be developed in surprising new ways. The new Chapter 26 describes these developments in more detail than before.

The first edition of this book included many short features with applications of physics to modern technology. For this edition, I have updated most of those features and added new ones describing recent technologies. I have also updated the content of the book to reflect the new fundamental discoveries that have taken place in recent years.

Finally, I have added in Appendix E a list of questions that you can use to test what you learn as you read the book. These questions are, for the most part, entertaining puzzles and do not require memorization of specific facts.

Acknowledgments

I wish to thank first my wife, Dr. Luz Marina Calle, a fellow NASA research scientist and my invaluable support throughout the many years that writing this book took. She witnessed all the ups and downs, the difficulties, the setbacks, and the slow progress in the long project. She read the entire manuscript and offered many suggestions for clarification, especially in the chapters where, as a physical chemist, she is an expert.

No book can be written without the important peer review process. The criticisms, corrections, and, sometimes, praise made the completion of this book possible. Over a dozen university physics professors reviewed this book during the different stages of its development. I wish to thank them for their invaluable advice. The work of two reviewers was particularly important in the development of the book. I appreciate the comprehensive reviews of Professor Michael J. Hones at Villanova University, who reviewed the manuscript four times, offering criticism and advice every time. Professor Kirby W. Kemper at Florida State University reviewed the book several times and suggested changes, corrections, and better ways to describe or explain a concept. The book is better because of them.

I would also wish to thank Professors Karen Parshall, George H. Lenz, Scott D. Hyman, Joseph Giammarco, Robert L. Chase, Graham Saxby, and Albert Chen, who read all or part of the manuscript and offered many comments. I am grateful to Karla Faulconer for many of the illustrations that appear in the book. I would especially like to acknowledge the invaluable help of my son Daniel, now a software engineer at MITRE, who read the entire manuscript, made many important suggestions and was my early test for the readability of many difficult sections. Finally, I wish to thank Dr. John Navas at Taylor & Francis for his kind support through all the phases of the preparation of this edition.

Carlos I. Calle
Kennedy Space Center, Florida

Author

Carlos I. Calle is a senior research physicist at NASA Kennedy Space Center and the founder and director of NASA's Electrostatics and Surface Physics Laboratory. He received his PhD in theoretical nuclear physics from Ohio University. He is the recipient of numerous NASA awards, including the Spaceflight Awareness Award in 2003 for exceptional contributions to the space program and the Silver Snoopy Award in 2007 for outstanding support of the Space Shuttle Program. With his laboratory staff, he has developed technologies for NASA's lunar and Martian exploration programs. He has also developed new testing techniques for several Space Shuttle systems and for the thermal shrouds on the International Space Station modules.

Dr. Calle has been working on the physical properties of the lunar and Martian soil and is currently designing and building instrumentation for future planetary exploration missions. As one of the world's experts on the electrostatic properties of lunar and Martian dust, he has been developing technologies to keep dust from the spacesuits and equipment being planned for the lunar exploration missions.

His earlier research work involved the development of a theoretical model for a microscopic treatment of particle scattering. He also introduced one-particle excitation operators in a separable particle-hole Hamiltonian for the calculation of particle excitations. As a professor of physics during the early part of his career, he taught the whole range of college physics courses. He has published over 150 scientific papers and is the author of *Einstein for Dummies*, *Coffee with Einstein*, and *The Universe—Order without Design*.

Part I

Introductory Concepts

1 Physics: The Fundamental Science

WHAT IS PHYSICS?

Physics deals with the way the universe works at the most fundamental level. The same basic laws apply to the motion of a falling snowflake, the eruption of a volcano, the explosion of a distant star, the flight of a butterfly, or the formation of the early universe (Figure 1.1).

It is not difficult to imagine that, some 30,000 years ago during a cold, dark spring night, a young child, moved perhaps by the pristine beauty of the starry sky, looked at his mother and, in a language incomprehensible to any of us today, asked: "Mother, who made the world?"

To wonder how things come about is, of course, a universal human quality. As near as we can tell, human beings have been preoccupied with the origin and nature of the world for as long as we have been human. Each of us echoes the words of the great Austrian physicist Erwin Schrödinger, "I know not whence I came nor whither I go nor who I am," and seeks the answers.

Here lies the excitement that this quest for answers brings to our minds. Today, scientists have been able to pierce a few of the veils that cloud the fundamental questions that whisper in our minds with a new and wonderful way of thinking that is firmly anchored in the works of Galileo, Newton, Einstein, Bohr, Schrödinger, Heisenberg, Dirac, and many others whom we shall meet in our journey into the world of physics.

Physics, then, attempts to describe the way the universe works at the most basic level. Although it deals with a great variety of phenomena of nature, physics strives for explanations with as few laws as possible. Let us, through a few examples, taste some of the flavor of physics.

We all know that if we drop a sugar cube in water, the sugar dissolves in the water and as a result the water becomes thicker, denser—that is, more viscous. We, however, are not likely to pay a great deal of attention to this well-known phenomenon. One inquisitive mind did.

One year after graduating from college, the young Albert Einstein considered the same phenomenon and did, indeed, pay attention to it. Owing to his rebellious character, Einstein had been unable to find a university position as he had wanted, and was supporting himself with temporary jobs as a tutor or a substitute teacher. While substituting for a mathematics teacher in the Technical School in Winterthur, near Zurich, from May to July 1901, Einstein started thinking about the sweetened water problem. "The idea ... may well have come to Einstein as he was having tea," writes a former collaborator of Einstein.

FIGURE 1.1   The laws of physics apply to (a) a falling snowflake. (Courtesy of W. P. Wirgin.) (b) The explosion of a star, and (c) the eruption of a volcano. (Courtesy of NASA.)

Einstein simplified the problem by considering the sugar molecules to be small hard bodies swimming in a structureless fluid. This simplification allowed him to perform calculations that had been impossible until then that explained how the sugar molecules would diffuse in the water, making the liquid more viscous.

This was not sufficient for the 22-year-old scientist. He looked up actual values of viscosities of different solutions of sugar in water, put these numbers into his theory, and obtained from his equations the size of sugar molecules! He also found a value for the number of molecules in a certain mass of any substance (Avogadro's number). With this number, he could calculate the mass of any atom. Einstein wrote a scientific paper about his theory, entitled "A New Determination of the Sizes of Molecules."

Albert Einstein. (Terracota sculpture by the author.)

On the heels of this paper, Einstein submitted for publication another important paper on molecular motion, in which he explained the erratic, zigzag motion of individual particles of smoke. Again, always seeking the fundamental, Einstein was able to show that this chaotic motion gives direct evidence of the existence of molecules and atoms. "My main aim," he wrote later, "was to find facts that would guarantee as far as possible the existence of atoms of definite finite size."

Almost a century earlier, Joseph von Fraunhöfer, an illustrious German physicist, discovered that the apparent continuity of the sun's spectrum is actually an illusion. This seemingly unrelated discovery was actually the beginning of the long and tortuous road toward the understanding of the atom. The 11th and youngest child of a glazier, Fraunhöfer became apprenticed to a glassmaker at the age of 12. Three years later, a freak accident turned the young lad's life around; the rickety boarding house he was living in collapsed and he was the only survivor. Maximilian I, the elector of Bavaria, rushed to the scene and took pity on the poor boy. He gave the young man 18 ducats. With this small capital, Fraunhöfer was able to buy books on optics and a few machines with which he started his own glass-working shop. While testing high-quality prisms Fraunhöfer found that the spectrum formed by sunlight after it passed through one of his prisms was missing some colors; it was crossed by numerous minuscule black lines, as in Figure 1.2. Fraunhöfer, intrigued, continued studying the phenomenon, measuring the position of several hundred lines. He placed a prism behind the eyepiece

FIGURE 1.2   (See color insert following page 268.) The solar spectrum with some of the dark lines discovered by Fraunhöfer.

of a telescope and discovered that the dark lines in the spectrum formed by the light from the stars did not have quite the same pattern as that of sunlight. He later discovered that looking at the light from a hot gas through a prism produced a set of bright lines similar to the pattern of dark lines in the solar spectrum.

Today we know that the gaps in the spectrum that Fraunhöfer discovered are a manifestation of the interaction between light and matter. The missing colors in the spectrum are determined by the atoms that make up the body emitting the light.

In the spring of 1925 a 24-four-year-old physicist named Werner Heisenberg, suffering from severe hay fever, decided to take a two-week vacation on a small island in the North Sea, away from the flowers and the pollen. During the previous year Heisenberg had been trying to understand this interaction between light and matter, looking for a mathematical expression for the lines in the spectrum. He had decided that the problem of the relationship between these lines and the atoms could be analyzed in a simple manner by considering the atom as if it were an oscillating pendulum. In the peace and tranquility of the island, Heisenberg was able to work out his solution, inventing the mechanics of the atom. Heisenberg's new theory turned out to be extremely powerful, reaching beyond the original purpose of obtaining a mathematical expression for the spectral lines.

In 1984, this idea of thinking about the atom as oscillations took a new turn. John Schwarz of the California Institute of Technology and Michael B. Green of the University of London proposed that the fundamental particles that make up the atom are actually oscillating strings. The different particles that scientists detect are actually different types or *modes* of oscillation of these strings, much like the different ways in which a guitar string vibrates. This clever idea, which was incredibly difficult to implement, produced a theory of enormous beauty and power, which explains and solves many of the difficulties that previous theories had encountered. The current version of the theory, called *superstring theory*—which we will study in more detail in Chapter 26—promises to unify all of physics and help us understand the first moments in the life of the universe. Still far from complete, superstring theory is one of the most active areas of research in physics at the present time.

In all these cases, the scientists considered a phenomenon of nature, simplified its description, constructed a theory of its behavior based on the knowledge acquired by other scientists in the past, and used the new theory not only to explain the phenomenon, but also to predict new phenomena. This is the way physics is done. This book shows how we can also do physics, and share in its excitement.

THE SCIENTIFIC METHOD: LEARNING FROM OUR MISTAKES

In contrast to that of many other professionals, the work of a scientist is not to produce a finished product. No scientific theory will ever be a *correct*, finished result. "There could be no fairer destiny for any ... theory," wrote Albert Einstein, "than that it should point the way to a more comprehensive theory in which it lives on, as a limiting case."

Science is distinguished from other human endeavor by its *empirical method*, which proceeds from observation or experiment. Distinguished philosopher of science Karl R. Popper said that the real basis of science is the possibility of empirical disproof. A scientific theory cannot be proved correct. It can, however, be disproved.

According to the scientific method, a scientist formulates a theory inspired by the existing knowledge. The scientist uses this new theory to make predictions of the results of future experiments. If when these experiments are carried out the predictions disagree with the results of the experiments the theory is disproved; we know it is incorrect. If, however, the results agree with the forecasts of the theory, it is the task of the scientists to draw additional predictions from the theory, which can be tested by future experiments. No test can prove a theory, but any single test can disprove it.

In the 1950s, a great variety of unpredicted subatomic particles discovered in laboratories around the world left physicists bewildered. The picture that scientists had of the structure of matter up to the 1940s—as we will learn in more detail in Chapters 7 and 8—was relatively simple and fairly easy to understand: matter was made of atoms, which were each composed of a tiny nucleus surrounded by a cloud of electrons. The nucleus, in turn, was made up of two kinds of particles, protons and neutrons. The new particles being discovered did not fit this simple scheme. Two theories were formulated to explain their existence. The first one proposed a "particle democracy," in which no particle was any more fundamental than any other. This theory was so well received by the scientific community in the United Sates that one of the proponents of the second theory, Murray Gell-Mann of the California Institute of Technology, decided to publish his paper in a European journal, where he felt the opposition to his new ideas would not be so great. Gell-Mann and (independently) George Zweig, also of Caltech, proposed that many of the growing number of particles and in particular the proton and the neutron were actually made up of smaller, indivisible particles, which Gell-Mann called quarks. Different combinations of quarks, in groups of two or three, were responsible for many of these particles. According to their theory, the growing number of new particles being discovered was not a problem anymore. What mattered was that the objects of which these particles were made were simple and small in number.

Which theory was correct? In 1959 Stanford University built a large particle accelerator which, among other things, could determine whether or not quarks existed. Seven years later, experiments carried out at the Stanford Linear Accelerator Laboratory, SLAC, allowed physicists to determine the presence of the quarks inside protons and neutrons. Since then, many experiments have corroborated the Stanford results; the quark is accepted today as one of the fundamental constituents of matter and the particle democracy theory is no longer viable. We shall see in the final chapters of this book that these new theories of matter are far from complete. Nevertheless, the knowledge obtained from these theories has given us not only a better understanding of the universe we live in, but has also produced the modern technological world based largely on the computer chip.

We can summarize the scientific method by saying that *we can learn from our mistakes*. Scientific knowledge progresses by guesses—by conjectures which are controlled by criticism and by critical tests. These conjectures or guesses may survive the tests; but they can never be established as true. "The very refutation of a theory," writes Popper, "is always a step forward that takes us nearer to the truth. And this is how we learn from our mistakes."

PHYSICS AND OTHER SCIENCES

Physicists often become interested in phenomena normally studied by scientists in other scientific disciplines, and apply their knowledge of physics to these problems with great success. The recent formulation of the impact theory of mass extinctions is a good illustration of physicists becoming involved in other scientific fields and of the way working scientists apply the scientific method to their work.

In 1980, the Nobel Prize-winning physicist Luis Alvarez and his son Walter, a professor of geology at the University of California at Berkeley, reported in a paper published in the journal *Science* that some 65 million years ago a giant meteorite crashed into the earth and caused the extinction of most species. The dinosaurs were the most famous casualties. Alvarez and his collaborators based their theory on their study of the geological record. Walter Alvarez had told his father that the 1-cm-thick clay layer that separates the Italian limestone deposits of the Cretaceous period—the last period of the age of reptiles—from those of the Tertiary period—the first period of the age of mammals, was laid down during precisely the time when the great majority of the small swimming animals in the marine waters of that region had disappeared. What made it even more exciting was the fact that this time also coincided with the disappearance of the dinosaurs.

The layer of clay—observed worldwide and known as the K–T boundary layer—contains an unusually high concentration of the element iridium (Figure 1.3). This element is present in very small amounts in the earth's crust but is much more abundant in meteorites. The father and son team thought that by measuring the amount of iridium present in the clay they could determine how long the layer had taken to form. They assumed that iridium could have rained down on the earth from meteoritic dust at a fairly steady rate during the thousand years that it took to form. If that were the case, they could measure the amount of iridium in the clay and in the rocks above the clay (formed later) and below (formed earlier) and determine the time it had taken for the iridium to accumulate. To that effect, they enlisted the help of Frank Asaro and Helen Michel, nuclear chemists at the Lawrence Berkeley Laboratory. Asaro and Michel showed that the clay layer contains three hundred times the amount of iridium as the layers above and below.

The source of this unusual amount of iridium had to be extraterrestrial, Luis Alvarez reasoned. Meteorites, which are extraterrestrial, have fallen on the earth since its formation. If the iridium came from the meteorites, why this sudden increase

"YOU'RE BEING RECALLED. HE'S
GOING TO TRY MAMMALS."

An unorthodox theory of the extinction of the dinosaurs. (Courtesy of Sidney Harris.)

FIGURE 1.3 K–T boundary layer with a high concentration of iridium. (Courtesy of Alessandro Montanari.)

in the meteorite rate during this particular time and why did it decrease again to previous levels? What was so special about this particular time in the history of the earth? More importantly, why did it coincide with the extinction of about 50% of the species in existence at the time?

The Alvarez team first proposed that the iridium could have come from the explosion of a supernova near the solar system. Astrophysicists had proposed that the mass extinctions could have been caused by such an explosion. Since these tremendous explosions produce heavy elements, Luis Alvarez proposed analyzing the samples taken from the clay for their presence. Detailed measurements revealed no heavy elements, however, and the supernova idea had to be abandoned.

While Walter Alvarez returned to Italy to collect more clay samples, his father worked on the theory, inventing "a new scheme every week for six weeks and [shooting] them down one by one," as he wrote later. Luis Alvarez then considered the possibility of an asteroid or a comet passing through the atmosphere, breaking up into dust which would eventually fall to the ground, like the comet that broke up over Tunguska, Siberia, in 1908. Calculations showed him that a larger asteroid, ten kilometers in diameter for example, would not break up into pieces. The Tunguska comet was smaller.

Alvarez then concluded that some 65 million years ago, a ten-kilometer comet or asteroid struck the earth, disintegrated and threw dust into the atmosphere. The dust remained in the atmosphere for several years, blocking sunlight, turning day into night, and preventing photosynthesis (the process by which, in the presence of light, plants convert water, carbon dioxide, and minerals into oxygen and other compounds). Without plants to eat, animals starved to death. We see the remnants of dust today as the global K–T boundary layer between the Cretaceous and Tertiary layers. Alvarez calculated the diameter of the object from the known concentration of iridium in meteorites and his group's data on the iridium content of the Italian clay samples.

Other scientists proposed the idea that intense volcanic eruptions could account for the mass extinctions. These scientists found high levels of iridium in tiny airborne particles released by the Kilauea volcano in Hawaii and concluded that iridium from the inner earth can reach the surface. For a few years after they were proposed, both ideas could be used to explain the K–T extinctions. However, different predictions could be drawn from the two competing ideas and scientists scurried to find new evidence in support of the different predictions. Recent findings, however, appear to confirm the predictions of the impact theory.

According to the scientific method, however, no theory can ever be proved correct. One of the theories will eventually be shown to be incorrect, leaving the remaining theories stronger, but not proven. "You will never convince some [scientists] that an impact killed the dinosaurs unless you find a dinosaur skeleton with a crushed skull and a ring of iridium around the hole," joked a scientist at a conference on the subject.

SIZES OF THINGS: MEASUREMENT

Most work in physics depends upon observation and measurement. To describe the phenomena encountered in nature and to be able to make observations, physicists must agree on a consistent set of *units*.

Throughout history, several different systems of units were developed. It began with the Babylonians and the Egyptians, thousands of years ago. The earliest recorded unit of measurement, the *cubit*, based on the length of the arm, appeared in Egyptian papyrus texts. According to Genesis, Noah's Ark was 300 cubits long (about 150 m). Because the length of the arm varies from person to person, so did the cubits used among various civilizations. The Egyptians used a short cubit of 0.45 m and a royal cubit of 0.524 m. The ancient Romans used the *mille passus*, 1000 double steps by a Roman legionary, which was equal to 5000 Roman *feet*. In the fifteenth century, Queen Bess of England added 280 feet to the mile to make it eight "furrow-longs" or furlongs.

In 1790, Thomas Jefferson proposed a system based on units of 10, in which 10 feet would equal a decad, 10 decads a road, 10 roads a furlong, and 10 furlongs a mile. Congress did not approve Jefferson's system.

In France, however, the French Revolution brought an interest in science, and another base 10 system—the metric system—was born. This system, based on the *meter*, from the Greek *metron* meaning measure, was more scientific. Instead of using human anatomy, the meter, as approved by the French National Convention in 1795, was defined as 1/10,000,000 of the length of earth's meridian between the Equator and the North Pole (Figure 1.4).

Once the meter was defined, a unit of volume, the *liter*, could be defined by cubing a tenth of a meter. From the liter, the *kilogram* as a unit of mass was derived. Multiples of 10 provided larger units indicated by Greek prefixes, and for smaller units Latin prefixes were used.

Due to the consistency and uniformity of the system and the easiness of defining new units merely by adding a Greek or a Latin prefix, the metric system was adopted in Europe in the nineteenth century. Today, an expanded version of the system, SI units, for *Le Système International d'Unités*, is used by 95% of the world's population and is the official system in science. In Table 1.1 we list the standard prefixes used in the SI system.

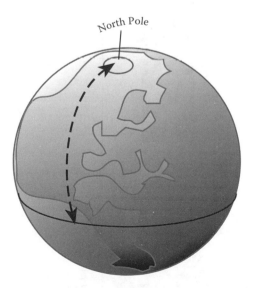

FIGURE 1.4   The meter was originally defined as 1/10,000,000 of the length of earth's meridian from the North Pole to the Equator.

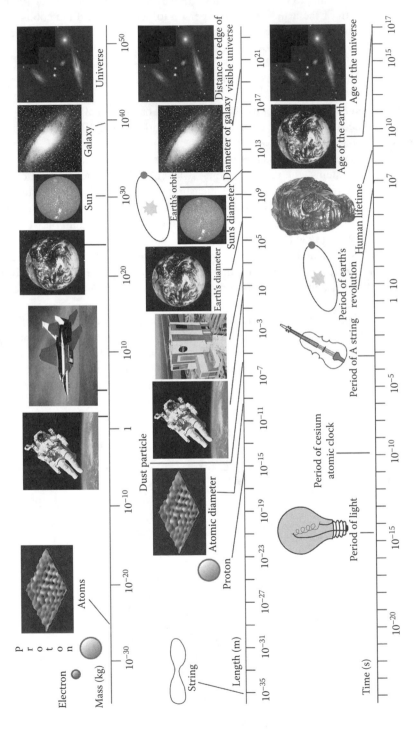

Range of masses, lengths, and time intervals found in the universe.

TABLE 1.1
Powers of Ten Prefixes

Value	Symbol	Prefix
$10^{18} = 1\ 000\ 000\ 000\ 000\ 000\ 000$	E	exa
$10^{15} = 1\ 000\ 000\ 000\ 000\ 000$	P	peta
$10^{12} = 1\ 000\ 000\ 000\ 000$	T	tera
$10^{9} = 1\ 000\ 000\ 000$	G	giga
$10^{6} = 1\ 000\ 000$	M	mega
$10^{3} = 1000$	k	kilo
$10^{2} = 100$	h	hecto
$10^{1} = 10$	da	deca
$10^{-1} = 0.1$	d	deci
$10^{-2} = 0.01$	c	centi
$10^{-3} = 0.001$	m	milli
$10^{-6} = 0.000\ 001$	m	micro
$10^{-9} = 0.000\ 000\ 001$	n	nano
$10^{-12} = 0.000\ 000\ 000\ 001$	p	pico
$10^{-15} = 0.000\ 000\ 000\ 000\ 001$	f	femto
$10^{-18} = 0.000\ 000\ 000\ 000\ 000\ 001$	a	atto

Notice in Table 1.1 that for large and small numbers it is easier to use scientific notation. In the scientific notation, numbers are written as a number between 1 and 10 multiplied by a power of 10. The radius of the earth, for example, which is 6380 km, can be written in scientific notation as $6.380\ 10^3$ km. To see why, note that we can write the number 1000 as follows:

$$1000 = 10 \times 10 \times 10 = 10^3.$$

The radius of the earth is, then,

$$6380 \text{ km} = 6.38 \times 1000 \text{ km} = 6.38 \times 10^3 \text{ km}.$$

FUNDAMENTAL UNITS

All mechanical properties can be expressed in terms of three fundamental physical quantities: length, mass, and time. The SI *fundamental units* are:

Quantity	Fundamental Unit	Symbol
Length	meter	m
Mass	kilogram	kg
Time	second	s

The General Conference on Weights and Measures held in Paris in 1983 defined the meter as the distance traveled by light through space in 1/299,792,458 of a second. Notice that the unit of length is defined with such high precision in terms of the unit of time. This is possible because the second is known to better than one part in ten trillion. The 1967 General Conference on Weights and Measures defined the second as the duration of 9,192,631,770 periods (durations of one oscillation) of a particular radiation emitted by the cesium atom. The device that permits this measurement is the cesium clock, an instrument of such high precision that it would lose or gain only three seconds in one million years (Figure 1.5).

The last fundamental mechanical quantity is *mass*. Mass is a measure of the resistance that an object offers to a change in its condition of motion. In the case of an object at rest with respect to us, mass is a measure of the amount of matter present in the object. The standard unit of mass is the standard kilogram, a solid platinum–iridium cylinder carefully preserved at the Bureau of Weights and Measures in Sèvres, near Paris. The kilogram is now derived from the meter, which is derived from the second. A copy of the standard kilogram, the Prototype Kilogram No 20, is kept at the National Bureau of Standards in Washington, DC. A high precision balance, especially designed for the National Bureau of Standards, allows the comparison of the masses of other bodies within a few parts in a billion.

The mass of an atom cannot be measured by comparison with the standard kilogram with such a high degree of precision. However, the masses of atoms can be compared against each other with high accuracy. For this reason, the mass of an atom is given in

FIGURE 1.5   A cesium atomic clock at the National Institute of Standards and Technology in Washington, DC. (Courtesy of National Institute of Standards and Technology.)

"MY GOODNESS, IT'S 12:15:0936420175. TIME FOR LUNCH."

(Courtesy of Sidney Harris.)

atomic mass units (amu). The mass of carbon in these units has been assigned a value of 12 atomic mass units. In kilograms, an atomic mass unit is

$$1 \text{ amu} = 1.6605402 \times 10^{27} \text{ kg.}$$

PHYSICS AND MATHEMATICS

Physics and mathematics are closely intertwined. Mathematics is an invention of the human mind inspired by our capacity to deal with abstract ideas; physics deals with the real material world. Yet, mathematical concepts invented by mathematicians who did not foresee their applications outside the abstract world of mathematics have been applied by physicists to describe natural phenomena. "It is a mystery to me," wrote the Nobel Prize-winning physicist Sheldon Glashow, "that the concepts of mathematics (things like the real and complex numbers, the calculus and group theory), which are purely inventions of the human imagination, turn out to be essential for the description of the real world."

Physicists, on the other hand, have invented powerful mathematical techniques in their search to understand the physical world. Newton developed the calculus to solve the problem of the attraction that the earth exerts on all objects on its surface. Mathematicians later continued the development of calculus into what it is today.

Mathematics is then the instrument of physics; the only language in which the nature of the world can be understood. Nonetheless, in this book we are interested in the *concepts* of physics. These concepts can usually be described with words and examples. In some instances, however, there is no substitute for the elegance and conciseness of a simple formula. In these cases, we shall consider such a formula to see how it explains new concepts and how they can be linked to other concepts already learned. The reader should always keep in mind that our purpose is to understand the physical phenomenon, not the mathematics that describes it.

THE FRONTIERS OF PHYSICS: VERY SMALL NUMBERS

What does a mass like $1.6605402 \times 10^{-27}$ kg mean? Suppose that you start with one grain of salt, which has a mass of about one ten-thousandth of a gram, and with a very precise instrument you divide it into ten equal parts, take each one of the tenths, divide them into ten new equal parts, and so on. You will not arrive at single electrons this way because, as we shall see in Chapters 7 and 8, the electron is one of the several constituents of atoms. Although atoms can be split, you cannot do it with a cutting instrument.

Suppose, however, that we divide the grain of salt into the smallest amounts of salt possible, single molecules of salt. One single molecule of table salt has a mass of about 9×10^{-23} g. Let's round this number up to 10^{-22} g. If your instrument takes one second, say, to take each piece of salt and divide it into ten equal parts, how long would it take to end up with individual molecules of salt? The answer is 3×10^{10} years.

Astrophysicists estimate that the age of the universe is of the order of 10^{10} years. It would take our hypothetical instrument roughly the age of the universe to arrive at a single molecule of salt!

Grains of salt, magnified 100 times. (Courtesy of V. Ward, NASA.)

PIONEERS OF PHYSICS: MEASURING THE
CIRCUMFERENCE OF THE EARTH

The meter, as we saw, was defined in 1975 as 1/10,000,000 of the length of the earth's meridian from the Equator to the North Pole. For that definition to make sense, an accurate knowledge of the earth's dimension was needed. That is, the actual length of the meridian from the Equator to the North Pole had to be known with good precision. How did we come to know the earth's dimensions before the advent of twentieth-century technology?

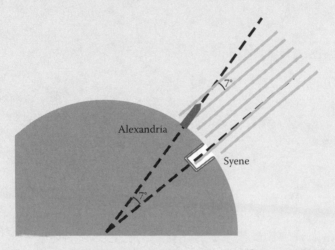

FIGURE 1.6 Eratosthenes' method to measure the circumference of the earth.

The dimensions of the earth have been known since the time of the ancient Greeks. The Greek astronomer Eratosthenes, who lived in the third century BC in Alexandria (Egypt), came up with a very clever method for obtaining the circumference of the earth. Eratosthenes had heard that in the city of Syene, an ancient city on the Nile, near today's Aswan, on the first day of summer the sun shone on the bottom of a vertical well at noon. However, in his native Alexandria, the sun's rays did not fall vertically down but at an angle of 7° to the vertical. This angle of 7° was about one-fiftieth of a circle and that meant that the distance between Syene and Alexandria must be one-fiftieth of the earth's circumference (Figure 1.6).

During Eratosthenes' time, the distance between these two cities was estimated to be 5000 stadia. So the circumference of the earth was 50 times this distance or 250,000 stadia. Although the exact length of that Greek unit is not known, we do know that the length of a Greek stadium varied between 154 and 215 meters. If we use an average value of 185 m, the result is only about 15% larger than modern measurements—a remarkable achievement.

Part II

The Laws of Mechanics

2 The Description of Motion

UNDERSTANDING MOTION

The understanding of motion is fundamental in our comprehension of nature. "To understand motion is to understand nature," wrote Leonardo da Vinci. If we understand how an object moves, we might be able to discover where it has been and predict where it will be at some time in the future, provided that the present conditions are maintained. In physics, we are interested in the description of the motion of the different bodies that we observe, such as automobiles, airplanes, basketballs, sound waves, electrons, planets, and stars.

To study how objects move, we need to begin by studying how a simple object moves. An object without moving parts, such as a ball or a block, is simpler than one with separate parts because we do not need to worry about the motions of the parts, and we can concentrate on how the object moves as a whole. A ball can roll and a block can slide on a surface. Which one is simpler? It would be easier for us if we did not have to decide beforehand either the shape of the object or its internal structure. Physicists simplify the problem by considering the motion of a point, an ideal object with no size, and therefore no internal structure and no shape.

We will consider first the motion of a point. However, in our illustrations and examples we might refer to the motion of real objects, like cars, baseballs, rockets, or people. If we do not consider the internal structure of the object, and do not allow it to rotate, this object behaves like a point for our purposes.

UNIFORM MOTION

"My purpose is to set forth a very new science dealing with a very ancient subject," wrote Galileo in his book *Two New Sciences*. He continued: "There is, in nature, perhaps nothing older than motion, concerning which the books written by philosophers are neither few nor small; nevertheless I have discovered by experiment some properties of it which are worth knowing and which have not hitherto been either observed or demonstrated."

Galileo, one of the first modern scientists and the first one to understand the nature of motion, was born in Pisa the same year that Shakespeare was born in England and three days before Michelangelo died. The year was 1564. His full name was Galileo Galilei, following a Tuscan custom of using a variation of the family name as the first name of the eldest son.

His father, a renowned musician, wanted his son to be a physician, a far more lucrative profession in those days. Thus, he entered the University of Pisa to study medicine. Upon hearing a lecture on geometry which encouraged him to study the work of Archimedes, the young medical student decided that science and mathematics seemed far more interesting than medicine. Galileo talked to his father about letting him switch. Fortunately for the world his father consented.

Galileo became well known throughout Italy for his scientific ability and at the age of 26 was appointed Professor of Mathematics at the University of Pisa. There he dug deeply into fundamental science. He also made some enemies, especially among the older and more respected professors, who did not like their opinions and views challenged by the young and tactless Galileo. Partly because of this, and partly because the Republic of Venice was, in 1600, the hub of the Mediterranean, which in turn was the center of the world, Galileo accepted a position as Professor of Mathematics at Padua, where he began the work in astronomy that was to bring him immortal fame.

Galileo's work on mechanics was published as *Discourses and Mathematical Demonstrations Concerning Two New Sciences Pertaining to Mechanics and Local Motion*, which appeared in 1638. In the chapter "De Motu Locali" (or "Change of Position") he writes:

> The discussion is divided into three parts; the first part deals with motion which is steady or uniform; the second treats of motion as we find it accelerated in nature; the third deals with the so-called violent motions and with projectiles.

Galileo then goes on to explain "motion which is steady or uniform:"

> By steady or uniform motion, I mean one in which the distances traversed by the moving particle during any equal interval of time, are themselves equal.

Figure 2.1 is an example of uniform motion; it shows several positions of an athlete running along a straight 100-meter track at a steady pace. The marks alongside the track show that the runner moves equal distances of 10 meters in equal intervals of six seconds.

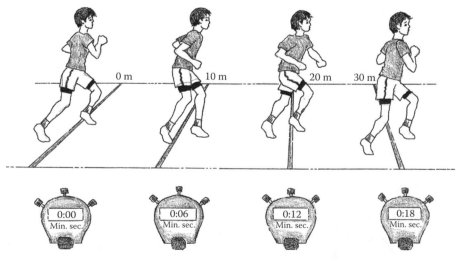

FIGURE 2.1   Several positions of a runner running along a straight track.

AVERAGE SPEED

The runner in Figure 2.1 travels 10 meters in six seconds or 100 meters in sixty seconds (one minute). We can say that the runner travels at 100 meters per minute. Average speed is defined as the total distance traveled divided by the time taken to travel this distance. If we use the letter d to indicate distance, and the letter t to indicate time, we can write the average speed $\bar{v}$ as

$$\bar{v} = \frac{\text{distance traveled}}{\text{time taken}} = \frac{d}{t}.$$

The bar above the letter v is used to indicate that this is the average value. The runner of our example travels a distance of 100 meters in one minute. The average speed of the runner, then, is

$$\bar{v} = \frac{d}{t} = \frac{100 \text{ m}}{1 \text{ min}} = 100 \text{m/min}.$$

Figure 2.2 shows a multiple-exposure photograph of a disk of solid carbon dioxide (dry ice) in uniform motion. The disk, resting on the polished surface of a table, is given a gentle push. With the room darkened, the shutter of a camera set on a tripod is kept open while at equal intervals of time a strobe is fired. Since the only source of light comes from the strobe which for this experiment was fired at 0.10 second intervals—the film records the position of the disk as it slides on the table. The meter rule shows that the disk moves 13 cm between flashes. The disk, therefore, traverses equal distances of 13 cm in equal intervals of 0.10, s or 130 cm in 1.0 second. We can say that the disk travels at an average speed $\bar{v}$ of 130 cm per second.

The *units* of speed are units of distance divided by units of time. Speed can thus be given in miles per hour, kilometers per hour, meters per second, feet per minute, etc. The SI unit of speed is the meter per second (m/s).

In both of those cases, the speed did not change. The runner and the disk were moving at a uniform or constant speed, at least for the intervals that were considered. However, few motions are uniform. The most common situation is that of variable

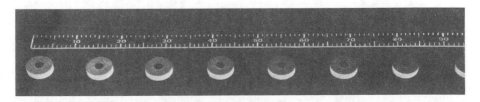

FIGURE 2.2   Multiple-exposure photograph of a disk of dry ice moving on a smooth surface. (Illustration from PROJECT PHYSICS, copyright ©1981 by Holt, Rinehart and Winston, Inc. Reprinted with permission from the publisher.)

speed. If you drive from your dorm to the movies, you start from rest, speed up to 30 miles per hour and probably drive at that speed for a few minutes, until you have to slow down to make a turn or come to a stop at a traffic light. The speed at which you drive will change many times throughout your trip. We can obtain the average speed of the motion by dividing the total distance traveled by the time it took to cover that distance.

THE FRONTIERS OF PHYSICS: FRICTION

The disk of dry ice shown in the multiple-exposure photograph of Figure 2.2 moves an equal distance in equal time because friction between the puck and the smooth surface is negligible. When two surfaces rub together, the atoms that make up the two surfaces interact in ways that depend on the atomic composition of the substances, making them stick to each other. Although the general mechanism is well understood, the details of how friction appears are only now beginning to become clear. Recently, scientists at Georgia Institute of Technology used an atomic force microscope, which measures the forces between two objects separated by less than ten nanometers, to examine the tip of a tiny nickel probe moving on a gold surface (Figure 2.3).

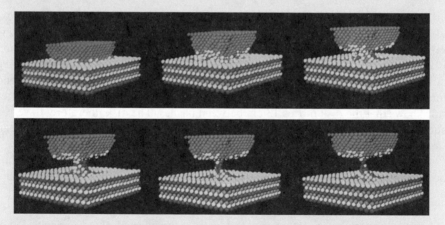

FIGURE 2.3   Friction seen at the atomic level between a nickel tip and a gold surface. (Courtesy of Prof. Uzi Landman.)

When the nickel tip was pulled back slightly after it had made contact with the surface, a connective "neck" of atoms developed between the two surfaces, a sort of bridge at the atomic scale. After the tip was pulled far enough, the neck snapped, leaving the tip covered with gold atoms. Why would gold atoms move over to the nickel tip instead of the other way around? Gold, it turns out, requires less energy to have one of its surface atoms removed than nickel. The researchers believe that these differences in energy account for the differences in friction between different substances.

Let's consider a numerical example. A boy takes 15 minutes to ride his bicycle to his friend's house, which is 2 km away. He talks to his friend for 20 minutes and then continues toward his grandparents' home, 4 additional km. He arrives there 25 minutes after leaving his friend's house.

Since the total distance traveled is 6 km (2 km to the friend's house and 4 km to the grandparents' home) and it took the boy a total of 60 minutes to get there, the average speed is

$$\bar{v} = \frac{d}{t} = \frac{6 \text{ km}}{60 \text{ min}} = 6 \text{ km/h.}$$

Notice that we have included the time the boy spends at his friend's house in our calculation of the total time taken for the trip.

INSTANTANEOUS SPEED

Average speed is useful information. When we are traveling, we can calculate the average speed for a section of the trip and use it to estimate how long it will take us to complete the trip, provided we continue driving under similar conditions. However, in some cases, we might be interested in obtaining more information.

It would take you about 8 hours to drive from Washington, DC to Charlotte, North Carolina, a distance of 400 miles. Although the average speed in this case is 50 mph, you know that at times you would drive at a higher speed, whereas heavy traffic or lower speed limits through certain parts would force you to drive at a lower speed. Knowing that you can average 50 mph for this trip does not provide information about how fast you actually traveled or whether you stopped at all along the way. The *instantaneous speed*, when we can obtain it, will give us information about the detail of the trip. Instantaneous speed is the speed given by a car's speedometer, the speed at a given instant.

If your car speedometer fails before you complete a 60-mile trip that usually takes you one hour, you cannot be sure that you did not break the 65 mph speed limit at any time during your trip even if it still took you one hour to arrive at your destination. However, if you time yourself between successive mile posts and it takes you about one minute to travel one mile, you know that you are maintaining a speed close to 60 mph. To measure your speed more accurately you would need to reduce the time intervals to one second, perhaps—during which you would travel only about 90 feet—or, with precision equipment, even to 1/10 second. Yet, this very small interval of time would still not give you the instantaneous speed. You would need to reduce that interval to an *instant*.

Modern instrumentation allows measurements of the speed of an object at intervals small enough to provide us with excellent approximations to the instantaneous speed. Mathematically, it is possible to obtain the exact value of the instantaneous speed by the use of calculus, a mathematical technique developed over three hundred years ago by Isaac Newton.

VELOCITY: SPEED AND DIRECTION

In some cases the *direction* in which we are moving is also important information. A pilot needs to know how fast the wind is blowing and in what direction. The pilot needs to know the wind velocity. Velocity gives the speed and the direction of motion.

The wind velocity of a 50-mph wind blowing east would push off course a small airplane heading north at 80 mph. If the pilot did not correct the airplane's heading it would end up flying in a northeasterly direction.

Velocity, and other quantities that require a magnitude and a direction, are called *vector* quantities. In the next section we shall study some properties of these new quantities. Quantities that do not require a direction are said to be *scalar* quantities. Speed is a scalar quantity.

VECTORS

As we have said, vector quantities are those that possess both magnitude and a direction. We represent vectors as arrows with length proportional to the magnitude and with a direction indicating the direction of the vector quantity. We use bold face letters, (**v**, **V**) to completely represent vectors, and standard letters (*v*, *V*) to indicate their magnitudes.

To illustrate some properties of vectors, let's consider the following situation. A woman wants to buy a paperback she has heard about recently. She walks 3 km east to her friend's house and then both walk together to the nearest drugstore, 4 km from her friend's house and in the same direction (Figure 2.4a).

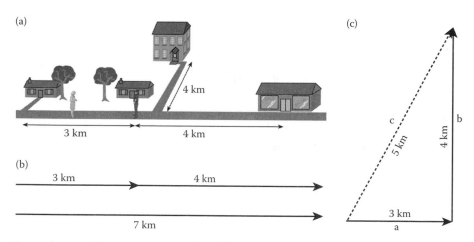

FIGURE 2.4   (a) A woman walks 3 km to her friend's house and then 4 km to the drugstore. The man is 7 km away from home. If instead she walks 4 km from the friend's house to the library, the woman is only 5 km from home. (b) The two individual trips of 3 and 4 km are equal to a single trip of 7 km. (c) Walking east 3 km (labeled with vector a) and then north 4 km (vector b) is equivalent to walking across in the direction shown by vector c.

Obviously, the total distance traveled by the woman is 7 km. Since the two trips take place in the same direction, the woman finds herself 7 km away from home. Graphically, we can illustrate this situation as in Figure 2.4b.

Let's suppose now that the woman wants to borrow the book from the local library instead of buying it at the drugstore (Figure 2.4a). If the library is also 4 km from her friend's house, but north instead of east, as illustrated in Figure 2.4b, the woman is now only 5 km from home. Using Pythagoras' theorem we get a displacement of $\sqrt{(4^2 + 3^2)}$ km = 5 km. If she had wanted, she could have cut across the field straight from her home to the library, walking only the 5 km.

If vector **a** represents the 3-km walk east, and **b** the 4-km walk north, vector **c** represents the straight walk across the field from the house to the library. Vector **c** is equivalent to the two vectors **a** and **b** together. In other words, walking east for 3 km to the friend's house and then north for 4 km to the library is the same as walking 5 km across the field from the house to the library. We call vector **c** the *resultant* of vectors **a** and **b**, and we call the process the *addition of vectors*.

An alternative method, the so-called *parallelogram method* for the addition of vectors, consists of placing the vectors to be added tail-to-tail instead of head-to-tail, keeping their orientations fixed. The resultant is obtained by completing the diagonal of the parallelogram. In Figure 2.5 we illustrate the addition of horizontal vector **a** and vertical vector **b**, producing resultant vector **c**. (In this case the parallelogram is a rectangle.)

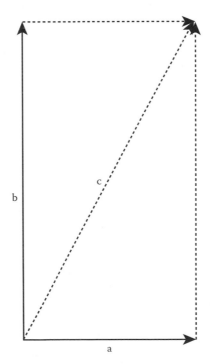

FIGURE 2.5   Addition of two vectors using the parallelogram method.

Just as horizontal vector **a** and vertical vector **b** can be combined to form the resultant vector **c**, any vector can be viewed as the resultant of two vectors that are perpendicular to each other. These new vectors are called the *components* of the original vector.

ACCELERATION

"The properties belonging to uniform motion have been discussed in the preceding section; but accelerated motion remains to be considered," wrote Galileo in the chapter named "De Motu Locali" of his book *Two New Sciences*, where he discussed his discoveries on uniform motion. He continued:

> And first of all it seems desirable to find and explain a definition best fitting natural phenomena. This we readily understand when we consider the intimate relationship between time and motion...

We can illustrate this relationship between time and motion with the following situation. Laurie and her friend Matthew are discussing the responsiveness of Laurie's recent investment: a nice used hybrid car she bought with the money she earned last summer. Laurie says the dealer assured her that the car can make 0 to 50 miles per hour in less than 10 seconds. Matthew does not think that a three-year-old hybrid subcompact can do that and suspects that the speedometer might have been altered. To settle the argument Laurie asks Matthew to ride with her while she accelerates and to record the speed of the car every second. Since this is difficult to do, they also convince ten of their friends to line up at regular intervals along a straight and flat section of an infrequently traveled road, to time the car as Laurie starts accelerating from rest. The ten friends are separated by intervals of 10 meters each and start their digital timers as soon as Laurie steps on the gas pedal. Laurie and Matthew collect their friends' data and tabulate it as follows:

x (m)	0	10	20	30	40	50	60	70	80	90	100
t (s)	0	3.2	4.4	5.5	6.3	7.1	7.7	8.4	8.9	9.5	10

Since the car is accelerating, the speed is not constant. This can be seen by studying the data. Laurie covers the first 10 meters in 3.2 seconds, but takes only 1.2 seconds to cover the second interval of 10 meters. The last 10 meters are traveled in half a second!

According to the car's speedometer, Laurie was traveling at 45 miles per hour at the end of the 10 seconds. To check this, Laurie calculates the average speed during the last interval, which is equal to the 10 meters traveled (100 m − 90 m) divided by the 0.5 seconds it took, or

$$\overline{v} = \frac{10 \text{ km}}{0.5 \text{ s}} = 20 \text{ m/s}$$

or 45 miles per hour, confirming Matthew's data, and assuring Laurie that the speedometer had not been tampered with. Since 45 miles per hour is the average speed during the last 10 meters, the actual speed at the 100-m mark is slightly higher (48 mph in this case). However, since this was only one trial, Laurie is satisfied that the 0–50 in about 10 seconds may be possible.

When we say that the car makes "0 to 50 in 10 seconds" we obviously mean that the car accelerates from rest to 50 miles per hour in 10 seconds. Or in the case of Laurie's car, 45 miles per hour in 10 seconds. Acceleration is the rate at which velocity changes and is given in units of distance per time squared (miles per hour per second, for example). Thus, average acceleration is given by

$$\bar{\mathbf{a}} = \frac{\Delta \mathbf{v}}{t}$$

where $\Delta \mathbf{v}$ is the change in velocity that takes place during the time interval t. (The Greek capital delta is used in this fashion to indicate *change* in some quantity, here the change in velocity.) The SI unit for acceleration is the meter per second squared (m/s^2). Since velocity is a vector quantity, acceleration is also a vector quantity.

UNIFORMLY ACCELERATED MOTION

In the previous example, the acceleration of the car is constant. In other words, the velocity increased at a constant rate. In *Two New Sciences,* Galileo called motion with constant acceleration uniformly accelerated motion. He wrote:

> [F]or just as uniformity of motion is defined by and conceived through equal times and equal spaces (thus we call a motion uniform when equal distances are traversed during equal time intervals), so we may, in a similar manner, through equal time-intervals, conceive additions of speed as taking place without complication...
> Hence the definition of [uniformly accelerated] motion...may be stated as follows:
> A motion is said to be uniformly accelerated when, starting from rest, it acquires during equal time-intervals, equal increments of speed.

Uniformly accelerated motion is a special case of accelerated motion, since acceleration does not have to be always constant. Here we shall consider only this special case. When the acceleration is constant, $a = \bar{a}$. Thus, we can write,

$$a = \bar{a} = \frac{\Delta v}{t}.$$

Let's consider a numerical example. A driver enters a straight highway at 30 km/h and accelerates to 55 km/h in 5 seconds. What is the acceleration? Calling the initial and final velocities v_i and v_f, we can calculate the change in velocity is $\Delta v = v_i - v_f = $ 55 km/h − 30 km/h = 25 km/h. The driver accelerates at

$$a = \frac{\Delta v}{t} = \frac{25 \text{ km/h}}{5 \text{ s}} = 5 \text{ km/h/s}.$$

That is, the driver speeds up at a rate of 5 km/h every second.

If we know the acceleration of an object moving with uniformly accelerated motion and are interested in calculating the change in the velocity Δv after certain time t, we can turn our equation around to obtain

$$\Delta v = at.$$

We can use this equation to compute, for example, the takeoff speed of a jet airplane. Suppose an airliner increases its speed by 10km/h every second; that is, its takeoff acceleration is 10 km/h/s. If the plane takes off 36 seconds after it first began accelerating, the takeoff speed can be obtained from the previous equation by realizing that the change in speed during these 36 seconds is simply the takeoff speed v since the initial speed is 0. The takeoff speed of a jet airliner is then,

$$v = at = \frac{10 \text{ km/h}}{\text{s}} \times 36 \text{ s} = 360 \text{ km/h}.$$

FALLING BODIES

One important example of uniformly accelerated motion is the vertical motion of an object falling. The acceleration g is caused by the gravitational attraction of the earth upon the object. The magnitude of g at the earth's surface is 9.8 m/s². This value varies slightly with altitude and latitude and with the different geological features near the earth's surface. In the following chapter we shall see how Galileo first calculated this value. Here we merely use the fact that, neglecting air resistance, all objects near the surface of the earth fall with an acceleration $g = 8$m/s² as an example of motion with constant acceleration. Since for this case, $a = g$, the last equation should be written as

$$\Delta v = gt.$$

A common situation is the fall of an object from rest. In this case, in a certain time t, the velocity changes from zero to the value of the instantaneous velocity, v. The change in velocity after a time t is the instantaneous velocity, and we can write

$$v = gt.$$

Thus, after one second, the stone that the boy in Figure 2.6 drops from a bridge is moving with a velocity of $v = 9.8$ m/s² × 1 s = 9.8 m/s towards the water. After 5 seconds, the velocity has increased to $v = 9.8$m/s 5s² = 49 m/s. In Table 2.1, we show the instantaneous speed of an object falling at intervals of one second.

FIGURE 2.6 After a time *t* the instantaneous velocity of the stone that the boy drops from the bridge is $v = gt$.

TABLE 2.1
Instantaneous Speed of an Object
Falling from Rest

Time of Fall (s)	Velocity (m/s)
0	0
1	9.8
2	19.6
3	29.4
4	39.2
5	49.0
6	58.8
7	68.6
8	78.4
9	88.2
10	98.0

Table 2.1 illustrates Galileo's definition of uniformly accelerated motion as that which acquires equal increments of speed during equal time intervals. We see that for each time interval of 1 second, the speed increases by 9.8 m/s. This result is in agreement with our commonsense idea of acceleration. When we say that an object is accelerating, we seem to imply that *the longer the elapsed time, the faster it goes.* However, we could also imply that *the farther the object goes, the faster it goes.* In fact, Galileo wrote that at one time he thought it would be more useful to define uniform acceleration in terms of the increase in speed and distance traveled.

PIONEERS OF PHYSICS: GALILEO'S METHOD

Galileo arrived at the expression for computing the distance traveled by a falling object guided by his desire to find quantities that he could measure directly. A motion is uniformly accelerated (he had written) when, starting from rest, it acquires equal increments of speed during equal time intervals. Measuring increments of speed at equal time intervals for a falling object was not practicable with the rudimentary clocks that were available to him.

Galileo had discovered, however, that an object falling from rest increases its speed uniformly from an initial value of 0 to a final value v_{final}. He realized that for a quantity that changes uniformly, the average value is halfway between the initial and final values. Since the initial value of the speed of an object falling from rest is zero, the average value of the speed is then $\frac{1}{2} v_{final}$ (halfway between 0 and v_{final}). Using the definition of average speed, $\bar{v} = d/t$, one can obtain the distance d traveled in terms of the average speed and the time; that is $d = \bar{v} t$. Thus, the distance traveled by an object moving with an average speed $\frac{1}{2} v_{final}$ is

$$d = \frac{1}{2} v_{final} \times t.$$

Galileo could not test this expression directly either, since he would have had to measure the speed of the falling object just before it hit the ground, a task that is difficult even today.

The final speed of an object falling from rest is, as we saw above, $v_{final} = gt$. This expression for v_{final} can be substituted into the expression for the distance d:

$$d = \frac{1}{2} gt \times t = \frac{1}{2} gt^2.$$

This is the expression that Galileo was seeking. It relates the total distance that the object falls from rest to the total time, quantities that were easier to measure. However, the timing devices that Galileo had at his disposal only allowed him to test this expression in an indirect way.

We already know how to compute the change in speed with time elapsed. How do we find the distance traveled as the speed increases? Galileo, guided by a need to find an expression that he could test with the limited instruments of his time, was able to show from the definitions of average speed and acceleration that the distance traveled by a falling object that starts from rest is

$$d = \frac{1}{2} g t^2.$$

Notice that since the acceleration due to gravity g is constant, the distance traveled is proportional to the square of the time elapsed. The stone that the boy in Figure 2.5 drops moves 4.9 m in the first second. In two seconds, *twice* the time, it moves *four* times the distance, or 19.6 meters; and in *three* seconds it moves *nine* times as far, or 44 meters.

THE MOTION OF PROJECTILES

The third and last part into which Galileo divided his study of motion is the motion of projectiles. "I now propose to set forth those properties which belong to a body whose motion is compounded of two other motions, namely, one uniform and one naturally accelerated. This is the kind of motion seen in a moving projectile." Galileo further explains:

> Imagine any particle projected along a horizontal plane without friction; then we know, from what has been more fully explained in the preceding pages, that this particle will move along this same plane with a motion which is uniform and perpetual, provided the plane has no limits. But if the plane is limited and elevated, then the moving particle, which we imagine to be a heavy one, will on passing over the edge of the plane, acquire, in addition to its previous uniform motion, a downward propensity due to its own weight; so that the resulting motion, which I call projection, is compounded of one which is uniform and horizontal and of another which is vertical and naturally accelerated.

The strobe photograph in Figure 2.7 illustrates Galileo's experiment. Two balls were released simultaneously from the same height. One ball was simply dropped while the other was thrown horizontally. The equally spaced horizontal lines in the photograph show us that both balls keep pace as they fall, accelerating toward the ground at the same rate. This acceleration is the acceleration due to gravity, g. The initial horizontal velocity given to one of the balls does not affect its vertical motion. Careful examination of the photograph tells us that the horizontal distance between positions of the ball that was given an initial horizontal velocity are all equal. The horizontal component of the motion is uniform motion with a constant horizontal velocity component. Therefore, the vertical motion of the ball does not affect its horizontal motion.

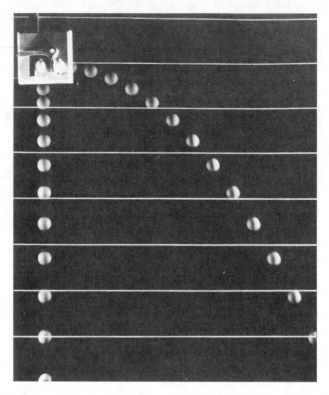

FIGURE 2.7   Stroboscopic photograph of two balls released simultaneously from the same height. One of the balls is given an initial horizontal velocity and moves off to the side as it falls. The horizontal lines in the photograph help us see that both balls hit the ground simultaneously. (From *PSSC Physics Seventh Edition*, by Haber-Schaim, Dodge, Gardner, and Shore. With permission from Kendall/Hunt Publishing Company, Dubuque, IA, 1991.)

The independence of horizontal and vertical motions allows us to predict the position and velocity of projectiles at any time during their flight merely by applying what we have learned about motion to the two independent components. The horizontal motion is uniform, at a constant speed v_x, and the vertical motion is uniformly accelerated motion, with a constant acceleration g.

3 The Laws of Mechanics: Newton's Laws of Motion

THE CONCEPT OF FORCE

Central to the laws of mechanics is the concept of *force*. Our idea of force is closely related to muscular activity. When we push or pull on an object, we exert a force on it (Figure 3.1). When we push a lawn mower across a yard, pull a hand truck loaded with boxes, push against the arms of a chair to get up from it, or when we turn the ignition key with our index finger and thumb to get the car started, we are applying a force. These forces associated with muscular activity are not the only ones that exist in nature. When you bring a small magnet near a nail, a *magnetic force* pulls the nail toward the magnet; and a *gravitational force* keeps the moon orbiting around the earth and the earth around the sun, and keeps us attached to the ground.

The concept of force is directly involved in the formulation of the laws of motion. The discovery of these laws marks the birth of our modern understanding of the universe.

THE ANCIENT IDEA OF MOTION

We all know today that, neglecting the very small effect of air resistance, an object falling toward the ground experiences a constant acceleration caused by the gravitational attraction of the earth upon the object, and that all falling objects experience this acceleration. This was not known before the early 1600s. Until then, it was believed that heavier objects would fall toward the ground faster than lighter ones. This idea was based on the teachings of Aristotle, the greatest scientific authority of antiquity.

Born in the Greek province of Macedonia in the year 384 BC, Aristotle was raised by a friend of the family, having lost both parents while still a child. At the age of 17 he went to Athens for his advanced education and later joined Plato's Academy, becoming "the intelligence of the school," as Plato himself called him. It was in Athens, many years later, that Aristotle founded the Lyceum, a school so named because it was near the temple to Apollo, also known as Lykaios or Lyceius (apparently because he protected the flocks from wolves (*lykoi*)). This was the famous "peripatetic (walk about) school" where Aristotle would sometimes lecture while strolling in the school's garden.

Aristotle's lectures were collected in some 150 volumes, of which only about 50 have survived. Aristotle's writings remained largely forgotten until the thirteenth century AD. Throughout the Middle Ages, Aristotle became one of the most important influences and perhaps the greatest philosopher.

According to Aristotle, there was a sharp distinction between heaven and earth, with different sets of natural laws for each region. The boundary between these two

FIGURE 3.1　The mother pushing the stroller exerts a force on it. There are other forces in nature.

regions was the sphere of the moon, above which all motion was perpetual, circular, and uniform. This was the region of no-change, the home of the *æther*, where things were eternal and unchanging. In contrast, in the region below the sphere of the moon, all motion was along straight lines. Things constantly changed due to the interplay between hot and cold, dry and moist. The four combinations of these opposites produced the four elements: *earth*, *water*, *air*, and *fire*. These four elements had their own natural place, and motion was an attempt to reach that place. Since the earth was at the center, an object composed mostly of earth, like a rock, would fall toward the ground, its natural place.

Thus, the fall of an object toward the earth is an example of natural motion. Moreover, since a heavy object contains more earth than a light one, it could have a stronger tendency to fall toward its natural place. According to Aristotle, heavier objects fall faster than lighter objects. It took nineteen centuries and the genius of Galileo for this error to be corrected.

THE BIRTH OF MODERN SCIENCE

In *Two New Sciences*, Galileo presents his theories in the form of a dialog among three people: Simplicio, who represents the views of Aristotle; Salviati, who represents Galileo; and Sagredo, who represents the intelligent layman. At one point, after discussing whether Aristotle had ever tested by experiment whether a heavier stone would fall to the ground faster than a lighter stone, Simplicio and Salviati continue:

> Salviati [Galileo]: If then we take two bodies whose natural speeds are different, it is clear that on uniting the two, the more rapid one will be partly retarded by the slower, and the slower will be somewhat hastened by the swifter. Do you not agree with me in this opinion?
> Simplicio [Aristotle]: You are unquestionably right.

Salviati [Galileo]: But if this is true, and if a large stone moves with a speed of, say, eight, while a smaller moves with a speed of four, then when they are united, the system will move with a speed less than eight; but the two stones when tied together make a stone larger than that which before moved with a speed of eight. Hence the heavier body moves with less speed than the lighter one; an effect which is contrary to your supposition. Thus you see now, from your assumption that the heavier body moves more rapidly than the lighter one, I infer that the heavier body moves more slowly.

Simplicio [Aristotle]: I am all at sea... This is, indeed, quite beyond my comprehension...

Galileo had actually proved theoretically in 1604 that falling bodies are accelerated toward the ground at a constant rate. An object falling to the ground experiences what came to be known as uniformly accelerated motion. In fact, Galileo actually defined uniform acceleration in terms of the behavior of falling bodies. However, with the clocks and instruments available to him at the time, he could not directly test whether his theoretical predictions were correct. The legend that he dropped weights from the leaning tower of Pisa is very likely false.

Galileo realized that there was an indirect way of testing his theory. If an object is falling to the ground at a *slower* rate, as when a ball rolls down a smooth inclined plane, it can be timed with good accuracy. Galileo actually constructed such a plane and left detailed notes on his experiment (Figure 3.2). In *Two New Sciences* it is Salviati who describes in great detail how the hundreds of experiments were performed. By changing the angle of inclination and determining the acceleration of the ball as it rolled down, Galileo was able to infer that in the limiting case, when the angle was 90°, the acceleration, having been constant for all the other angles, had to be constant too. And 90° was, of course, free fall.

FIGURE 3.2 Picture painted in 1841 by G. Bezzuoli. Galileo (the tall man in the center, pointing with his right hand at the open book) demonstrates one of his experiments with the ball rolling down an inclined plane. Changing the angle of inclination of the plane allowed Galileo to infer that when the angle was 90° (vertical plane) the acceleration of the ball was also constant. He could not perform this last experiment because he lacked a timing device accurate enough to time the rapidly falling ball. (From painting in 1841 by G. Bezzuoli. Used with permission of Alinari/Art Resource, New York.)

Thus, we see how Galileo not only was able to argue against the Aristotelian approach with his mathematical rationalism, but in the process he established the modern scientific method from observation to hypothesis. From the mathematical analysis of the hypothesis, predictions are drawn, and these can in turn be tested by experimental observation. Galileo was aware of his having founded the experimental method. In *Two New Sciences* he writes:

Salviati: … we may say the door is now open, for the first time, to a new method fraught with numerous and wonderful results which in future years will command the attention of other minds.

PIONEERS OF PHYSICS: GALILEO'S DIALOG WITH ARISTOTLE

Simplicio: Your discussion is really admirable; yet I do not find it easy to believe that a bird-shot falls as swiftly as a cannon ball.

Salviati: Why not say a grain of sand as rapidly as a grindstone? But Simplicio, I trust you will not follow the example of many others who divert the discussion from its main intent and fasten upon some statement of mine which lacks a hair's-breadth of truth and, under this hair, hide the fault of another which is as big as a ship's cable. Aristotle says that "an iron ball of one hundred pounds falling from a height of one hundred cubits reaches the ground before a one-pound ball has fallen a single cubit." I say that they arrive at the same time. You find, on making the experiment, that the larger outstrips the smaller by two finger-breaths; now you would not hide behind these two fingers the ninety-nine cubits of Aristotle, nor would you mention my small error and at the same time pass over in silence his very large one.

Galileo Galilei, *Dialog Concerning Two New Sciences*, 1638, translated by Henry Crew and Alfonso de Salvio (Macmillan, New York, 1914; reprinted by Dover, New York. With permission), pp. 64–65.

GALILEO FORMULATES THE LAW OF INERTIA

In one of Galileo's experiments, he drops two balls, "one, say, of lead, the other of oak, both descending from a height of 150 or 200 *braccia* [yards]." "Experience shows us," writes Galileo in *Two New Sciences*, "that [the] two balls … arrive at the earth with very little difference in speed." Galileo shows that the opposition presented by the air when the object is moving with great speed is not much larger than when the object is moving at a lower speed. He further explains:

As to speed, the greater this is, the greater will be the opposition made to it by the air, which will also impede bodies the more, the less heavy they are. Thus the falling heavy thing ought to go on accelerating in the squared ratio of the duration of its motion; yet, however heavy the movable might be, when it falls through very great heights the impediment of the air will take away the power of increasing its speed further, and will reduce it to uniform and equable motion. And this equilibration will occur more quickly and at lesser heights as the movable shall be less heavy.

Galileo affirms here that a body falling through a great distance will be slowed down by increasing air resistance until this resistance equals the weight of the falling object. At this moment, the object is prevented from any increase in its speed and will continue falling with a constant velocity due to the cancellation of the forces acting on the object. This is contrary to the Aristotelian idea that to keep an object in motion you would need to supply a force, and that if the motive force is balanced by the resistance, the speed becomes zero. Thus, according to Galileo, an object moving at constant speed along a straight line will continue to do so in spite of the fact that it has lost all contact with the source of that motion (Figure 3.3). This can be interpreted as the tendency of a body to resist any change in its state of motion, a property that Newton would later call *inertia*. In *Two New Sciences*, Galileo writes:

> I mentally conceive of some movable projected on a horizontal plane, all impediments being put aside. Now it is evident from what has been said elsewhere at greater length that equable [uniform] motion on this plane would be perpetual if the plane were of infinite extent.

FIGURE 3.3 An ice skater moving on perfectly smooth ice at a constant speed along a straight line would not slow down. Air resistance and some friction (since even Olympic ice rinks are not perfectly smooth) actually slow down the ice skater.

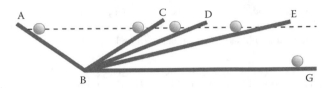

FIGURE 3.4 Galileo's clever design to prove that, neglecting friction, an object moving along a horizontal plane will continue moving forever at a constant velocity.

Galileo arrived at this important conclusion in a very clever way. "Let us suppose," he wrote, "that the descent [of a body] has been made along the downward sloping plane AB, from which the body is deflected so as to continue its motion along the upward sloping plane BC." In the same way that a pendulum bob reaches nearly the same height from where it was released after one swing, the rolling body in Galileo's experiment reaches the same height when rolling upwards on plane BC. "And this is true whether the inclinations of the planes are the same or different, as in the case of the planes AB and BD." Thus, if the inclination of the plane on the right in Figure 3.4 is changed from C to D to E, the body must roll farther in each case to reach the same height. When the second plane is horizontal, the rolling body never reaches its original height and must "maintain a uniform velocity equal to that which it had acquired at B after fall from A."

There is still some disagreement among historians of science as to whether Galileo believed that a plane of infinite extent could actually exist, and some historians interpret Galileo's phrase "rolling on forever" as meaning staying at a constant height above the earth, not moving along a straight line through space. It is, however,

PHYSICS IN OUR WORLD: THE LEANING TOWER OF PISA

The Leaning Tower of Pisa has only been straight for a short period during its 800-year history. Right after its construction in the twelfth century it tilted north. It straightened up a few years later and stood nearly upright for almost a century, only to start tilting at an alarming rate in a southward direction during the early part of the fourteenth century. During the early 1600s, when Galileo lived in Pisa, the tower had already leaned over 180 minutes of arc. Toward the end of the twentieth century, the tower tilted at an angle of 321 minutes of arc, making the top of the tower 5.227 meters off-center. The extent of this lean placed a very large stress on the southern wall of the tower and, without correction, might have actually caused it to collapse.

Construction of the tower began in 1173 in the Piazza dei Miracoli. The tower or campanile formed part of a grandiose and spectacular cathedral and baptistery built in white marble in a blend of Romanesque and Gothic styles. The soil under the piazza is composed of several layers of clay and sand. The area under the entire square has been sinking slowly and unevenly, with a few places where the soil is sinking faster than in the rest of the area. One of these places lies exactly below the tower.

Since 1911, when regular monitoring of the tower began, the top of the tower moved about 1.2 mm per year. Although several efforts to stop the tilting were initially undertaken, none of those early efforts succeeded. In 1990, the Italian government formed an international commission of experts to study methods to stabilize the tower. In 1992, the first level of the tower was strapped with steel bands to prevent the walls from breaking. In 1993 the commission proposed placing 750 tons of lead bricks on the northern side of the concrete ring that encircles the tower. These counterweights not only stopped the incline from increasing for nine months, but actually straightened it by about 2.5 cm.

In 1995, the lead bricks were replaced by a second ring anchored with steel cables to one of the deep clay layers. In 2008, after over ten years of work, engineers declared that the tower should remain stable for at least 200 years.

evident that he saw that a body could have a uniform, non-accelerated, constant velocity, and that Isaac Newton would use Galileo's ideas in the formulation of his laws of motion.

NEWTON'S FIRST LAW: THE LAW OF INERTIA

Isaac Newton was born in Woolsthorpe, in the county of Lincolnshire, on Christmas Day 1642, the same year that Galileo died.* His father, Isaac, had died three months before he was born. Of his early education little is known, except that he attended two small schools in villages near Woolsthorpe. At the age of 12, Newton enrolled in King's School at Grantham, a few miles north of Woolsthorpe, where his intellectual interests were awakened. He began building ingenious mechanical toys and clocks and became interested in space and time and in the motion of objects.

In 1660, Newton entered Trinity College and five years later he took his Bachelor of Arts degree. That year the great plague broke out in London, and Newton went to his mother's farm in Woolsthorpe when the university closed in June 1665. He did not return to Cambridge until April 22, 1667. During the two years that Newton remained at his mother's farm, his mathematical genius flourished, and his studies reached a climax in October 1666. This period of creativity is usually known as the *annus mirabilis*, the year of wonders, telescoped for commemorative purposes into the single year of 1666. During that period he performed experiments to investigate the nature of light, completing his theory of colors in the winter of 1666. He began his observations of comets, tracking them for nights on end until he fell ill from

---

* The Gregorian calendar was not in use in England at the time. On the Continent, however, where the Gregorian calendar had been in use since 1582, Newton's date of birth was ten days later: January 4, 1643. In addition, in England, which used the Old Style calendar, the new year started on March 25 rather than on January 1.

exhaustion, as he told his biographer. During that same period, Newton developed the main ideas of what he called his theory of *fluxions*, which we know today as calculus. It was also during that glorious period of creativity that Newton laid the foundations of his celestial mechanics with his discovery of the law of gravity. "All this was in the two plague years of 1665 & 1666," Newton wrote, "for in those days I was in the prime of my age for invention, & minded Mathematics & Philosophy [i.e. physics] more than at any time since."

It is not clear why Newton, after having conceived the idea of universal gravitation, did not attempt to publish anything about it for almost 20 years. When he finally did, at the instigation of one of his closest friends, the scientist Edmond Halley, he published what many consider the greatest scientific treatise ever written, the *Philosophiae Naturalis Principia Mathematica*. The physics of the *Principia* still guides our communication satellites, our space shuttles, and the spacecraft that we send to study the solar system (Figure 3.5).

The *Principia* is divided into three books or parts. In Book One, The Motion of Bodies, Newton develops the physics of moving bodies. This first part is preceded by "axioms, or laws of motion:"

Law I: Every body continues in its state of rest, or of uniform motion in a right line, unless it is compelled to change that state by forces impressed upon it.

"Projectiles continue in their motions," explained Newton after stating his first law, "so far as they are not retarded by the resistance of the air, or impelled downwards by the force of gravity." Therefore, as long as there are no forces pushing or pulling on an object, it will not change its state of motion. This state of motion could be the absence of motion in relation to an observer, that is, a body at rest. Newton is stating very clearly in his first law what Galileo hinted at, that an object moving along a straight line with a constant velocity (uniform motion) will continue moving along that line with the same velocity unless a force is applied to the object so as to change its velocity.

FIGURE 3.5 The physics of the *Principia* still guides our space vehicles. (Courtesy of NASA.)

PHYSICS IN OUR WORLD: CAR SEAT BELT

A car seat belt moves freely when it is pulled and locks when the car suddenly stops. The mechanism that makes a seat belt work is based on the concept of inertia.

The belt wraps around a belt shaft. Attached to the shaft, a toothed plate rotates freely during normal operation (Figure 3.6). When the car suddenly stops, however, the heavier end of an elongated clutch moves forward because of its inertia, engaging the toothed plate with its own inner teeth. The clutch itself is prevented from rotating further by a small pawl which engages and locks a ratchet, also attached to the belt shaft. The pawl is attached firmly to the car body.

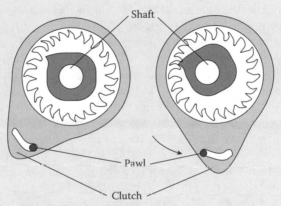

FIGURE 3.6 Car seat belt mechanism.

Regardless of how fast you pull on the belt, you will not be able to lock it, since the clutch is not engaged. When the driver suddenly slams on the brakes or the car collides, there is a rapid decrease in the car's speed. You, whatever your have loose in your car, and, more importantly, the elongated clutch, continue moving with the velocity that the car had. The objects will probably fall on the floor, but you will be prevented from hitting the windshield when the moving clutch engages the toothed wheel.

The tendency of objects to maintain their state of motion (rest or motion with a constant velocity) is called the principle of inertia. Inertia is the resistance that all objects present to any attempt to change their state of motion. The inertia of an object is related to its mass. A massive object has a high inertia, which means it is hard to move or to stop once you get it moving. Light objects can be moved about more easily. It is inertia that projects you out of your seat when you are in a car and it stops suddenly, and inertia is responsible for your difficulty in getting your car moving on an icy road. It is inertia that makes you hold on to a rotating merry-go-round so that you are not thrown off, and bursts a flywheel that is spinning too fast. Newton's first law is also called the law of inertia.

We have spoken of forces pushing and pulling on objects to change their states of motion. These forces caused the objects to move. However, we all know that we can push on a heavy piece of furniture, like a dresser for example, and fail to move it. In this case the force applied is balanced by the frictional force between the dresser and the floor. We would say that the net force acting on the dresser is zero. Forces are vector quantities and this knowledge helps us understand why we can have several forces applied to an object and still have a net or resultant force equal to zero. When the two teams participating in a tug-of-war contest are unable to drag each other across the center line (Figure 3.7), the net force on the rope is zero, even though the individual team members, each of different strengths, pull with different forces. The sum of all the individual forces pulling on the rope is zero and the rope does not move. In Figure 3.7, we have the three different forces F_1, F_2, and F_3 exerted by each one of the three women acting on the left side of the rope, and the three additional forces, F'_1, F'_2, and F'_3, exerted by the men, all different, and differing from the first ones, acting on the right side of the rope. In this vector diagram, the lengths of the vector forces are proportional to their magnitudes. Since the total length of the vectors on the left is equal to the total length of the three vectors acting on the right, the vector sum of these six forces is equal to zero. Thus the net force acting on the rope is zero and we say that the rope is in equilibrium.

Newton's first law says that, if the net force acting on an object is zero, the object will not change its state of motion. Thus, if the object is in equilibrium, its velocity remains constant. A constant velocity means that the velocity can take up any value that does not change, including zero, the case of an object at rest. This fact means that "motion with a constant velocity" and "a state of rest" are equivalent, according to Newton's first law. On close examination, a velocity is always given in reference to some point. If you are traveling on an airplane, the book you might be reading is at rest with respect to you, but is traveling, along with the other passengers and the entire contents of the plane, at the cruising speed of the plane, about 1000 km/h

FIGURE 3.7 Equilibrium: The net force on the rope is zero if neither team is able to drag the other. Although the forces acting on the rope are all different, the net force is equal to zero.

with respect to the ground. A flight attendant walking up the aisle at 2 km/h with respect to the plane would be moving at 1002 km/h with respect to the ground. Whether an object is at rest or moving with a constant velocity depends on the frame of reference to which it is referred. These special reference frames, in which the law of inertia is valid, are called *inertial frames of reference*. Thus, motion is entirely relative, an idea that is central to Newton's laws and was recognized by Newton himself.

NEWTON'S SECOND LAW: THE LAW OF FORCE

Newton's first law tells us what happens to an object in the absence of a net force. What happens if the net force acting on an object is not zero? According to the first law, the object will not move with a constant velocity. This means, of course, that the object will experience acceleration in the direction of the applied net force. Newton explained it in Book One of the *Principia*:

> **Law II:** The change of motion is proportional to the motive force impressed; and is made in the direction of the right line in which that force is impressed.

According to Newton's second law, when an instantaneous force acts on a body, as when a baseball bat strikes a ball, a change in the body's motion takes place which is proportional to the "force impressed." If the force acts on the body continuously rather than instantaneously, as when we push on an object for some time, acceleration is produced that is proportional to the applied force.

As an illustration of the second law, suppose that your sports car has stalled and you decide to push it (Figure 3.8a). If your car has a mass m of 1000 kg, you would apply a force F for 10 seconds to accelerate it from rest to a speed of 5 km/h. When two of your friends decide to join you and help you push the car (Figure 3.8b), a larger force of about $3F$ applied to the car of mass m for the same 10 seconds would bring the car to a speed of 15 km/h, producing acceleration three times larger and perhaps getting your car started. Thus, for a constant mass, we can write

$$F \text{ is proportional to } a \text{ or } F \propto a.$$

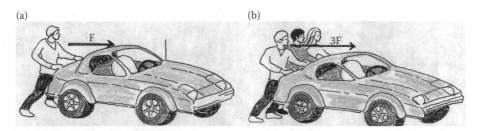

FIGURE 3.8 (a) An applied force F on the car of mass m, produces an acceleration a as the car increases its speed from zero to a final v during a time t. (b) A force of $3F$ produces an acceleration $3a$, thus increasing the car's speed from zero to a final $3v$ during the same time t.

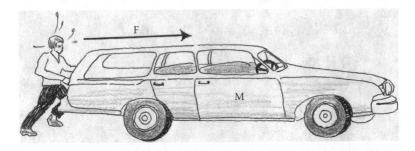

FIGURE 3.9   A larger mass *M* requires a larger force *F* to produce a given acceleration.

When your 2000-kg wagon stalls you know that you have to push much harder, with a larger force, to accelerate it from rest to 5 km/h in 10 seconds (Figure 3.9). In fact, you have to apply a force twice as large as you did for your sports car to produce the same acceleration, since the mass of the wagon is twice as large. We should write then, for a fixed acceleration,

$$F \propto m.$$

In the first case, for a constant mass (the sports car), the applied force is proportional to the acceleration produced. In the second case, for a given acceleration (0 to 5 km/h), the applied force is proportional to the mass of the object, larger for the larger mass of the wagon and smaller for the smaller mass of the sports car. We can combine these two proportionality expressions into a single equation:

$$\mathbf{F} = m\mathbf{a}$$

which is the mathematical expression for Newton's second law. In converting from proportionality to equality, a constant is usually introduced. In this case, however, if SI units are used, the constant is unity. Notice that $\mathbf{F}$ and $\mathbf{a}$ are both vector quantities.

We can restate the second law as follows:

The acceleration of an object is directly proportional to the net force acting on it and inversely proportional to its mass.

The SI unit of force is the *newton* (N), defined as the force that produces an acceleration of 1 m/s² when acting on a 1-kg mass. From this definition we can write:

$$1\text{N} = 1\text{kg m/s}^2.$$

In Newton's second law, it is the net force $\mathbf{F}$ acting on an object of mass *m* that produces an acceleration $\mathbf{a}$.

PUSHING SINGLE ATOMS

It takes 17 piconewtons to push a cobalt atom along a copper surface and 219 pN to push it over a platinum surface. Researchers at IBM's Almaden Research Center and at the University of Regensburg in Germany used an atomic force microscope to measure for the first time the force required to push atoms across different surfaces. Their goal is to learn about the way atoms interact with the surfaces that they rest on. The scientists hope that this knowledge can lead to an increase in the miniaturization of computer chips. This research can help IBM researchers experimenting with ways to arrange strands of carbon atoms that conduct electricity—carbon nanotubes—into arrays with DNA molecules. After the nanotube array is constructed, the DNA molecules can be removed. This ordered nanotube grid could work as a data storage device.

NEWTON'S THIRD LAW: THE LAW OF ACTION AND REACTION

Newton's second law allows us to calculate the average force required to accelerate a certain sports car from 0 to 60 mph in 9.9 seconds, for example, by determining the car's mass and multiplying it by the acceleration, which we would of course obtain from the initial and final speeds, and the time taken to accelerate the car between those two speeds. However, where does this force come from? What is the source for this force? The car's engine, which would perhaps come to mind as the answer, only makes the wheels spin; it does not make the car go! Moreover, the spinning wheels act on the pavement, not on the car. Newton gave us the answer to this problem in the form of his third law:

> **Law III**: To every action there is always an equal reaction: or, the mutual actions of two bodies upon each other are always equal, and directed to contrary parts.

"Whatever draws or presses another is as much drawn or pressed by that other," he explained. Thus, the car's wheels push on the pavement with a force produced by the engine. This push or action is matched by a force from the pavement on the wheels, which makes the car go. These two equal forces, the force of the wheels on the pavement and the force or reaction of the pavement on the wheels act, as we can see, on *different* bodies, the pavement and the wheels.

Newton's third law tells us that forces always come in pairs, acting on different bodies. A single isolated force acting on a body without another equal force acting somewhere else cannot exist.

According to the third law, the force that keeps the moon orbiting around the earth, F_{EM}, is equal and opposite to the force that attracts the earth towards the moon, F_{ME}, (Figure 3.10). The force of attraction between the earth and any object is called the *weight* of the object, w. If we drop a ball near the surface of the earth and neglect air resistance, the force of attraction between the earth and the ball makes

FIGURE 3.10   The earth and the moon attract one another with equal and opposite forces.

FIGURE 3.11   The earth pulls on the falling ball with a force that we call the ball's weight. The ball pulls on the earth with an equal and opposite force.

the ball fall towards the center of the earth with an acceleration of 9.8 m/s², which is, as we learned in Chapter 2, the acceleration due to gravity, g. Newton's second law tells us that this force due to gravity or weight (w), is given by

$$w = mg$$

where m is the object's mass. The third law tells us that the object pulls on the earth with an equal and opposite force. Thus, the ball pulls on the earth with a force equal in magnitude to its weight (Figure 3.11). Newton's second law also explains why the acceleration at which the earth "falls" towards the ball is negligibly small, since

FIGURE 3.12   The skaters pull on each other's arms with equal and opposite forces.

Frank and Ernest

(Courtesy of Bob Thaves.)

the force exerted by the ball on the earth is equal in magnitude to the ball's weight. Therefore,

$$F = w = M_E\, a = mg.$$

Since M_E is much greater than m, a must be much smaller than g.

Any two objects interacting with each other obey Newton's third law. The two skaters of Figure 3.12 pull on each other's arms with equal and opposite forces acting on two different bodies, namely the two skaters.

4 Energy

WHAT IS ENERGY?

Energy is one of the more important concepts in science. It appears in many forms, including mechanical energy, thermal energy, electromagnetic energy, chemical energy, and nuclear energy. Wherever and whenever anything happens, like the explosion of a distant sun or the falling of a golden leaf from a tree in autumn, a change in some form of energy is involved. In spite of our familiarity with the concept of energy, few of us can define it properly. What is energy? Can we measure it? Can we touch it?

Energy is an abstract concept introduced by physicists in order to better understand how nature operates. Since it is an abstract idea, we cannot form a concrete picture of it in our minds, and we find it very difficult to define it in simple terms. But we can perhaps understand what it can do. Energy is the ability to do work. Therefore, before we can fully understand what this definition of energy means we need to know what we mean in physics by work, a word that we use in our everyday language.

THE CONCEPT OF WORK

Work involves an effort directed toward the production of something. In physics, what we understand for work differs somewhat from our everyday meaning of the word, and we must be careful to distinguish between the two meanings. A few examples should illustrate what work means in physics. Suppose your car's battery dies and you decide to push the car to the gas station 500 feet along the road. When you push the car you exert a force on it, even if the car does not move. However, if the car begins to move as you push it, then you are doing work on it. The man pushing the stalled car (Figure 4.1a) performs work only if he is able to move it through some distance. If he attempts to push a bulldozer, he probably will not be able to move it. In this case the work done is zero, even if the man gets very tired (Figure 4.1b).

Suppose now that a child is pulling a wagon (Figure 4.2a). The wagon rolls along the horizontal surface of the sidewalk as the child pulls on the handle. We know from our own experience that if the direction in which the handle is pulled is too close to the vertical, the child's effort to move the wagon is not as effective. If the wagon is extremely heavy, the child must lean forward and pull on the handle in a horizontal direction. In fact, the horizontal direction makes the force most effective in this case (although this position is probably uncomfortable for the child; as seen in Figure 4.2b). If the child were to pull on the wagon in a direction perpendicular to the direction of motion (Figure 4.2c), his effort would be

(a)   (b)

FIGURE 4.1 (a) Work is done when the force applied by the man pushing makes the car move some distance. (b) The man pushing the bulldozer will not be able to move it. In this case, he performs no work on the bulldozer.

(a)   (b)

(c)

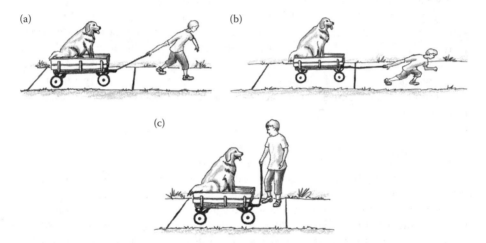

FIGURE 4.2 (a) Pulling a moderately heavy wagon. (b) To pull effectively on an extremely heavy wagon the child must lean forward and pull in a horizontal direction. (c) Pulling the strap in a vertical direction makes it impossible to move the wagon.

wasted; the force that he would apply in this case would do nothing for the motion of the wagon along the floor.

If the child pulls with the same force in all cases described in the previous situation, the work done on the wagon will be maximum when the applied force is in the direction of motion of the wagon, zero when the force is at right angles to the direction of motion, and an intermediate value when the force is along any other direction. The closer the applied force gets to the direction of motion, the more effective it becomes in producing work. We can say that work is a measure of the productivity of a force. For the simple case of a constant force acting on an object along the direction of motion of the object, as in Figures 4.2b and 4.3, the work done on the object is the product of the force and the distance the object moves, or

$$\text{Work} = \text{force} \times \text{distance or } W = F \times d$$

FIGURE 4.3 The force exerted by the horse is in the same direction as that of the resultant motion. In this case, the work done by the horse is equal to the product of the magnitude of the force and the distance traveled.

When the applied force is not along the direction of motion, the force can be resolved into two components, one parallel to the direction of motion and the other perpendicular to it. As we saw from our discussion, only the parallel component does work.

It is interesting to notice that if a person walks on a horizontal surface at a steady pace while carrying a suitcase, the force that he exerts on the suitcase to prevent it from falling to the ground does not produce work on the suitcase. However, if the suitcase is heavy and the person has to walk a long distance, he begins to sweat. Why should he sweat if he is doing no work? Even if he is standing still while holding the suitcase, he will get tired. The fact is that in this case there actually is motion inside the man's arm and work is being done on the muscle fibers. While the man is holding the suitcase, nerve impulses are continuously reaching the muscles in his arm. When these nerve impulses reach a muscle fiber, the fiber lurches for an instant and then relaxes. At any one time, large numbers of fibers are tightening up while the rest are relaxing. Since we were analyzing the motion of the suitcase held by the man, a situation external to the activity of the muscle fibers inside the man's arm, we concluded correctly that no work was done on the suitcase.

UNITS OF WORK AND ENERGY

The SI unit of energy (and work, since energy is the ability to do work, and thus must have the same units) is the *joule* (J). This unit is named in honor of the English physicist James Prescott Joule (1818–1889) whose work clarified the concepts of work and energy. A joule combines the units of force and distance:

$$1 \text{ J} = 1 \text{ Nm}$$

We can illustrate this unit with two examples. It requires about one joule of work to lift a baseball from the ground to your chest, whereas it takes about ten joules of energy to pick up an average physics textbook from the bottom shelf of a bookcase and stand up to read it.

When dealing with the energies of atoms or electrons, though, the joule is too large a unit. For these purposes, another unit, the electron volt (eV), is used. The conversion factor between electron volts and joules is

$$1 \text{ eV} = 1.602 \times 10^{-19} \text{ J}.$$

A frequently used multiple of the eV is the MeV which equals one million eV.

THE CONCEPT OF ENERGY

The concept of work is very useful in understanding the concept of energy. As we stated at the beginning of the chapter, energy is the capacity to do work; that is, energy allows us to perform tasks, to do work. We can also think of energy as the result of doing work. It is the chemical energy stored in the man's body in Figure 4.4 that enables him to do the work on the car as he pushes it, converting chemical energy into energy of motion of the car and into heat (thermal energy) as the tires rub against the pavement.

This idea of energy as something stored that can do work was called first *vis viva* (Latin for "living force") by the German philosopher Gottfried W. Leibniz (1646–1716), because he thought that only living things could have the capacity to do work. The English scientist Thomas Young (1773–1829) realized that inanimate objects, like the wind, can do work—by moving a windmill or a ship, for example. He proposed the name *energy*, a name he fashioned from Greek words meaning "work within," for this work stored in bodies.

Of the various types of energy listed at the beginning of the chapter, we will only consider mechanical energy for the moment. Later in the book, most of the other kinds of energy will be studied in some detail. An object may have mechanical energy by virtue of its state of motion, its location in space, or its internal structure.

FIGURE 4.4   Energy is the capacity to do work. The chemical energy in the man's body enables him to do work on the car.

PIONEERS OF PHYSICS: JAMES PRESCOTT JOULE

The second son of a wealthy brewer, James Joule had a good early education. As a young man, he was taught by the renowned chemist John Dalton and showed a talent for science. At 19, he did several experiments investigating the nature of electromagnets, which resulted in a published paper.

Born in 1818 near Manchester, England, Joule developed an early interest for the machines in his father's large brewery. This interest made him proficient at designing experiments and building the machines required to run them. He soon developed an almost fanatical zeal for accurate measurements. Such was his dedication that he even took time during his honeymoon to design a special thermometer with which to measure the temperature difference between the top and the bottom of a waterfall that he and his bride visited.

When Joule was 15 his father became ill and retired. Although the young James had to spend time running the brewery, he continued his scientific endeavors. At 22, he calculated the amount of energy produced by an electric current and went on to spend the next 10 years devising experiments to measure energy in every conceivable way.

The initial report of his experiments was met with skepticism and even rejection. The Royal Society did not accept his original paper and Joule was forced to present his results at a public lecture. His report was finally published in the Manchester newspaper at the instigation of his brother, who was the paper's music critic.

Eventually, his work caught the attention of other scientists and Joule gained the recognition he deserved. He was elected to the Royal Society in 1850 and years later became president of the British Association for the Advancement of Science. In 1854 his wife died after only six years of marriage and Joule, deeply distressed, retreated to his work. In 1875, he began to have financial difficulties and Queen Victoria granted him a pension. Toward the end of his life, he became concerned and disturbed about the applications of his work to warfare. He died in 1889, at the age of 71, after a long illness.

ENERGY OF MOTION

As we have seen, work involves forces and motion. An object in motion has the capacity to do work: running water can turn a millstone, a gusty wind sets a windmill in motion and drives a sailing ship, and a truck ramming into the rear of a small car at a traffic light will surely move it some distance!

Thus, an *object in motion* has energy. It is the motion of the object that causes it to contain energy. Still water does not turn a millstone and still air does not drive a sailing ship. We call the energy of an object in motion *kinetic energy*. The word kinetic was first introduced by the English physicist Lord Kelvin in 1856 and comes from a Greek word that means motion. The amount of kinetic energy that an object has depends on its mass and on its speed. Thus, a large truck traveling at the same speed as a small sports car would have more kinetic energy due to its larger mass. Likewise, a runner would have more kinetic energy than a person of similar weight walking along the same path, due to the runner's greater speed.

Kinetic energy—the energy that an object has by virtue of its motion—is thus proportional to the mass of the object and to its speed. It is equal to one-half the product of the mass m and the square of the speed v^2:

$$\text{Kinetic Energy } (KE) = \frac{1}{2}mv^2.$$

ENERGY OF POSITION

A snowball at rest at the top of a hill has no kinetic energy since it is not moving. However, it is *potentially* capable of doing work on a snowman at the bottom of the cliff, if it is set into motion by the boy (Figure 4.5). This type of energy, which we call gravitational potential energy, is due to the object's separation from the earth. It is called *gravitational* because the gravitational force of attraction of the earth does work on the object as it falls toward the ground, and it is called *potential* because energy has been stored for later use. Notice that unless the boy pushes it, the snowball will not reach the ground below where it can harm the snowman. If the snowball is never pushed, it will never become separated from the earth, remaining on the ground at the top of the cliff. When it is pushed, it acquires a separation from the earth equal to the height of the cliff. What we call ground, then, is actually the lowest position the object can reach in a particular situation. This lowest position or ground

FIGURE 4.5 The snowball has gravitational potential energy by virtue of its position with respect to the ground. If pushed off the cliff, its potential energy will allow it to do work on the snowman below.

is the *reference level* from which the position of the object is measured. We are free to choose an arbitrary reference level that better suits our particular situation.

The gravitational potential energy depends also on the object's mass. Thus, a tree falling on a house after a storm does much more damage (more work!) than a walnut falling from a tree standing on the roof of the same house. If an object's height changes, so does its potential energy because the distance to the ground increases. If you lift a box full of books from the ground and place it on a chair, the box acquires potential energy because of its height with respect to the ground. This potential energy comes from the work done by your muscles in lifting the box. If you now decide to place the box on the table, the potential energy of the box increases by an amount proportional to the increase in height. The increase in potential energy results from the additional work that you have to do to lift the box from the chair up to the table.

We can summarize the previous discussion as follows:

> Gravitational potential energy is the energy that an object has by virtue of its separation from the earth's surface. Potential energy is proportional to the mass of the object and to the height above an arbitrary reference level.

It is not difficult to obtain the exact expression for the gravitational potential energy. Consider the boy lifting the baseball of mass *m* in Figure 4.6. The potential energy of the ball before the boy picks it up is zero, if we choose the floor as our reference level. As the boy slowly lifts the ball, the potential energy increases until the ball reaches the table. The work done by the boy in lifting the ball slowly, without acceleration, is equal to the force applied by the boy multiplied by the distance traveled by the ball, which is equal to the height of the table *h*. The magnitude of the

FIGURE 4.6   As the boy slowly lifts the ball, the ball gains gravitational potential energy.

force applied by the boy equals the ball's weight mg. The force is in the same direction as that of the motion. Therefore,

$$W = Fh = mgh.$$

Thus, $W = mgh$ is the work done by the boy on the ball. When the ball is resting on the table, it has no speed, and therefore no kinetic energy. It, however, has stored energy by virtue of being at a height h above the floor; in other words, by being at a distance h above the reference level. If the boy decides to push the ball off the table, it would, of course, fall down and gain kinetic energy, which would have come from the stored energy when it was resting on the table. This stored energy is what we call potential energy. We can write for potential energy, then,

$$PE_{grav} = mgh.$$

There are other kinds of potential energy (Figure 4.7). When we push two magnets together with their north poles facing, there is an increase in *magnetic* potential energy. If you compress a spring by holding it between the palms of your hands and pushing, the *elastic* potential energy of the spring increases.

In all these cases, the potential energy is stored in the entire system of interacting bodies. After we bring the two magnets together, for example, we could hold either one of the two magnets in place and let the other move away. Since either magnet can be released and allowed to move away from the other magnet, making use of the available potential energy, this magnetic potential energy must reside in the system of the two interacting magnets. When we lift a baseball up to a certain height and then release it so that it falls toward the earth, it seems as if the potential energy belongs only to the baseball. However, if we could devise a method to secure the baseball in space with respect to the sun, the earth would "fall" toward the baseball. The gravitational potential energy in this case resides in the earth-baseball system.

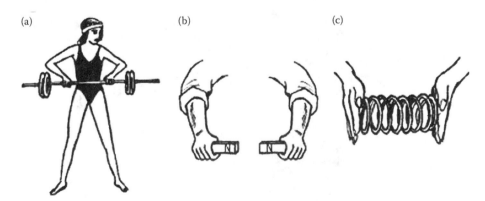

FIGURE 4.7 Different kinds of potential energy. (a) The gravitational potential energy of the weights increases as the weights are lifted. (b) The magnetic potential energy increases as the two magnets are brought closer together with their north poles facing. (c) The elastic potential energy of the spring increases as it is compressed.

ELASTIC POTENTIAL ENERGY

The stretched spring in Figure 4.8 has stored energy. We call this elastic potential energy. If we pull on the spring with a force **F**, the increase in length x is proportional to the stretching force (as long as the spring is not stretched too much). We can write for the force exerted by the person pulling on the spring:

$$F = kx$$

where k is called the *force constant* of the spring. This force law is known as Hooke's law. By Newton's third law, the spring exerts an equal and opposite force on the person pulling, or

$$F = -kx.$$

The applied force increases from 0 when the spring is not stretched, to kx when the spring has been stretched a distance x. Thus the average force is the sum of these two values divided by two, or

$$\bar{F} = \frac{0 + kx}{2} = \frac{1}{2}kx$$

The work done in stretching the spring from the equilibrium position out to a distance x is the product of this average force $\bar{F}$ and the displacement x:

$$W = \bar{F}x = \left(\frac{1}{2}kx\right)x = \frac{1}{2}kx^2.$$

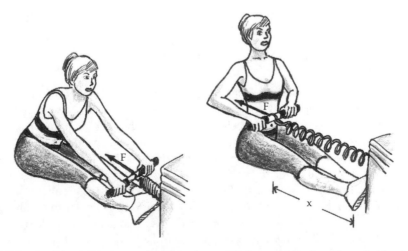

FIGURE 4.8 The spring is stretched from its equilibrium position by a force that is proportional to the displacement x.

This work is converted into elastic potential energy in the spring:

$$PE_{\text{elastic}} = \frac{1}{2}kx^2.$$

THE WORK–ENERGY THEOREM

When you throw a bowling ball you do work on it as you push the ball through some distance. The bowling ball gains speed and its kinetic energy increases. After the ball leaves your hand, it travels along the lane and hits the pins, pushing them down, thereby doing work on them. The kinetic energy that the ball acquired came from the work done on it. The work that the ball does on the pins comes at the expense of some of its kinetic energy: the ball slows down after it hits the pins. In this case, work is being converted into kinetic energy (and some of that kinetic energy is being converted back into work).

We had observed when discussing gravitational energy that the work done by the boy of Figure 4.6 in lifting the ball up to a certain height h was $W = mgh$. The potential energy acquired by the ball when lifted to this height comes from the work that the boy does on it. If the boy lifts the ball without accelerating it, all the work done on the ball is converted into potential energy.

In general, work can be converted into both kinetic and potential energy. This statement is what we call the *work–energy theorem*. When a basketball player shoots a basket, the work done on the ball is changed into an increase in the kinetic energy of the ball, as it accelerates in the player's hands, and into potential energy as the ball gains height.

CONSERVATIVE AND NONCONSERVATIVE FORCES

When we lift an object of mass m, initially at rest on the ground, slowly up to a height h, the work that we do on the object is

$$W = mgh.$$

If the object is a book with a mass of 1 kg and we lift it to a height of 0.5 m, the work done on the book would be $W = 1\ \text{kg} \times 9.8\ \text{m/s}^2 \times 0.5\ \text{m} = 4.9\ \text{J}$. Now, suppose that we pick up the same book from a shelf that is 1 meter above the floor and place it on another shelf 1.5 m high (see Figure 4.9). The change in height would be $h = 1.5\ \text{m} - 1\ \text{m} = 0.5\ \text{m}$ and the work done by us in moving the book is the same 4.9 J. This tells us that the work done against the force of gravity on the book of mass m depends only on the difference in heights as we lift it from 0.5 m to 1 m above the ground; that is, the work depends on the initial and final heights. Motion perpendicular to the direction of the force of gravity contributes nothing to the work done on the book. The work done does not depend on the path through which the book moved between the two points. We could move it following a straight path between the two end points (path 1 in Figure 4.9) or we could move it to the side first, then up, then to the other side so that it ends at the same end point (path 2); we could, in fact, move the book following any trajectory that begins and ends at the same points (as in path 3); the work done will always be the same. A force with the property to produce work that is independent of the path, such as the gravitational force, is called a *conservative force*.

If, on the other hand, you reach for a book that lies on your desk at the other end of where you are and slide it toward you following a straight path (Figure 4.10), the

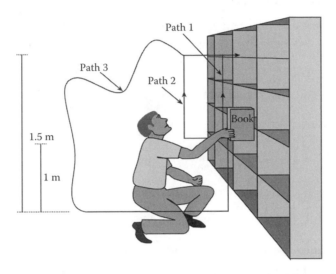

FIGURE 4.9   Three different paths that can be followed to move a book from a shelf 1 m high to a second shelf 1.5 m high. The work done against gravity is the same in all three cases.

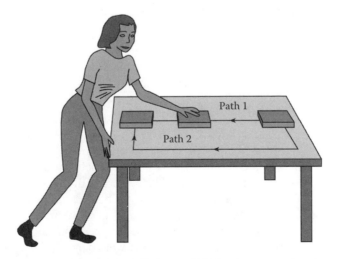

FIGURE 4.10   The work done against the force of friction is greater when the woman slides the book along the longer path 2 than along the straight path 1. The force of friction is a non-conservative force.

work done against the force of friction that exists between the book and the surface of the desk is less than if you decide to slide the book between the same initial and final points following some other path longer than the previous straight path. In this case, the work done against the force of friction does depend on the path taken. We call these forces, like the force of friction, *nonconservative forces*.

5 Conservation of Energy and Momentum

TRANSFORMATION OF ENERGY

As we have learned in the previous chapter, the amount of work done on an object equals the energy transformed from one form to another. The chemical energy in our bodies is transformed into potential energy as we lift a baseball, and this energy is further transformed into kinetic energy as we drop the ball and it falls with increasing speed toward the ground. Energy, then, can be transformed from one form to another. Before the boulder in Figure 5.1 starts rolling down the cliff, its energy is potential. As the boulder falls, its potential energy is continuously converted into kinetic energy. When it reaches the ground below, the boulder has no potential energy.

We can better understand how energy transforms from one type to another with the example of a girl on a swing (Figure 5.2). When the swing is at H it momentarily stops to reverse direction, so its kinetic energy is zero there, but its potential energy is maximum because it has reached its maximum height. At point L the situation is reversed: the swing flies pass that point at its maximum speed (maximum kinetic energy) but its potential energy is zero because it cannot fall any further. This maximum kinetic energy begins to decrease as the swing moves away from the lowest position L. At the same time, the potential energy increases as the swing approaches H' (which is at the same height as H) where it momentarily stops. At this point, the kinetic energy is again zero.

Notice that we have chosen the lowest position of the swing (point L) as the point at which the potential energy is zero. However, we could have chosen the ground as the point of zero potential energy. In this case, the potential energy at point L has a minimum value which is not zero. Since the swing never reaches the ground, the potential energy decreases from its maximum value at H to its minimum at L, while the kinetic energy changes from zero at H to its maximum value at L. It is then the *difference* in the potential energy at the highest point and at the lowest point that is important. As we stated in the previous chapter, the reference level at which the potential energy is zero is arbitrary.

In most cases, it is more convenient to use the lowest position that the object under consideration can take as the reference point; that is the point of zero potential energy.

THE PRINCIPLE OF CONSERVATION OF ENERGY

The example of the girl on the swing described in the previous section is, of course, an ideal situation. We know from experience that the swing will not reach the same height unless the girl "pumps." In the ideal situation, however, assuming that there is

FIGURE 5.1　The potential energy of the boulder will be converted into kinetic energy when the boulder rolls down the cliff.

no friction, the girl on the swing will continue oscillating between points H and H', forever exchanging kinetic energy and potential energy and vice versa. As one form of energy decreases, the other increases with the total amount remaining constant. This constant value is called the total mechanical energy (E) of the system. Clearly, at any point, the sum of the two energies has to be the same. We can express this relation as *kinetic energy + potential energy = total mechanical energy = constant*.

This, the *principle of conservation of mechanical energy*, states that the sum of the kinetic energy (KE) and potential energy (PE) of an isolated system remains constant. An isolated system is one that experiences no external forces and into or out of which there is no flow of energy. In the case of our idealized swing—that is, one which neither interacts with the air nor has friction in the supports—the isolated system consists of the swing, the girl, and the earth.

In the ideal situations described so far, the forces acting on the system are conservative forces. In general, however, both conservative and nonconservative forces (like friction) act on a system. In reality, as we have stated before, when the girl on the swing of the earlier example swings back and forth, she pushes the air aside, losing some of the energy. Even if we replace the girl with a mannequin, enclose the entire swing in a big container, and evacuate the air, the swing will still not reach the same height at every swing. The metal hooks holding the swing rub against the bar, however well lubricated they might be, producing *heat*, which is a form of energy, in the same way that rubbing your hands together when you are cold makes your hands warmer. This heat or *thermal energy* is dissipated into the environment, producing an energy loss. As the girl swings, mechanical energy is continuously transformed

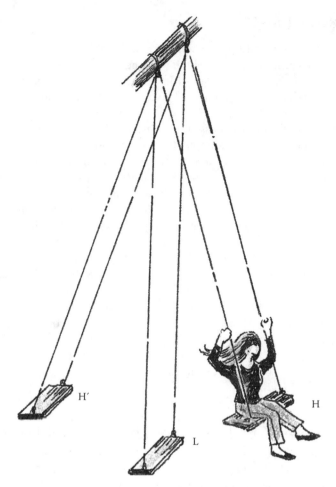

FIGURE 5.2 The girl on the swing sways from a maximum height *H* to a lowest point *L*. The potential energy at *H* is transformed into kinetic energy, which reaches a maximum value at *L*.

into thermal energy; this results in a decrease in height with each swing. This rubbing of parts of the system against each other is what we call *friction*. Likewise, when the child playing in a park slides down a slide some of her kinetic energy is converted into heat or thermal energy. These situations do not violate the principle of conservation of mechanical energy as stated above, however, since thermal energy is not a form of mechanical energy.

Could we include nonconservative forces and still have conservation of energy? If we consider all forms of energy in a system, we can expand the principle of conservation of mechanical energy to a more general *principle of conservation of energy*, which can be stated as follows:

Energy is neither created nor destroyed; it only changes from one kind to another.

THE ENERGY OF MASS

In 1905, Einstein extended the principle of conservation of energy still further to include mass. In a beautiful paper written when he was 26 years old—the fifth scientific paper that he published that year—Einstein deduced that mass and energy were equivalent. The famous formula $E = mc^2$ gives the energy equivalent of a mass m; c is the speed of light.

According to Einstein, the mass of an object is a form of energy. Conversely, energy is a form of mass. For example, the combination of one pound of hydrogen with four pounds of oxygen to form water releases enough energy to run a hair dryer for about ten hours. If we had an extremely precise balance, we would discover that the mass of the water formed is less than the total mass of the oxygen and hydrogen used by about one part per billion. The mass loss is exactly equivalent to the energy released in the process. A more dramatic example—as we shall see in Chapter 23—is the release of energy in nuclear reactions. If the same amount of hydrogen were to be used in a nuclear reaction, we could obtain about ten million times more energy. As in the chemical process, the end product would weigh less than the original material. The difference in mass is converted into energy.

Einstein's extension of the principle of conservation of energy is a profound generalization. In everyday life, however, the limited principle of conservation of energy is sufficient. A flying bird has more energy than a bird standing on the branch of a tree, but the increase in the bird's mass due to its greater energy when flying is so small that it cannot be measured by any experiment.

PIONEERS OF PHYSICS: THE PHYSICISTS' LETTERS

"I have heard rumors that you are on the war path and wanting to upset Conservation of Energy, both microscopically and macroscopically. I will wait and see before expressing an opinion, but I always feel 'there are more things in heaven and earth than are dreamed of in our Philosophy.'" Thus wrote the New Zealand-born physicist Ernest Rutherford in a letter to the great Danish physicist Niels Bohr in November 1929.

Rutherford's letter was referring to a discovery that had been made a few years before concerning the behavior of some subatomic particles. It had been observed that the kinetic energies of the particles emitted in certain radioactive processes were not in accordance with the laws governing the motion of subatomic particles. It appeared as if some energy was not accounted for in spite of the best efforts of the experimental physicists. Bohr, a towering figure in twentieth-century physics (whom we will meet in Chapter 7), began to doubt the validity of the principle of conservation of energy in these processes and the letter from Rutherford (whom we will also meet in Chapter 7) showed his great concern.

In February of that same year the German theoretical physicist Wolfgang Pauli had written in a letter to his friend the physicist Oscar Klein that "with his consideration about a violation of the energy law, Bohr is on a completely wrong track." By December of the following year Pauli would write a letter to the physicists attending a conference in Tübingen that began: "Dear radioactive ladies and gentlemen, I have come upon a desperate way out regarding the [problem of the energy violation]. To wit, the possibility that there could exist ... neutral particles which I shall call neutrons. The [energy violation problem] would then become understandable from the assumption that ... a neutron is emitted along with the [other particles] in such a way that the sum of the energies ... is constant."

"For the time being," Pauli continued in his letter, "I dare not publish anything about this idea and address myself confidentially first to you, dear radioactive ones, with the question how it would be with the experimental proof of such a neutron..." He closes his letter with the following: "I admit that my way out may not seem very probable *a priori* since one would probably have seen the neutrons a long time ago if they exist. But only he who dares wins."

Pauli won. His neutron, rechristened the *neutrino*, for "little neutral one" by the Italian physicist Enrico Fermi, was discovered 25 years later.

Physicists have great faith in the principle of conservation of energy. Whenever a new phenomenon seems to violate this principle, physicists invariably look for some hidden object or particle that could account for the missing energy rather than accept a violation of this principle. In addition to the neutrino, many other particles have been discovered this way.

EFFICIENCY

The principle of conservation of energy tells us that energy can change its form but it never disappears. In any process that involves friction, however, we lose control of some energy; it dissipates into the environment. The moving parts of an automobile engine, for example, require a lubricant to minimize friction. When we fail to add oil to the engine, it overheats and the heat produced by friction dissipates into the environment. Even the most efficient engine cannot regain energy lost through friction.

According to the principle of conservation of energy, if there are any energy losses, the input energy must equal the output work plus the energy losses. Of course, the smaller the energy losses are, the greater the output work becomes. Thus, if we minimize the energy losses, the output work increases, approaching the input energy. Since a real machine always has energy losses, the output work is always less than the input energy. The ratio of the output work to the input energy can give us a way of comparing how efficient different machines are. We define the *efficiency* of a machine by the relationship

$$\text{efficiency} = \frac{\text{work or energy out}}{\text{work or energy in}}$$

TABLE 5.1

Efficiencies of Some Energy Conversion Devices

	Efficiency, ε (%)
Incandescent lamp	4
Fluorescent lamp	20
Solar panel	20
Automobile engine (gasoline)	25
Automobile engine (diesel)	35
Home oil furnace	65
Electric motor	95
Electric generator	99

or, in symbols

$$\varepsilon = \frac{W_{OUT}}{E_{IN}} < 1.$$

ε, the Greek letter epsilon, is the standard symbol for efficiency.

Efficiency is often expressed as a percentage by multiplying it by 100. Table 5.1 shows the efficiency of some energy conversion devices.

POWER

The Great Pyramid at Giza in Egypt was constructed by thousands of Egyptian peasants, who worked for years transporting, preparing, and laying about 2,300,000 blocks of an average weight of 2½ tons. If a similar project were to be undertaken today using the same materials but modern equipment and techniques, the task could be accomplished in a much shorter time but the total work done would be the same, since the same 2,300,000 blocks would have to be lifted to the same heights. What we call *power*, however, has changed by using different methods of construction that allow the project to be completed in less time. Power is the rate at which work is done. If we use P to indicate power, we can write

$$P = \frac{W}{t}$$

and since energy is the ability to do work, power can also be expressed in terms of the energy used per unit time, or

$$P = \frac{E}{t}.$$

It was James Watt (1736–1819), a Scottish engineer, who during his pioneering experiments to improve the steam engine first attempted to describe the power of an engine by comparing it with the power exerted by an average horse. He determined that a horse could lift a 550-pound weight slightly less than four feet in 4 seconds. This led him to assume that an average horse could lift 550 pounds up to a height of one foot in one second. He defined this rate of performing work as 1 *horsepower* (hp). Thus

$$1\text{hp} = 550 \text{ ft lb/s}$$

The SI unit of power is the *watt* (W), named in honor of James Watt. 1 watt is 1 joule per second, that is, 1 W = 1 J/s. This unit is the same unit as we encounter in the description of electrical appliances. A 1500-watt electric heater uses 1500 watts of power; this means that it uses 1500 joules of electrical energy every second. A 1000-watt hair drier uses 1000 joules of electrical energy per second. The conversion factor from horsepower to watts is

$$1\text{hp} = 746 \text{ W}.$$

There is a unit of energy that is derived from the expression of energy in terms of power. From our definition of power, we can write for energy

$$E = Pt.$$

If we operate a 1000-watt electric heater for one hour, the total energy used is 1 kilowatt-hour:

$$E = 1000 \text{ W} \times 1\text{h} = 1\text{kWh}.$$

This unit may be familiar to the reader, as it is used by power companies in their monthly statements to indicate the amount of electrical energy used.

We can calculate the power required to move an object at a constant speed v against a constant force F, as when pushing it up a ramp, by expressing the work done as

$$W = Fd$$

where d is the displacement of the object caused by the applied force F. Since d/t is the speed v, power is then

$$P = \frac{W}{t} = \frac{Fd}{t} = Fv.$$

PHYSICS IN OUR WORLD: AUTOMOBILE EFFICIENCY

Today, a fleet of 800 million vehicles travel the world's roads. These vehicles consume half of the world's oil and produce carbon monoxide and other gases that are harmful to the global environment. Most of the industrialized nations have implemented policies to reduce automobile emissions and conserve energy.

Most energy-saving procedures involve new engine and body designs to make them more efficient. These designs have reduced the average consumption of gasoline in the United States during the last 15 years by about half.

Energy losses occur in the engine itself. Friction among the different engine components converts some of the available energy into heat, which escapes into the environment. Additional energy losses take place in the transmission, where friction removes some of the energy. Finally, friction between the tires and the pavement and the air resistance that appears as the vehicle moves (aerodynamic drag) produce additional losses in efficiency.

Most of the current advances in engine design are in hybrid electric vehicles, which contain both an electric motor and an internal combustion gasoline engine. However, advances in the plain internal combustion engine are also taking place. Manufacturers in recent years have experimented with engine designs that reduce energy losses. Some of these engines involve new light-weight materials and new computerized fuel injection designs. At MIT's Sloan Automobile Laboratory, researchers are developing an engine that uses direct fuel injection instead of the port injection system used in most cars today. Although direct fuel injection was developed earlier, its combination with turbocharging, another preexisting technology, has yielded impressive increases in the performance of internal combustion engines.

When used separately, turbocharging and direct fuel injection yield modest improvements in efficiency. Turbocharging compresses the air just before it enters the combustion chamber, thus increasing the amount of oxygen in the chamber. The additional oxygen burns more fuel in the chamber, providing a boost to the engine. But this boost is limited because too much oxygen in the chamber can cause the fuel to explode out of step with the spark, what is commonly known as engine knock.

Alternative fuels, such as ethanol, which contains less energy per volume than gasoline, prevents knock, but at the expense of fuel efficiency. However, the MIT researchers realized that since ethanol produces an appreciative cooling effect when it vaporizes, it can counterbalance the rise in temperature in the combustion chamber due to turbocharging. They have designed a gasoline engine with an additional direct ethanol injection system that introduces ethanol when turbocharging begins to generate the increase in temperature in the chamber. Increases in efficiency of 25% or more can be achieved with this new system.

Finally, reducing the vehicle's weight with plastics, aluminum, and high-grade steel reduces the rolling resistance of the tires. Aerodynamic styling also improves the appearance of the new automobile designs and at the same time reduces drag.

IMPULSE AND MOMENTUM

You are freewheeling down a small hill on your bicycle and when the road starts going uphill, you decide that you are too tired to pedal, so you coast. As we all know from experience, the bicycle does not stop moving the very moment one stops pedaling. It keeps going, moving against the force of friction that exists between the tires and the pavement and the much greater friction from the resistance of the air. The force that overcame the force of friction is no longer there. However, the bicycle keeps going because it has *momentum*. In fact, on flat ground, if we were to remove all frictional forces, the bicycle would keep moving forever. Newton's first law tells us that this is the case: unless there is a force, an object continues moving in the same direction and with the same speed. As we shall see, the concept of momentum helps us quantify what we already know from Newton's first law.

Let us return to the example with the bicycle. Suppose now that a 50-kg boy is riding his bicycle on flat ground. As soon as the boy reaches a speed of 20 km/h, he stops pedaling and coasts the rest of the way. After he returns, his friend, who weighs 60 kg, decides to see how far she is able to coast using her friend's bicycle. When she reaches the same speed of 20 km/h she also stops pedaling, and she coasts for a longer distance, as we would have been able to predict even without being present at this experiment. The girl, by virtue of her larger mass, has a larger *momentum*.

In a second phase of this experiment, the boy asks his brother, who is a year older but weighs the same 50 kg, to see how far he can coast. His brother is stronger, however, and is able to reach a speed of 30 km/h before he stops pedaling. Again, we would agree that the brother will coast for a longer distance, simply because he is moving with a greater speed before starting to coast. For this second experiment, we can say that the brother has a larger momentum.

Momentum depends on both of these quantities. It is proportional to the product of the mass of the object and the object's speed. If we use the letter p to indicate momentum, we can write,

$$p = mv.$$

The units of momentum are the units of mass times the units of speed.

Momentum is a vector quantity. In vector form, it is proportional to the velocity of the body and is defined as

$$\mathbf{p} = m\mathbf{v}.$$

Since momentum is a vector quantity, an object could be moving with a constant speed and still change its momentum due to the fact that velocity, being a vector, can change direction even if its magnitude (the speed) remains constant. An object moving in circles at a constant rate is one case where the speed is constant but there is a change in momentum.

If we push a lawn mower of mass m across a yard, the lawn mower accelerates. Since momentum is the product of mass and velocity, a changing velocity implies a changing momentum. Thus, if we apply a force to an object for any length of time, the momentum of the object changes. Push a stalled car for a few seconds and the momentum of the car changes by a small amount. Push with the same force for one whole minute, and the car's momentum increases by a greater amount. The product of the applied force and the time during which the force is applied, $F \times t$, is called impulse I, and this impulse is equal to the change in momentum of the body. In symbols:

$$I = Ft = \Delta p.$$

The units of impulse are units of force multiplied by units of time. The SI unit of impulse is the Newton second (N s). According to our previous equation, this is also the unit of momentum.

It is because impulse is equal to the change in momentum that we can see why it is advantageous to bend our knees to absorb the shock when jumping (or falling!) from a certain height. In this particular case the change in momentum is fixed; we arrive at the ground with a momentum that is determined by the height from where we jumped and we must stop. The momentum must change from the value just before hitting the ground to zero. How fast this change takes place determines the magnitude of the force that we must exert with our legs. The longer we take to stop by flexing our knees, the smaller the average force.

PHYSICS IN OUR WORLD: AIR BAGS

Air bag systems that protect automobile drivers and passengers are probably the most effective of the automatic or passive safety devices currently used in automobiles. When a sudden decrease in speed is detected by an electronic sensor an electric contact is closed which causes a small explosive to rapidly release nitrogen gas into the bag. (Nitrogen comprises 78% of the air we breathe.)

Air bags are designed to pop out only in a frontal crash that is equivalent to hitting a wall at 10 to 15 miles per hour (16 to 24 kilometers per hour). At that speed, the force that the wall exerts on the automobile to stop it would cause severe injury or death to the driver in only 1/8 s. Since the air bag pops out in just 1/25 s, the driver hits the cushion of nitrogen gas in the air bag. Since the bag deforms, the driver continues moving into the bag until finally stopped.

Although the change in momentum is the same since the car is stopped completely, the time during which the force acts on the driver is longer. The change in momentum equals the impulse, which in turn is the product of the net force F multiplied by the time t during which the force acts. Since this product Ft is the same with or without the air bag (it equals the change in momentum), a larger value of t implies a smaller value of the force F. This smaller force should not injure the driver.

One second after the impact, the nitrogen escapes through tiny holes in the fabric of the bag and the bag rapidly deflates to avoid suffocating the driver. The powder released from the airbag after deployment is either cornstarch or talcum powder, which is used to keep the bag lubricated while stowed.

CONSERVATION OF MOMENTUM

When we look at the world around us, we notice that all objects that are moving eventually stop, that even the most modern machines cannot run forever. Skylab, the first space laboratory, came down to earth in fragments after circumnavigating the earth more than 31,000 times. Even motion in the heavens is not forever. Stars explode and disappear, together with their possible planetary systems. In a few billion years, the time will come for our own sun to brighten until it swells and engulfs Mercury and Venus, disturbing the motions of the remaining planets.

An object set in motion with respect to an observer will eventually stop owing to its interaction with other objects in the universe. When objects interact, however, there is still a quantity, related to the motion of the objects, which does not change. This *quantity of motion* is the total momentum of the interacting bodies. As an illustration of this, let's consider two identical carts with well-greased wheels moving

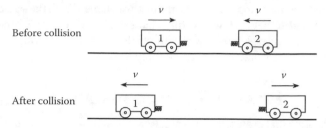

FIGURE 5.3 Two equal carts approach each other with the same speed v. After the collision they recede from each other at the same speed, if we neglect friction.

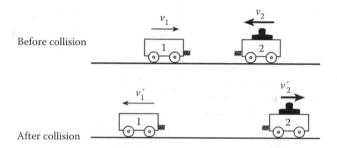

FIGURE 5.4 Adding a weight to cart 2 to increase its mass allows us to see that the velocity is not the conserved quantity. We need to take into account the different masses. It is the total momentum of the two carts that is conserved.

toward each other at the same speed (Figure 5.3). If the cart bumpers were to be fitted with perfectly elastic springs, you would notice that after the collision takes place, the two carts would be moving in opposite directions at the *same* speed. In fact if you were to measure the speeds of each cart before and after the collision with an instrument, you would find that they are almost exactly equal, the friction in the wheels accounting for the small difference. If the friction is reduced, so is the difference in the speeds before and after. Considering the velocities to include also the direction of motion of the carts, we can easily see that the sum of the velocity $\mathbf{v}_1$ of one cart and the velocity $\mathbf{v}_2$ of the other cart does not change with the collision of the carts. We can then say that, neglecting friction, velocity is a *conserved quantity* for this particular case.

In a second part of this experiment, we add weights to one of the carts so as to double its mass and set them in motion so that, again, the two carts approach each other with the same speed (Figure 5.4). In this case, however, the two carts do not maintain their speeds after the collision. Rather, the light cart bounces back with a larger speed than that of the cart carrying the big mass. Certainly, velocity is not a conserved quantity for this more general case. The different masses of the carts prevented the vector sum of the velocities from remaining unchanged. Could momentum be conserved, then? Multiplying the velocity times its mass, we obtain the momentum of each cart, and adding them together, we discover that the *total* momentum of the carts is conserved. Representing the momenta of the two carts after the collision with primes, we have

$$p_1 + p_2 = p'_1 + p'_2$$

or
$$p_{\text{before}} = p_{\text{after}}$$

This is a very important law in physics. It is called the *principle of conservation of momentum*, and can be stated as follows: *If no net external force acts on a system, the total momentum of the system is conserved.* This law is general and universal, which means that it applies to any system of bodies anywhere in the universe on which no external forces are acting. It holds regardless of the type of force that the bodies exert on each other.

The principle of conservation of momentum was clearly stated first by the Dutch physicist Christiaan Huygens, one of the most gifted of Newton's contemporaries. Employing kinematical analyses and the methods of ancient geometers rather than the modern analytical methods of dynamics (at the time known only to Newton), Huygens extended the work of John Wallis and Robert Boyle who, responding to a challenge by the Royal Society to investigate the behavior of colliding bodies, suggested in 1668 that the product *mv* was conserved in collisions.

ELASTIC AND INELASTIC COLLISIONS

You might have seen displayed in novelty shops a small device consisting of five or six shiny steel balls attached by means of two threads to two parallel, horizontal rods (see Figure 5.5). The device is often called *Newton's cradle*. When one of the two

(a)   (b)

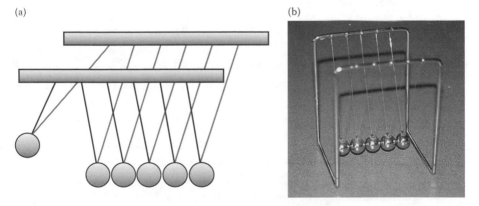

FIGURE 5.5 (a) After the swinging ball collides with the first stationary ball, the momentum of the swinging ball is transmitted to this stationary ball which collides with the next, until finally the last ball swings up to nearly the same height. Conservation of momentum alone does not explain completely the motion of these balls. (b) An actual Newton's cradle.

end balls is raised up to a certain height and allowed to swing down to collide with the next ball, which is at rest, the momentum of the first ball is transmitted through all the other stationary balls to the last one, which swings up to nearly the same height. This motion continues for several swings.

In 1666, Christiaan Huygens saw a demonstration before the recently founded Royal Society of London of a similar experiment, with two balls. For the two following years, the members of the Royal Society argued as to why the struck ball would swing up to nearly the same height while the first ball would stop completely after the collision. According to the principle of conservation of momentum, the momentum of the swinging ball before the collision must equal the momentum of all the moving balls after the collision, without the need to specify how the balls must move.

In 1668, Huygens explained to the Royal Society that there was another conservation principle involved in this process. In addition to momentum, the product mv^2 was also a conserved quantity. In other words, according to Huygens, the sum of the products mv^2 for all the moving objects before the collision must equal the sum of the products mv^2 after the collision. This product (divided by 2) is what we call today *kinetic energy*.

Thus, in addition to momentum, *kinetic energy is also conserved* in this type of collision. Collisions where kinetic energy is also a conserved quantity are called *elastic collisions*. Collisions involving objects that do not lose their shape or heat up in any way are elastic collisions. The perfectly elastic bumpers are restored to their original shapes immediately after the collision, and two "perfectly hard" spheres would collide without any distortion of their shapes. Of course "perfectly elastic" springs and "perfectly hard" balls do not exist, which is why the second ball in the Royal Society demonstration rose to *nearly* but not *exactly* the same height as that from which the first ball was released.

Collisions where the kinetic energy is not conserved are called *inelastic collisions*. The collision of two automobiles is an example of an inelastic collision. All types of collision conserve momentum, however.

CANNONS AND ROCKETS

Let's apply the law of conservation of momentum to the firing of a cannon and a rocket. Consider first the cannon of Figure 5.6, where we have called M the

FIGURE 5.6 After firing, the cannonball moves to the right with a velocity V and the cannon recoils with a smaller velocity v, due to its larger mass. That is why you feel a kick on firing a gun.

mass of the cannon and m the mass of the cannonball. Obviously, as the cannon and cannonball are at rest before firing, the total momentum of our system, cannon plus cannonball, is zero. Conservation of momentum tells us that the total momentum of this system must remain zero after the cannonball is fired. After firing, the cannonball acquires a velocity V and a momentum $p_b = mV$ and the cannon recoils with a momentum $p_c = Mv$. The total momentum after firing is equal to zero, i.e.

$$p_b = p_c = 0.$$

If we call the direction in which the cannonball moves positive, the magnitude of its momentum is mV, whereas the magnitude of the cannonball's momentum, moving in the opposite (negative) direction would be $M(-v)$. We can use our expressions for momentum above to write for the magnitude of the momentum of each of the two components of this system:

$$mV + M(-v) = 0$$

or

$$mV = Mv.$$

This relation tells us how the total momentum remains zero after firing. The large mass of the cannon moves with a small velocity in the opposite direction to the motion of the cannonball, which, because of its smaller mass, moves with a large velocity. The momentum of the cannon and of the cannonball is equal in magnitude and opposite in direction.

Rockets work on the same basic principle. Burning expanding gases, which are produced at high pressures in a combustion chamber, escape at large velocities through a constricted nozzle propelling the rocket in the opposite direction (Figure 5.7). As was the case with the cannon, the magnitude of the momentum of the gas exiting through the rear of the rocket must equal the momentum of the rocket in the opposite direction. In symbols:

$$M_{rocket} \, v_{rocket} = m_{gas} \, V_{gas}.$$

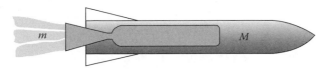

FIGURE 5.7 As the gases are ejected through the nozzle with a momentum mV, the rocket moves in the opposite direction with a momentum Mv.

This expression is actually an approximation since we have neglected the loss in mass due to the gases ejected. It would be a good approximation for a rocket out in space, away from the gravitational pull of the earth, as it undergoes a small change of course, for example. In this case, the loss of mass due to the ejection of gases is very small compared to the mass of the rocket itself.

6 Rotation and the Universal Law of Gravitation

ROTATIONAL MOTION

Our lives are spent moving in circles. The earth spins on its axis once every day and revolves around the sun once a year (Figure 6.1a). The solar system revolves, together with billions of other stars, around the center of the Milky Way Galaxy, which forms part of a small conglomerate of galaxies called the Local Group; this also rotates in space (Figure 6.1b). Up to this point, we have confined our study of motion to motion in one dimension, in which the entire object is displaced from one point to another following a straight path, without changing the orientation in space of the object. This type of motion is called *translational motion*. In this chapter we will extend our study to motion in a circle, which will help us understand Newton's law of universal gravitation.

In studying circular motion we should distinguish between *rotation* and *revolution*. A turntable *rotates* or *spins* around its axis and the earth revolves around the sun. The rotating turntable does not travel anywhere, it merely spins in place. On the other hand, as the earth *revolves* around the sun its spatial location changes while it rotates on its axis. The motion of the turntable is an example of pure *rotational motion*, whereas that of the earth spinning and revolving combines rotational and translational motion.

Consider the motion of a merry-go-round in an amusement park and assume that it is rotating at a constant rate. In Figure 6.2, we have depicted the merry-go-round as seen from above and rotating counter-clockwise. A child riding a horse on the rotating platform at a distance r from the center would describe a circular path. As you have probably experienced, the farther away you are from the center of the merry-go-round, the faster you move. In other words, the speed increases in proportion to your distance from the center. This phenomenon can be dramatically observed in ice skating shows. As the skaters begin forming a spinning line, the ones near the center of rotation barely move, whereas the skaters at the two ends of the line have to work hard to keep up their speeds and maintain the line straight.

As the merry-go-round of our example rotates in relation to the ground, the line joining the center with the place where the child is riding sweeps out an angle θ (the Greek letter theta) in the time t that it takes the child to travel the length s (Figure 6.2). The rate at which this angle increases with respect to time as the merry-go-round rotates is the same regardless of where the child rides. We call this rate of change the *angular velocity* of the merry-go-round. It is standard practice to use

(a)   (b)

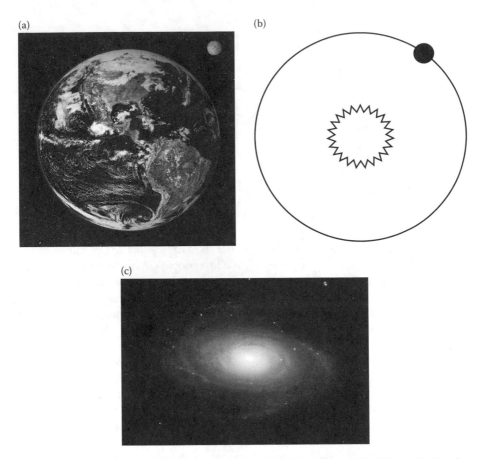

(c)

FIGURE 6.1 (**For Figure 6.1c, see color insert following page 268.**) The earth (a) spins around its axis and revolves around the sun. (Courtesy of NASA.) (b) The Galaxy itself is spinning around its center. (c) This is a photograph of the spiral galaxy M-81, believed to have a similar structure to that of the Milky Way. (Courtesy of NASA, Space Telescope Science Institute.)

Greek symbols to represent angular quantities, and we use a lowercase omega (ω) for angular velocity. If it took 18 seconds for the merry-go-round to complete a turn, we could say that its angular velocity was 20°/s. The earth, on the other hand, rotates around its axis at the rate of 15°/h (360°/24h).

As we have just seen, the speed v at each point on the platform depends on the distance to the center of rotation, whereas we can say that the entire merry-go-round rotates with the same angular velocity ω. Angular velocity, then, is a more useful quantity for rotational motion. The units of angular velocity that are commonly used are radians per second. The conversion from degrees to radians is $360° = 2\pi$ radians. (In this book, however, we will use degrees per unit time for angular velocity.) We can relate the angular velocity of the rotating platform

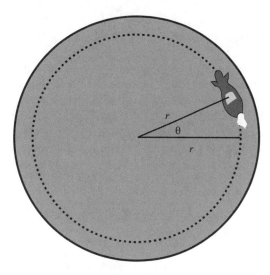

FIGURE 6.2 A merry-go-round rotating counter-clockwise, as seen from above.

to the linear speed of a particle on the platform with the expression $v = r\omega$, with ω in radians.

When the rotating platform completes one revolution, the angle θ becomes equal to 360°. It is customary to call the time taken in completing one revolution the period T of the motion. For one revolution the distance traveled is the length of the circumference or $2\pi r$. The speed can thus be written as

$$v = \frac{2\pi r}{T}.$$

THE FRONTIERS OF PHYSICS: DVD PLAYERS

DVDs and CDs transfer data at different rates, depending on the technology used by the manufacturer. A DVD is made of several polymer layers each containing microscopic bumps that form a spiral track of data some 12 km long. A laser beam focusing on a particular layer decodes the digital information stored in the bumps. A tracking system moves the laser from the center of the disc outward, as the laser reads the first layer, and from the outside toward the center, when the laser is focused on the second layer.

Standard DVDs maintain their angular velocity constant, like hard drives and magneto-optical drives. This technology, appropriately called constant angular velocity (CAV), implies a variable transfer rate, since the linear speed at which the tracks move relative to the head increases with increasing distance from the center.

Blu-ray discs, on the other hand, maintain a constant data flow by keeping the linear speed at which the tracks move past the head constant, regardless of the position of the head. That means that the disc spins faster when the head is near the center and slower when farther away. Thus the angular velocity is not constant; it decreases with increasing radius. This technology is called constant linear velocity (CLV), although it is the linear speed, and not the linear velocity, that remains constant.

TORQUE AND ANGULAR MOMENTUM

Suppose that you need to loosen up a tight bolt and are using a medium size adjustable wrench without success. You know from your past experience that you probably will have better luck if you switch to a larger wrench (Figure 6.3). Although you might apply the same force with either wrench, the larger one will have a longer handle, allowing you to apply the force at a larger distance from the wrench. The same force is more effective in rotating the stuck bolt if you apply it at a larger distance from the bolt.

To push open a heavy metal door in a building, the farther away from the hinges you push the easier it is to open it. If you push from the wrong end you might still be able to open it but, as you know, it will be much more difficult.

From these examples we can see that the ability to rotate an object depends on the applied force *and* the point of application of the force. In Figure 6.3, the wrench rotates around the center of the bolt O, and the force is applied at point P. The distance from the center of rotation O and the point P, where the force is applied, is called the lever arm r. We call *torque*, τ (the Greek letter tau), the quantity that depends on both force and lever arm. Torque is defined as the product of the lever arm and the applied force:

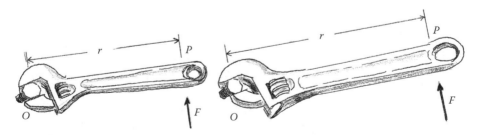

FIGURE 6.3 A larger wrench facilitates loosening up a tight bolt because of its longer handle.

$$\text{torque} = \text{lever arm} \times \text{force.}$$
$$\tau = rF.$$

Thus, to be able to start an object rotating you need to apply a torque to it. The disk in Figure 6.4 will start rotating only if the force is applied to a position other than A, such as B or C. At position A the force will not produce any rotation; at this point the torque is zero because the lever arm is zero.

In Chapter 5 we learned that if the net force on an object is zero, the momentum of the object is constant. For a rotating object, however, torque is what causes the object to change its state of rotation. In other words, if the object is not rotating, a torque is required to set it into rotation; if it is rotating at a constant rate, a torque is needed to change its rate of rotation to a faster or slower rate; and, finally, a torque is needed to stop a rotating object. In this sense, torque is the rotational analog of force, since in linear motion force is what causes an object to change its state of motion. If the net torque acting on an object is zero, we say that the *angular momentum* of the object is constant. Angular momentum is a measure of the rotation of an object; it is the rotational analog of linear momentum, which, as we remember, is a measure of the linear motion of an object.

Since angular momentum is the rotational analog of momentum, we can express it in terms of analog quantities. We expressed momentum as the product of mass times velocity or $p = mv$. Mass is the measure of the inertia of a body, the resistance to a change in the state of motion of the body. Angular momentum can be expressed as the product of the rotational inertia of the body and the angular velocity. Rotational inertia or, more commonly, *moment of inertia I*, measures the resistance of a body to a change in its state of rotation; that is, the difficulty in starting or stopping a rotating

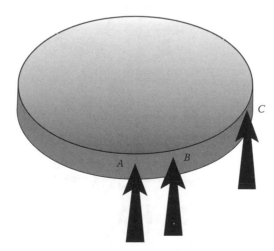

FIGURE 6.4 A force is applied at different distances from the rotating center of a disk. Any position other than A (B and C for example) will result in rotation.

object or changing its present rate of rotation. The moment of inertia of a body will in general change if its axis of rotation is changed. It is easier to spin a baton around its long axis than around an axis perpendicular to its center because in the latter situation most of the mass of the baton is not as close to the axis of rotation (Figure 6.5). The moment of inertia of a body changes if the mass distribution with respect of the axis of rotation changes, as when an ice skater spinning with her arms extended, draws them closer to her body (Figure 6.6). With her arms extended, the moment of inertia of the skater is larger because more of her mass is located farther away from the axis of rotation.

FIGURE 6.5 Spinning a baton around its long axis is easier than around an axis perpendicular to its axis because in the latter situation most of the mass is concentrated farther away from the center of rotation.

FIGURE 6.6   When the spinning skater draws in her arms, the moment of inertia of her body decreases and the angular velocity increases, keeping the angular momentum constant.

Using the letter L for angular momentum we can express it as:

$$L = I\omega.$$

We can think of angular momentum as the tendency of a rotating object to keep rotating because of its inertia.

The angular momentum of an object, as we have mentioned, is conserved or remains constant if the net external torque on the object is zero. This is a very important statement. It is called the *law of conservation of angular momentum*, and joins the other conservation laws that we have learned so far; the law of conservation of energy and the law of conservation of momentum. We can see the results of angular momentum conservation in an ice skater spinning on her toes (Figure 6.6). If she starts spinning with her arms extended and then draws them closer to her torso, the moment of inertia of her body decreases. Since friction is very small in this situation, the external torque on the skater is zero and angular momentum is conserved. Since angular momentum is the product of the moment of inertia and the angular velocity, a decrease in the moment of inertia of the skater must be compensated for by an increase in her angular velocity, which we observe in her faster spinning.

PHYSICS IN OUR WORLD: TWISTING CATS

One of the most spectacular maneuvers that a domestic animal is capable of is done by a cat. As many children know, when the cat is dropped upside down from a height of about 1 m, it is able to twist itself into the right position and land on its paws, unharmed. Not only is this feat dazzling, it also seems to violate the law of conservation of angular momentum. A more careful look at what the cat does in about one-third of a second helps us realize it is actually the conservation of angular momentum that allows the cat to turn.

Since the cat is dropped with no rotation to its body, the law of conservation of angular momentum tells us that, unless external torque was to act on the cat, it should fall without any rotation. Only two forces act on the cat as it falls: gravity, and friction with the air. The gravitational force acts on the entire body at once and provides no torque. As Newton discovered, the gravitational force of attraction of the earth upon any body can be considered as if it were acting on the center of mass of the body, and thus provides no torque. Although the frictional force due to the air does provide a torque, it is too small to cause the twist. How can the cat spin once released, then?

A careful look at the strobe photographs provides the clue. The second photograph shows the cat with its front paws brought in, closer to its body and its rear paws extended. This position has the effect of reducing the moment of inertia of the front of its body and increasing the moment of inertia of the lower part of the body. Since the body is bent in the middle, the front and rear parts of the body rotate about different axes. This second photograph also shows that the cat has twisted its upper body toward us, a 90° rotation, while twisting its lower body away from us. Since the lower body has a greater moment of inertia, a smaller rotation (in this case of about 10°) is equivalent to the larger rotation of the upper body. Since the two rotations are in opposite directions, the angular momentum of the front and rear cancel giving a zero net angular momentum.

In the third photograph, the situation is reversed. The front legs are extended, increasing the moment of inertia of the front, while the hind legs are pulled in, decreasing the moment of inertia of the rear. The rear end is now twisted by a large angle, which causes a small twist in the opposite direction in the front. The last three photographs show the cat, with zero angular momentum, on its way to a safe landing.

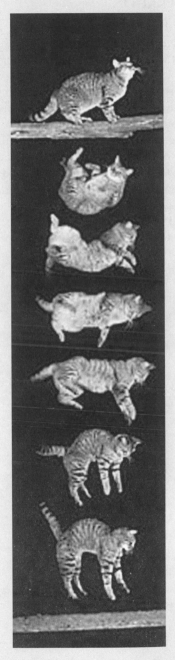

(From © Gerard Lacz/Natural History Photographic Agency. With permission.)

CENTRIPETAL ACCELERATION

Isaac Newton was one of the first scientists to recognize the importance of circular motion. Newton was able to show that an object moving in circles needs an unbalanced force to maintain this circular motion. This unbalanced force produces an acceleration on the object which we call *centripetal acceleration.*

If we twirl a ball on a string, the ball will keep moving in a circle for as long as we keep the tension on the string. If we let the string loose or the string snaps, the ball will go off in a straight line and will continue to move in the direction the velocity vector had at the moment the string was cut (neglecting gravity). The string then exerts a force on the ball which acts toward the center of the circle and prevents the ball from moving along a straight line. This force is called *centripetal force,* literally "the force that seeks the center," from the Latin *centripetus.*

The centripetal force causes the ball to accelerate toward the center of the circle. If we increase the speed of the twirling ball, we feel a greater pull on our hand. On the other hand, if we keep the ball moving at the same speed but increase the length of the string, the pull on our hand decreases. We conclude that the centripetal acceleration increases with increasing speed (it actually increases with the *square* of the speed) and decreases with increasing radius. Therefore, an object moving with a constant speed v in a circular path of radius r (Figure 6.7) has an acceleration directed toward the center of the circle called centripetal acceleration, a_c, which is directly proportional to the square of the speed and inversely proportional to the radius:

$$a_c = \frac{v^2}{r}.$$

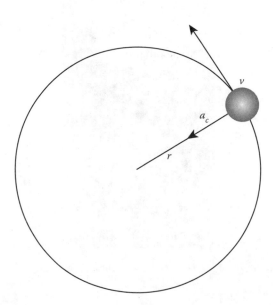

FIGURE 6.7   The centripetal acceleration of an object moving in a circular path of radius r with a constant speed v is $a_c = v^2/r$.

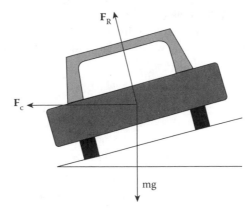

FIGURE 6.8   In a banked race track, the horizontal component of the reaction force $\mathbf{F}_R$ between track and car, which is perpendicular to the track, is equal to the centripetal force $\mathbf{F}_c$.

The centripetal force causing the ball to accelerate toward the center of the circle is given by Newton's second law. If the mass of the ball is m, the centripetal force is $F_c = ma_c$, or

$$F_c = \frac{mv^2}{r}.$$

This equation gives the force required to maintain an object moving about a fixed center with a constant speed. For example, when a car takes a curve, the friction between the pavement and the tires provides this force. If the pavement is wet or is covered with snow or ice, the friction force might not be large enough and the car is unable to negotiate the curve. On the other hand, even on dry pavement, if the speed of the car is too fast, the centripetal force required to keep the car moving in a circular path, being proportional to the square of the speed, could become too large and the car might skid. In older race tracks, where race cars take curves at high speeds, the tracks are sometimes banked steeply so that the horizontal component of the force that the track exerts on the car ($\mathbf{F}_R$), perpendicular to the banked track, is equal to the centripetal force (Figure 6.8).

SATELLITES

On July 26, 1963, the United States launched Syncom 2, the first communications satellite in geosynchronous orbit (i.e., it stays above the same spot on the earth) at an average height of 35,725 km. Syncom 2 completes an orbit in the same time that the earth completes a revolution, and because it revolves around the earth at the same rate as the earth rotates, the satellite moves with uniform circular motion. With this information we can calculate the orbital speed of this satellite by dividing the distance traveled by the time taken. The distance traveled in the 24-hour time interval or period is the length of the circumference, $2\pi r$, with r being the radius of the orbit, which is the distance between the center of the earth and the satellite; that is, r = (radius of the earth + height of satellite) = 6367 km + 35,725 km = 42,092 km, since the radius of the earth is known to be equal to 6367 km. We can calculate the linear speed of Syncom 2 as follows:

$$v = \frac{2\pi r}{T} = \frac{2 \times 3.14 \times (42,092) \times 10^3)}{24 \times 60 \times 60} \, \text{m/s}$$

which gives us a linear speed of 3,061 m/s or 11,020 km/h.

As we have learned, a centripetal force must exist that causes the satellite to orbit around the earth in an almost circular orbit. The centripetal acceleration produced by this force is $a_c = v^2 / r = (3,061 \text{ m/s})^2/(42,092 \times 10^3 \text{ m}) = (0.22 \text{ m/s})^2$. What is the source of this centripetal force? It is again due to the genius of Isaac Newton that we are able to understand that the force that keeps a satellite orbiting around the earth is also the force that keeps the moon going around the earth, the earth around the sun, and all the planets of the solar system moving around the sun; the force that we call *gravity*. The discovery of the universality of this force, one of the four forces known to us today, ranks as one of the major achievements of the human mind. We shall occupy ourselves with the discovery of this force in the following sections.

ORIGINS OF OUR VIEW OF THE UNIVERSE

To place Newton's discovery of the law of gravitation into historical perspective, we must briefly go back to the beginning, namely to the ancient Babylonians; in particular, the people of the Second Babylonian Empire, who lived in Mesopotamia, the region occupied today by Iraq. By 3000 BC the Babylonians had begun a systematic study of the sky out of their need to know the best harvesting times, since theirs was an agrarian culture. Modern astronomy can trace its roots back to their early discoveries. The Babylonians made deities out of the sun, the moon, and the five visible planets—Mercury, Venus, Mars, Jupiter, and Saturn; and this drove them to follow closely their motions across the sky. Their knowledge of the paths of the sun and moon was used in setting up a calendar. They also observed that the planets, unlike the sun and moon, did not follow simple paths across the sky, but would stop their eastward motion now and again, retracing part of their path and then stopping once more, before resuming their eastward motion.

Some of the knowledge acquired by the Babylonians reached the Greeks. From 600 BC to 400 AD, ancient science reached its highest peak with the Greek culture. The Greeks, unlike their predecessors, attempted to find rational explanations to the observed phenomena rather than accepting them as works of the gods. Thales of Miletus, born in 640 BC, one of the first known Greek thinkers, is unanimously acclaimed as the founder of Greek philosophy. Thales attempted to interpret the changing world in terms of physical processes.

Pythagoras, born in Samos in 560 BC, introduced the idea that the earth was spherical and held that the earth was at the center of the universe. This view was cast into a more complete theory by Aristotle, who thought that the earth was immovable and fixed at the center of the whirling heavens. Although this *geocentric* or earth-centered model, as it came to be known, was later expanded into an extremely complicated system by Ptolemy and accepted by nearly everybody for 18 centuries, certain thinkers, notably Aristarchus of Samos, held opposing views. Aristarchus, who lived between 310 and 230 BC, held that the sun was fixed at the center of the universe, and that the earth revolved around the sun in a circular orbit. He also held that the earth rotated on its axis as it revolved and that this axis was inclined with respect to the plane of the orbit.

There was one major obstacle that prevented the acceptance of Aristarchus' model: the earth seemed motionless. How could it rotate around the sun if no motion was detected? Moreover, if the earth rotated around the sun, there should have been an apparent shift in the position of the stars as the earth moves (Figure 6.9). This apparent shift in the position of the stars or parallax was never observed and the sun-centered or heliocentric model of Aristarchus was abandoned. We know today that, because the stars are so remote, none has a parallax that can be observed with the unaided eye. The first observation of the parallax of a star was made in 1838 by the German astronomer Friedrich Bessel, with the aid of a telescope.

The geocentric model developed by Claudius Ptolemy in the second century AD placed the earth in the middle of the rotating heavens. To explain the seemingly complicated motion of the planets in the sky that results, as we know today, from the combination of their own motions and the earth's motion around the sun (Figure 6.10a), Ptolemy had the planets moving in small circles call *epicycles* which in turn would rotate around the earth, following circular orbits called *deferents*. The combination of the motion of both these circles produced a cycloid pattern of the planet's motions. However, the center of the great circle, the deferent, was not at the earth's center. Instead, it was situated a short distance from it, at a different point for each planet. Moreover, the epicycle moved with a uniform rate not around the center of the large circle or deferent but around another point in space called the *equant* (Figure 6.10b). As more and more data on the planetary orbits were obtained, it became necessary to add

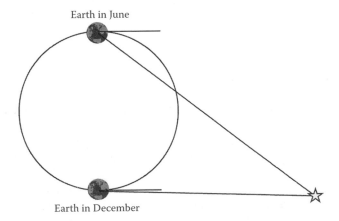

Earth in June

Earth in December

FIGURE 6.9   Apparent shift in the position of a star as the earth rotates around the sun.

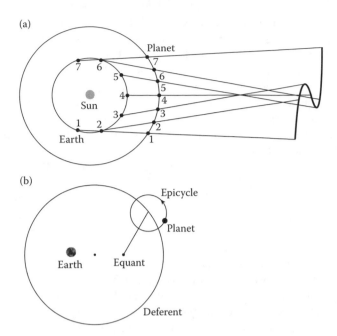

FIGURE 6.10 (a) Actual positions of the earth and another planet as both rotate around the sun. On the right of the diagram we see the apparent path of the planet in the sky as seen from the moving earth. (b) Ptolemy's system of the world.

more epicycles, giving epicycles moving on epicycles, with the planet moving on the last and smallest one. Eventually Ptolemy had a system of 40 epicycles that reproduced with fairly good accuracy the observations of his day. His system of the world was published in *The Mathematical Collection*, a treatise of 13 volumes that was passed on to the Arabs after the destruction of the famous library of Alexandria, and became known as *al Magiste* (The Greatest). The book was introduced in Europe by the Arabs where, known as the *Almagest*, it was famous for over a thousand years.

KEPLER'S LAWS OF PLANETARY MOTION

From Aristotle and Ptolemy until Copernicus, thirteen centuries later, no advance was made in man's understanding of motion in the heavens. Mikolaj Kopernigk (or Nicolaus Copernicus, his Latin name), born on February 19, 1473, in the little town of Torun in Poland, was the revolutionary who brought change to ideas that were firmly rooted in people's minds for more than a thousand years. He attended first the University of Cracow and later the University of Bologna in Italy where he became interested in astronomy.

The intellectual atmosphere that existed then in Italy facilitated the criticism of established ideas. Not only was the Ptolemaic system of the world cumbersome and inelegant, but recent astronomical data no longer supported the model regardless of the number of epicycles used. Copernicus especially disliked the idea of the equant, which had been introduced by Ptolemy in an effort to keep uniform circular motion in the description of the heavens. Copernicus realized that the motion was uniform about the equant and circular about the center of the deferent, two completely different points. It occurred to Copernicus that everything could be much simpler if the sun were at the center of the universe rather than the earth. In his model the earth and the five known planets—Mercury, Venus, Mars, Jupiter, and Saturn—would move around the sun with uniform circular motion.

The heliocentric model of Copernicus was considered heretical because it removed the earth from the center of Creation and for this reason Copernicus delayed publication of his theory until very late in his life. At the urging of some of his friends, Copernicus finally authorized the publication of his theory in a book. *De Revolutionibus* (*On Revolutions*) was published in 1543, a few weeks before Copernicus died at age 70.

The revolution that Copernicus started was not completely accepted until the early 1600s when Johannes Kepler, who had joined the superb observatory of the Danish astronomer Tycho Brahe as an assistant at the age of 30, made use of very precise data on the planet Mars to prove that the orbit of this planet was not a circle but an ellipse. Circular orbits, which were the basis for all the epicycles and deferents in the pre-Copernican theories, were not after all the actual paths of the planets.

Kepler was born at Weil der Stadt, in the wine country of Württemberg in southern Germany, on December 25, 1571, and attended the University of Tübingen where he graduated in 1588, having been converted to the Copernican theory by one of his professors. When he joined Tycho's observatory, Kepler had already published a book, *Mysterium Cosmographicum* (*Cosmic Mystery*), where he proposed, unsuccessfully, that there were six planets because there are only five perfect solids.* It was in search of better data to perfect his theory that he went to work in Tycho's observatory.

It took Kepler six years and thousands of pages of calculations to wrest the secret of the orbit of Mars. During his "battle with Mars," as he called his work, Kepler realized that the velocity of the planet should be related to its distance from the sun because it is the sun that provides the driving force. Since the orbit was not circular but elliptical, the distance from the sun varied. An ellipse, we might recall, is a very specific curve that can be obtained by placing two tacks some distance apart and loosely threading a loop of string through them. By moving a pencil in such a way as to keep the string taut we can draw an ellipse (Figure 6.11). The position of each tack is called a focus. Since the string is not elastic, the sum of the distances from any point on the ellipse to the two foci is always the same. Kepler placed the sun at a focus of the ellipse. His calculations showed him that a planet moved faster when it was nearer the sun and more slowly when it was farther away from the sun.

Kepler set out to calculate just how the speed of the planet changes as it orbits the sun. After many tedious and long calculations he found that the areas swept out in

---

* There are only five regular solids: the *triangular pyramid* (tetrahedron), whose faces are four equilateral triangles; the *cube*, which has six squares as faces; the *octahedron*, with eight equilateral triangles; the *dodecahedron*, with 12 regular pentagons; and the *icosahedron*, with twenty equilateral triangles as faces. Any regular solid may be inscribed in (have its vertices on) or circumscribed about (have its faces tangent to) a sphere.

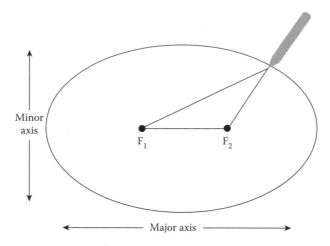

FIGURE 6.11 An ellipse.

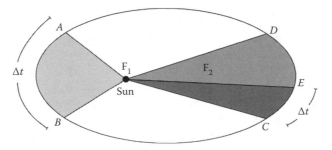

FIGURE 6.12 Kepler's second law: The law of areas. The planet moves from A to B in the same time interval Δt that it takes for it to move from C to E. The area swept by the line between the sun and the planet in moving from A to B is the same as the area swept out in moving from C to E.

equal times by the line joining the sun and the planet were equal. In uniform circular motion, equal angles are covered in equal times, so that it takes twice as long to go around one half of the circumference as it does to go any one quarter. For elliptical orbits, when the planet is close to the sun the area swept out by the line from the sun to the planet as it moves from A to B (Figure 6.12) is *smaller* than the corresponding area formed when the planet moves the same length, from C to D, at a farther distance from the sun.

Kepler found that since the planet moves at a slower speed when it is farther from the sun, the length of the orbit covered in a given time was smaller when the planet was farther from the sun. If it took a certain time Δt to move from A to B, the planet would move from C to E in the same time Δt. However, the area swept out in moving from A to B was the same as the area swept out in moving from C to E. Kepler published his two discoveries in a second book that he entitled *Astronomia Nova* (*New Astronomy*) in 1609. These two discoveries, known today as Kepler's first and second laws, lie at the foundation of modern astronomy.

Kepler did not rest on his discoveries of these laws. Convinced that God the Creator was also a musician, he set out to discover a relationship between the distance of a planet to the sun and its velocity that was similar to the musical ratios. After many efforts, he was forced to admit that "God the Creator did not wish to introduce harmonic proportions into the durations of the planetary years." His effort was not in vain, however, for Kepler was able to find a mathematical relationship between the planets' distances from the sun and their periods of revolution. He found that "the periodic times of any two planets are in the sesquilateral ratio to their mean distances," that is, the squares of the periods of any two planets (the time taken for a complete revolution around the sun) are proportional to the cubes of their mean distances from the sun. The mean distance is also the semi-major axis of the ellipse. If T is the period and r the mean distance or semi-major axis, Kepler's third law is

$$\frac{T_1^2}{T_2^2} = \frac{r_1^3}{r_2^3}.$$

The indices 1 and 2 indicate the period and average distance for each one of the two planets. This remarkable discovery, known today as Kepler's third law or the *harmonic law*, was contained in his monumental work *Harmonici Mundi* (*Harmonies of the World*), published in 1619.

We can state Kepler's three laws as follows:

1. Law of orbits: Each planet moves around the sun in an elliptical orbit, with the sun at one focus.
2. Law of areas: A planet moves around the sun at a rate such that the line from the sun to the planet sweeps out equal areas in equal intervals of time.
3. Harmonic law or law of periods: The squares of the periods of any two planets are proportional to the cubes of their average distances from the sun.

NEWTON'S LAW OF UNIVERSAL GRAVITATION

Kepler had deduced that a force or *anima motrix* was needed to keep the planets orbiting around the sun and that this force came from the sun. Galileo, who was a contemporary of Kepler, was trying to understand motion and had arrived at the conclusion that an object moving at a constant speed along a straight line will maintain that state of motion when there is no net force acting on the object. This idea was further developed by Newton into what became his first law of motion, the law of inertia, as we saw in Chapter 3. In his second law, Newton affirms that the only way to change the motion of an object is to apply a net force. As we learned earlier, to keep an object moving in circles we need to apply an unbalanced force to the object. This force, which we called the centripetal force, acts toward the center of the circle.

It was clear to Newton that to keep a planet orbiting around the sun, a force, directed exactly toward the sun, was required. What was the origin of this force? Newton hypothesized that there is a universal force of attraction between all bodies everywhere in the universe. By analyzing Kepler's third law, Newton was able to show that this force was inversely proportional to the square of the distance between the bodies. This was one of the problems that Newton considered while he was at his mother's farm in Woolsthorpe during the great plague years. It is not known whether or not Newton solved the entire problem then. Newton's biographer and contemporary, William Stuckey, who drew upon conversations with Newton and interviews with Newton's friends, writes that on April 15, 1726, "after dinner, the weather being warm, we went into the garden and drank tea, under the shade of some apple trees, only he and myself. Amidst other discourse, he told me, he was just in the same situation, as when formerly, the notion of gravitation came into his mind. It was occasion'd by the fall of an apple, as he sat in a contemplative mood."

What is clear is that Newton did not publish anything on what ranks as one of the greatest discoveries in the history of science for almost 20 years. And then only because two scientists of great renown, Edmond Halley and Robert Hooke, had also arrived at the mathematical form for the force of gravity, but had not been able to prove it. By January 1684, Halley had concluded that the force keeping the planets on their orbits "decreased in the proportion of the squares of the distances reciprocally." When Halley told Hooke at the Royal Society of his conclusion, Hooke boasted that he had arrived at exactly the same conclusion and had a mathematical proof of it but did not—and most probably could not—produce it. Halley then decided to consult with his dear friend Newton. Abraham de Moivre, a member of the Royal Society and friend of Newton and Halley, wrote of that visit:

> After they had been some time together, the Dr. [Halley] asked him what he thought the curve would be that would be described by the planets supposing the force of attraction towards the Sun to be reciprocal to the square of their distance from it. Sir Isaac replied immediately that it would be an ellipsis. The Doctor, struck with joy and amazement, asked him how he knew it. Why, saith he, I have calculated it.

What was Newton's method to calculate the force of gravity? It is not known in detail the technique that Newton used in developing his theory. However, in a book entitled A *View of Sir Isaac Newton's Philosophy* published in London in 1728, Henry Pemberton relates this account:

> As he sat alone in a garden, he fell into a speculation on the power of gravity: that as this power is not found sensibly diminished at the remotest distance from the center of the Earth, to which we can rise, neither at the tops of the loftiest buildings, nor even on the summits of the highest mountains; it appeared to him reasonable to conclude, that this power must extend much farther than was usually thought; why not as high as the Moon, said he to himself? and if so, her motion must be influenced by it; perhaps she is retained in her orbit thereby. However, though the power of gravity is not sensibly weakened in the little change of distance, at which we can place our selves from the center of the Earth; yet it is very possible, that so high as the Moon this power may differ much in strength from what it is here. To make an estimate, what might be the degree of this diminution, he considered with himself, that if the Moon be retained in her orbit by the force of gravity, no doubt the primary planets are carried round the Sun by the like power. And by comparing the periods of the several planets with their distances from

the Sun, he found, that if any power like gravity held them in their courses, its strength must decrease in the duplicate proportion of the increase of distance.

Newton himself mentions in a memorandum of about 1714 that he "began to think of gravity extending to the orb of the moon, and having found out how to estimate the force which [a] globe revolving within a sphere presses the surface of the sphere, from Kepler's rule of the periodical times of the planets... I deduced that the forces which keep the planets in their orbs must [be] reciprocally as the squares of their distances from the centers..."

We can use these accounts to reconstruct Newton's technique. Consider a planet of mass m revolving around the sun (Figure 6.13). If the planet moves with a speed v and orbits the sun in a near circular orbit of radius r, the centripetal force keeping the planet in its orbit is

$$F_c = \frac{mv^2}{r}.$$

If the period of the planet is T, then the speed can be written as the distance traveled in one period, which is the length of the circumference or $2\pi r$, divided by the period, or $v = 2\pi r/T$. Since we need the square of the speed in our expression for the centripetal force, that is, $v^2 = 4\pi^2 r^2/T^2$, we obtain after replacing v^2 in our expression for F_c:

$$F_c = \frac{mv^2}{r} = \frac{m(4\pi^2 r^2)}{r(T^2)}.$$

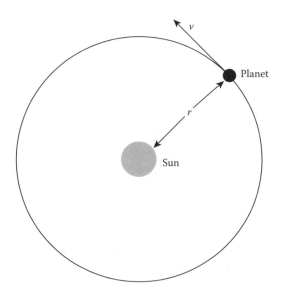

FIGURE 6.13   A planet revolving around the sun in an orbit of radius r.

Multiplying numerator and denominator by r and rearranging, we write

$$F_c = \frac{4\pi^2 m}{r^2} \times \left(\frac{r^3}{T^2} \right).$$

We notice that the expression in parenthesis, r^3/T^2, is a relationship between the square of the period and the cube of the distance between the planet and the sun. Kepler's third law tells us that this ratio is constant. If we call this ratio K, the centripetal force is

$$F_c = \frac{4\pi^2 m}{r^2} \times (K)$$

and if we multiply both numerator and denominator by the mass of the sun, it becomes

$$F_c = \left(\frac{4\pi^2 K}{M_S} \right) \frac{M_S m}{R^2}.$$

It is here where the genius of Newton made the incredible leap. Although this derivation was done for the case of a planet of mass m orbiting around the sun, Newton said that this same expression would apply to the force between the earth and the moon, between the earth and an apple falling from a tree, and between any two objects in the universe, by considering the expression in parenthesis to be a *universal* constant. This is Newton's law of universal gravitation. Any two objects of mass M and m, separated by a distance r, *anywhere in the universe*, will attract each other with a force given by

$$F = G\frac{Mm}{r^2}.$$

The value for the universal constant G was obtained in 1798 by Henry Cavendish by measuring the gravitational attraction between two pairs of lead spheres. The currently accepted value in SI units is

$$G = 6.67 \times 10^{11} \text{ N m}^2/\text{kg}^2.$$

Newton's law of universal gravitation means not only that the earth pulls on the moon with a gravitational force that keeps it in orbit or that the sun exerts a force of attraction on the earth and all the other planets in the solar system, but that this force is the same force that is responsible for making an apple fall from a tree. This universal law, which explains how the universe works, says that you pull the sun towards you with a force that can be calculated. And when we think about this, we begin to understand the scope of the mind of Newton. Since, as the distinguished historian of science I. Bernard Cohen writes, "[t]here is no mathematics—whether algebra, geometry, or the calculus—to justify this bold step. One can say of it only that it is one of those triumphs that humble ordinary men in the presence of genius."

Newton actually tested his great discovery. He realized that the moon is continuously *falling* toward the earth. We can understand this "falling" of the moon if we recognize that if we were to turn off the gravitational force, by Newton's first law, the moon would immediately go off in a straight line along the tangent. It is the gravitational force of attraction that makes it fall from the straight line into its circular orbit (Figure 6.14). Newton knew that the distance from the center of the earth to the center of the moon was about 60 times the radius of the earth. Since his gravitational law says that the force varies with the distance as $1/r^2$, the acceleration due to the gravitational attraction should be $1/60^2$ the acceleration due to gravity on the surface of the earth which in Newton's days was already known to be 9.80 m/s^2. According to this, the moon accelerates toward the earth at the rate of $(1/3600) \times 9.80$ m/s$^2 = 2.72$ mm/s^2.

To check his result, Newton calculated the centripetal acceleration of the moon knowing that its period was 27.32 days. As we have seen, the centripetal acceleration can be written as

$$a_c = \frac{v^2}{r} = \frac{4\pi^2 r}{T^2}.$$

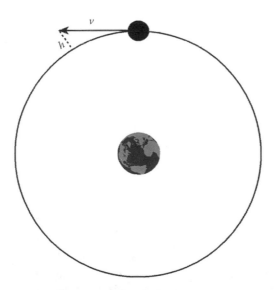

FIGURE 6.14 Acceleration of the moon toward the earth. By Newton's first law, if the gravitational force of attraction between the earth and the moon were to be eliminated, the moon would go off along a straight line with a speed v. The gravitational force pulls the moon toward the earth, making it "fall" a distance h into the circular orbit.

Newton obtained for this acceleration the value of 2.74 mm/s[2], in excellent agreement with the value predicted by his theory.

Some historians believe that Newton actually carried out a calculation like this during his *annus mirabilis* (year of wonders), as a young man of 23, although, due to incorrect data on the distance to the moon, his first attempt was off by a large amount. Only when, six years later, new measurements of the size of the earth yielded a more accurate value for the distance to the moon, could Newton make the calculation again with the agreement that we have just seen.

Newton's universal law of gravitation allows us to define the *weight* of an object more precisely than before as *the net gravitational force exerted on the object by all bodies in the universe*. Near the surface of the earth, the gravitational force of attraction of the earth is much larger than that of all other bodies in the universe and we can neglect their effect on the object; the weight of an object near the surface of the earth is then equal to the gravitational force of attraction of the earth.

The weight of an object of mass m near the surface of the earth is then

$$w = F_g = G\frac{M_E m}{r_E^2}.$$

As we learned in Chapter 3, Galileo discovered that the weight of an object is proportional to its mass:

$$w = mg$$

where g is the acceleration due to gravity. Therefore, from the last two expressions, we obtain

$$g = G\frac{M_E}{r_E^2}$$

where we can see that the acceleration due to gravity is the *same* for all bodies at the surface of the earth.

THE FRONTIERS OF PHYSICS: MEASURING THE DISTANCE TO THE MOON

To check his universal law, Newton estimated that the distance to the moon was about 60 times the radius of the earth. The method to calculate this distance was actually developed by the ancient Greeks. Around 280 BC, the Greek astronomer Aristarchus developed a procedure to measure the relative distances to the sun and the moon. He realized that during the moon's first and last quarter phases, the sun, the moon, and the earth formed a right triangle. He then carefully measured the angle between the moon and the sun, obtaining a value of 87°. With this information he estimated that the sun was about 20 times farther away than the moon (Figure 6.15).

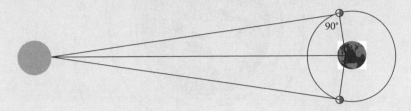

FIGURE 6.15   Aristarchus' method to measure the distance to the moon.

Aristarchus also obtained estimates for the relative sizes of the earth, the moon, and the sun. He measured the time it takes the moon to move through the earth's shadow during a lunar eclipse and from that value he calculated that the earth's diameter was three times greater than the moon's diameter. To calculate the sun's diameter, he used the fact that the apparent sizes of the moon and the sun, as seen from the earth, are nearly the same. Since he had calculated that the sun was 20 times farther away than the moon, he reasoned that the sun's actual diameter was 20 times that of the moon.

All that was needed to determine the actual distances and diameters was measuring the radius of the earth. Some 80 years later, the astronomer Eratosthenes devised a clever method to do just that. Aristarchus' measurements of the sun's size and of its distance were off due to the difficulty in determining the exact moment when the moon was at first or last quarter phase which introduced a large uncertainty in his measurement of the angles. However, his calculations of the size of the moon and its distance to the earth were incredibly close to modern measurements.

Today, we can measure the distance to the moon with a precision of 1 centimeter using lasers. From 1969 until 1971, the astronauts of the Apollo missions placed laser reflectors on the surface of the moon. Lunar laser ranging stations on the earth send laser pulses to these reflectors and detect the reflected light 2.6 seconds later. Since the speed of light is known with a high degree of precision, an extremely accurate determination of the distance to the moon and its variation as the moon orbits the earth is possible.

SPACECRAFT AND ORBITAL MOTION

Newton, as we have said before, realized that the moon is falling toward the earth. Any satellite in orbit around the earth is a free falling object. We can illustrate this in another way by using a diagram that appeared in Newton's *Principia* (Figure 6.16). If

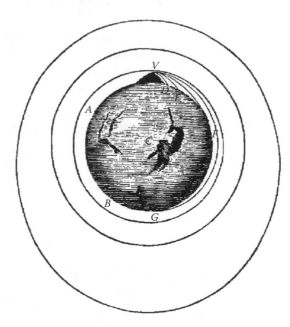

FIGURE 6.16 A projectile shot from a high mountain follows different curved paths. "[T]he greater the velocity ... with which it is projected, the farther it goes before it falls to the earth. We may therefore suppose the velocity to be so increased ... til at last, exceeding the limits of the earth, it should pass into space without touching it." (From Newton, Principia, Volume II: The *System of the World,* University of California Press, Berkeley, CA, 1934.)

we throw a stone horizontally from the tall mountain, it would fall to the ground following a path like the one that ends at point D. If we throw the stone harder (perhaps with the aid of some mechanical device), it would still fall to the ground, hitting at point E, farther away, due to the fact that the stone was thrown with a higher speed and also because the earth curves out from under it. At successively higher speeds the stone would hit the ground at points F and G, until finally, if thrown fast enough, its curvature would match that of the earth, and the stone would fall to the ground without ever reaching it. We would have placed the stone in orbit around the earth.

A satellite in orbit around the earth is, then, in free fall; it is continuously being accelerated toward the center of the earth. What is the value of this acceleration? We know that the force acting on the satellite is the gravitational force. Therefore, the satellite falls with the acceleration due to gravity. However, this acceleration due to gravity does not have the value of 9.80 m/s^2, but the value of the acceleration due to gravity at that particular distance from the center of the earth. Remember that 9.80 m/s^2 is the acceleration due to gravity at the surface of the earth, which is at a distance from the center of the earth equal to the earth's radius.

When the space shuttle is in orbit around the earth, the shuttle, the astronauts, and all the objects inside it appear weightless, since the shuttle in orbit is in free fall. As we all have seen many times on the television images broadcast from space,

FIGURE 6.17   The gravitational force acts on astronaut Wendy Lawrence (at work on the flight deck of the Space Shuttle Discovery) in such a way that she is kept moving in circles around the earth. (Courtesy of NASA.)

when an astronaut lets go of a pen, it continues going around the earth at the same speed as the rest of the shuttle and appears to be floating inside the spacecraft. Since the gravitational force is proportional to the mass, the magnitude of the gravitational force keeping the astronaut in orbit is larger than the gravitational force keeping the pen, with a smaller mass, in orbit. The gravitational force acts on the astronaut and the pen in the exact right proportion to keep them moving in circles together (Figure 6.17).

In Chapter 3 we said that mass was the measure of the inertia of a body. By Newton's second law, a net force acting on an object is proportional to the acceleration of the body and to the body's resistance to acceleration or its *inertial mass*. In this chapter, however, mass has been used as a quantity that is proportional to the gravitational force. We might call this mass *gravitational mass*. In 1922, an experiment performed in Budapest by Roland von Eötvös showed that the gravitational force acts in exactly the right proportions on objects of different masses. In other words, within the accuracy of his experiment, the gravitational and inertial masses were the same. A recent reanalysis of the data taken by Eötvös seemed to indicate a slight discrepancy implying that gravity does not act in the exact required proportions. If this were to be the case, an object with a larger gravitational mass would fall to the ground with a greater acceleration. Some scientists proposed the existence of a new force, the *fifth force*, which would explain the new findings. This fifth force is related to the total number of protons and neutrons in the nuclei of different substances. Recent experiments at Los Alamos National Laboratory and at the Joint Institute for Laboratory Astrophysics in Colorado have cast some doubts about the existence of such a fifth force. More sophisticated experiments are currently underway in an attempt to clarify the situation. We should point out, however, that these discrepancies are extremely small.

THE FRONTIERS OF PHYSICS: GLOBAL
POSITIONING SATELLITE SYSTEMS

In orbit at an altitude of about 20,000 km, an armada of 24 satellites broadcasting a continuous signal would potentially allow people to determine their location with millimeter precision. Although the satellites that form the U.S. Department of Defense's Global Positioning Satellite System (GPS) were placed in orbit in the late 1970s for military purposes, provisions were also made for civilian use.

The satellites broadcast high-frequency radio waves continuously. The signal propagates in all directions with the same speed (the speed of light). Since the speed of light is known to a high degree of accuracy, one would only need to know how long it took the signal to arrive at one's location to determine the distance to a particular satellite. The designers of the GPS system borrowed a technique used by astronomers to determine distances to other planets. This technique involves sending coded signals at predetermined known times. Since the code is known, comparing the signal from the satellite with the known coded signal shifted in time to various intervals allows the receiver to determine how long the signal has been traveling.

Knowing the distance to one satellite is not enough to determine one's location in space. If you were to determine that one of the satellites is 25,000 km away, you could be anywhere on a large sphere 25,000 km in radius centered on the satellite. Determining the distance to a second satellite would narrow your possible position to a circle (the intersection between the two spheres around each satellite). A third satellite will reduce the possible positions to two points (the intersection of this third sphere with the previous circle) and a fourth satellite will narrow this down to one unique point.

Since the GPS was intended for military purposes, the signals were originally encrypted to prevent potential adversaries from using the system to determine the position of their troops. The designers wanted to allow for limited civilian use and released slightly incorrect information about the timing of the signals. The designers estimated that this error would allow other parties to determine their position at best within 100 m. Soon, however, engineers were able to devise methods to circumvent the clock errors and managed to increase the accuracy to within a few millimeters. The intentional degrading of the signal has been discontinued, increasing the resolution of satellite navigation devices to a couple of meters.

U.S. and international organizations provide augmentations to the GPS system to meet specific user requirements for positioning, navigations, and timing. The Nationwide Differential GPS System (NDGPS), for example, is a ground-based system offered by the Federal Railroad Administrations, the U.S. Coast Guard, and the Federal Highway Administration to enhance GPS to within 10–15 cm accuracy. The Wide Area Augmentation System (WAAS) is a satellite-based system operated by the U.S. Federal Aviation Administration to provide GPS signals for aircraft navigation. This system is also used for ground applications. The Global Differential GPS is a high accuracy augmentation system developed by NASA's Jet Propulsion Laboratory to support NASA's space missions.

Many other countries around the world have implemented similar systems. The International Global Navigation Satellite Service (IGS) is a network of stations and organizations in 80 counties that provide accurate GPS data for scientific and educational purposes.

Part III

The Structure of Matter

7 Atoms: Building Blocks of the Universe

THE UNDERLYING STRUCTURE OF MATTER

The question of the nature of matter is perhaps as old as humanity. Twenty-five centuries ago, Thales of Miletus—the idealized Greek thinker credited by Aristotle as the founder of European philosophy—is believed to have posed that question. Clearly, discovering the hidden structure of nature constitutes one of the most basic reasons for our exploration of the world. Up to this point, we have occupied ourselves with the behavior of matter. We now turn our attention to the underlying structure of matter.

THE ATOMIC HYPOTHESIS

If some global catastrophe was to destroy civilization and you were one of the few survivors, what would be the best single statement that you could communicate to the next generations? The late Nobel Prize-winning physicist Richard Feynman had a good candidate for this statement: the *atomic hypothesis*. Feynman states this hypothesis as follows: "*All things are made up of atoms—little particles that move around in perpetual motion, attracting each other when they are a little distance apart, but repelling upon being squeezed into one another.*"

Why is this statement so important? Since it is the basis for modern physics, the physics that has made possible the computer chip, television and modern communications, and electronics. Modern physics began with the confirmation of the atomic hypothesis at the end of the nineteenth century.

EARLY CONCEPT OF THE ATOM

The Greek philosopher Leucippus (believed to have been born in Miletus about 490 BC) is sometimes credited as the originator of the atomic principle. However, it is his student Democritus, a Greek natural philosopher, who was responsible for its introduction into Greek thought.

Democritus said that matter was made up of small, indivisible particles, which he called atoms, from the Greek word for indivisible. As the scientific method (as we understand it today) did not exist in the time of Democritus, it did not occur to anyone to check ideas against experiment. People simply chose among these different ideas on the basis of taste and personal preference. And the views of Democritus were not widely accepted by most philosophers of his time.

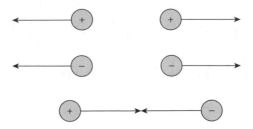

FIGURE 7.1 Like charges (all positive or all negative) repel each other, whereas unlike charges (a positive and a negative) attract each other.

The atomic concept lay dormant until the nineteenth century, when it was revived by the work of the English scientist John Dalton. Dalton proposed an *atomic theory* in which he stated that matter is composed of atoms, which he considered to be indivisible particles; that the atoms of a given element are identical; and that atoms are arranged, in chemical reactions, to form molecules.

By this time physicists had some understanding of electricity and had identified what the American physicist and statesman Benjamin Franklin called positive charge and negative charge. Experiments had shown that two positive charges repel each other; two negative charges also repel each other; but a positive charge and a negative charge attract each other (Figure 7.1). By studying how electrical currents decompose water and other chemical compounds, the English physicist Michael Faraday realized that electricity exists in discrete units or particles. In 1891 these particles of electricity were named *electrons.*

These two discoveries would eventually help explain a seemingly unrelated phenomenon that had baffled scientists for years. During the last decade of the nineteenth century, physicists tried to understand the nature of the strange and fascinating glow that appeared when an electric current passed through a wire inside a glass tube from which air had been evacuated. Two metal discs, or *electrodes*, sealed into the ends of a long glass tube, were connected to a high-voltage battery (Figure 7.2). When the battery was connected, a green glow became visible in the glass. Sometimes, investigators would place fluorescent materials inside the tube which would also glow when the electrical current was present. Since the rays seemed to emanate from the electrode connected to the negative end of the battery which was called the *cathode*, these rays were known as *cathode rays*, and the evacuated glass tube as a *cathode ray tube*. These tubes were the forerunners of today's television picture tubes. The modern computer monitor is still called a *CRT*, an abbreviation for cathode ray tube.

Sir Joseph John Thomson was the director of the Cavendish Laboratory in Cambridge, England. With a staff of 20, "J. J." had designed experiments with cathode ray tubes for more than a decade, attempting to decipher the nature of the glow. Obviously *something* was passing out of the cathode, through the vacuum, and colliding with the positive electrode. On Friday, April 29, 1897, Thomson announced that the glowing beam was not made up of light waves, as most physicists held. Rather, cathode rays were particles (*corpuscles*, as they were called then) carrying

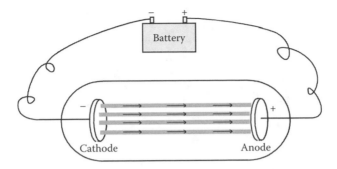

FIGURE 7.2   When a battery is connected across the two electrodes inside an evacuated glass tube or *cathode ray tube*, a green glow appears in the glass.

negative electric charge. These corpuscles were flying off the negatively charged cathode into the positive anode. Thomson was able to show that these corpuscles were electrons and was also able to determine the ratio of their mass to the electric charge. From this ratio he concluded that the mass of the electron was about two thousand times smaller than the mass of the lightest atom (hydrogen).

Thomson's discovery had tremendous implications. From Democritus to Dalton, atoms were thought to be the most basic constituents of matter. Yet Thomson had discovered something two thousand times smaller than an atom. From his experiments he was forced to conclude that atoms were not indivisible; they were made up of electrons. Electrons, he said, were the basic components of the atom.

FIRST MODELS OF THE ATOM

The idea that matter is made up of atoms was beginning to be accepted by the turn of the twentieth century. After Thomson discovered the electron in 1897 he and other physicists began to realize that the atom must be a complex structure. Thomson himself put forward a model for this complex atom. It has been called the "plum pudding" model because it represented the atom as a homogeneous sphere of positive electric charge in which the negatively charged electrons were embedded (Figure 7.3).

Things moved rapidly during the last decade of the nineteenth century. In 1895 Wilhelm Conrad Röntgen in Germany discovered a highly penetrating and invisible radiation that he called x-rays. A few weeks after Röntgen announced his discovery before the Physical Medical Society in Würzburg, Antoine Henri Becquerel in Paris discovered radioactivity by experimenting with uranium crystals. Working in Becquerel's laboratory, Pierre and Marie Curie discovered two other radioactive elements, which they named *polonium* and *radium*.

However, it was Ernest Rutherford, a New Zealander who studied under J. J. Thomson at England's Cavendish Laboratory, who showed that the rays given off by radium were of at least two different kinds. By placing thin sheets of aluminum in the path of the rays he stopped one kind of radiation, which he called *alpha rays*. The other kind, which penetrated thicker sheets, was named *beta rays*. A third type of radiation, *gamma rays*, was discovered a year later. The Curies and

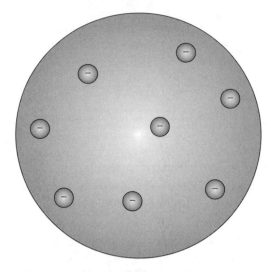

FIGURE 7.3   Thomson's model of the atom: a homogeneous sphere of positive charge with negative electrons embedded in it.

Henri Becquerel showed in 1900 that beta rays have negative electric charge. In 1907 Rutherford proved that the alpha particles were helium atoms with the electrons removed. Thus they had to be positively charged.

In 1908 Rutherford was awarded the Nobel Prize for his investigations on radioactivity. Most scientists do their most important work before they are honored with the Nobel Prize. This was not the case with Rutherford. In that very same year of 1908, Rutherford and his assistant Hans Geiger were looking for a way to detect alpha rays and discovered that a screen coated with zinc sulfide would flash where an alpha particle struck the screen. Thus, such a screen could serve as a detector of alpha particles. Rutherford decided to use his newly discovered detectors to investigate the structure of atoms.

This presented a problem, though. Rutherford knew that the atoms were too small to be seen. How could he investigate something that he could not even see? By firing a projectile at an unseen object and analyzing the projectile's behavior as it emerges, information about the nature of the interaction with the unseen object can be obtained—much in the same way as the sound of a stone hitting the surface of the water in a deep well can yield information about its depth (Figure 7.4). This is more or less what Rutherford decided to do. He set up an experiment to shoot alpha particles at a sheet of gold foil with a zinc sulfide screen placed at some distance from the gold foil (Figure 7.5). His collaborator Hans Geiger, who did most of the counting, observed that most of the alpha particles underwent very small deflections of only one degree or less. This was of course in accordance with Thomson's plum pudding model of the atom. A fast alpha particle—with positive electric charge—would pass through the electrons almost undeflected. The positive charge in Thomson's model was spread uniformly over the entire volume of the atom and thus was not concentrated, like that of the alpha particle. As soon as the alpha particle penetrated the

FIGURE 7.4 Although the water surface in a deep well might not be visible, we can find out how deep the well is by listening to the sound of a stone as it hits the water. Similarly, Rutherford's experiment allowed him to probe the atom, too small to be seen directly.

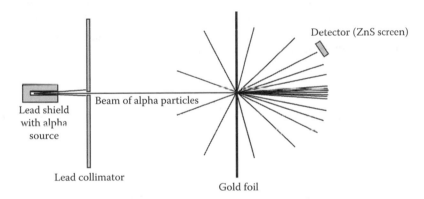

FIGURE 7.5 Schematic representation of Rutherford's experimental set up. The large angle deflections of the scattered particles forced Rutherford to conclude that the atom must have a positively charged nucleus.

atom, the positive charge of the atom would perhaps slow down the particle's motion but, since this charge was not localized at one point, it would not deflect the particle's trajectory. Rutherford, however, decided to ask a young undergraduate student named Ernest Marsden to look for alpha particles at angles up to 45°.

Marsden did find a few flashes at 45°. Encouraged by this, he decided to swing the telescope to larger angles, even past 90° and found flashes from the front side of the experiment. Rutherford was astonished. The deflections at such large angles forced Rutherford to conclude that the atom must have a positively charged nucleus. Some twenty years later he would comment, "It was quite the most incredible event that has ever happened to me in my life. It was almost as incredible as if you fired a 15-inch shell at a piece of tissue paper and it came back and hit you. On consideration, I realized that this scattering backward must be the result of a single collision, and when I made the calculations I saw that it was impossible to get anything of that order of magnitude unless you took a system in which the greater part of the atom was concentrated in a minute nucleus."

In 1911 Rutherford published his new model of the atom in which the positive electric charge is concentrated in a core that carries almost the entire mass of the atom, and the negatively charged electrons are uniformly distributed around the nucleus (Figure 7.6). Rutherford concluded that more than 99.9% of the mass of the atom was concentrated in the nucleus, which occupies about one part in 10^{14} of the volume of the atom.

The Japanese physicist Hantaro Nagaoka, one of Japan's first physicists, had proposed a model of the atom with a ring of electrons in orbit about a small nucleus. His conclusion did not come from experimental evidence, like Rutherford's. There is no evidence that Rutherford knew of Nagaoka's model.

Rutherford's nuclear model explained the observational evidence very well. There was a major problem, however. Physicists knew that if an electrically charged particle orbits around a center, it must radiate energy. The electron then must lose some of its energy with this radiation and should spiral down into the nucleus with the resulting collapse taking less than a microsecond. If this were really the case, atoms would not be stable and we would not be here to ponder these questions.

In 1913, the 28-year-old Danish physicist Niels Bohr, who had worked with Rutherford in England, proposed a solution to this problem, which involved a remarkable property of light that had been discovered only a few years before. Until then, light had been considered to be a wave phenomenon. This discovery would result in the awarding of the Nobel Prize to two of the most famous scientists that

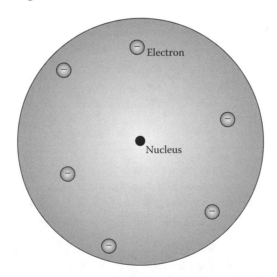

FIGURE 7.6 Rutherford's nuclear model of the atom. The positively charged nucleus is surrounded by a uniform spherical distribution of negative electrons.

ever lived: Albert Einstein and Max Planck. To understand the importance of this discovery, we need to become more familiar with the language of waves. (A more detailed treatment of waves is undertaken in Chapter 15.)

WAVES AND QUANTA

We are all familiar with water waves and waves on a string (Figure 7.7). The ripples that form on the surface of the water after a child throws a pebble in the water are one of the most common and familiar examples of what we call wave motion. The rapid up and down motion of a guitar string after a skilled player plucks it, displaces the air molecules around it, forcing them to collide with other nearby air molecules. The first molecules bounce back after the collisions only to collide with the returning string again, which sends them back toward the second set of molecules, themselves returning from other collisions. Very quickly, these molecular collisions spread out in all directions from the vibrating guitar string. Eventually, they reach our ears where, after traveling through the ear canal, they set the eardrum into similar oscillations. Hair cells a few thousandths of a millimeter thick begin then to move in response to these oscillations. Their motion is transformed into electrical impulses that travel to our brain at about one hundred miles per hour. There, in a process that physics cannot yet explain, we feel a sensation that we describe as pleasure.

These are examples of mechanical waves, and we know from experience that these waves propagate through some medium (the water, string, air, ear membrane, or hair cells in our examples). Although we cannot see sound waves, we know that they need some medium to travel through. The elasticity of the medium allows a disturbance created at some point in space to propagate through the medium. As a disturbance propagates, energy is transmitted through the medium without transferring matter. A wave, then, is a mechanism for the transmission of energy which does not require the actual translocation of matter. We can convince ourselves that this is the case by observing a piece of driftwood floating in the water. As the waves pass

FIGURE 7.7 Ripples formed on the surface of the water are one of the most familiar examples of wave motion.

by, we will see the wood rise and fall. The driftwood does not travel with the wave. The water moves up and down with the wood, while the wave travels from one point to another.

Light, on the other hand, is a wave that does not require a medium to propagate. The main evidence for the wave nature of light lies with the phenomenon of interference. If you drop two pebbles into a pond you will notice that at certain places the crests from the two waves meet and reinforce each other, while at other places, where a crest meets a trough, the two waves cancel each other out (Figure 7.8). When two crests or two troughs meet, the waves interfere constructively, and when a crest meets a trough, the waves interfere destructively. Combining two wave motions, then, can result in nothing. Two particles cannot annihilate each other like that.

We can draw an important conclusion from our brief look at wave phenomena. Waves behave very differently from particles. Particles travel in straight lines when free of external forces and they do not show interference. Waves do.

Although some Greek philosophers believed that light was composed of particles that travel in straight lines (even Isaac Newton in the seventeenth century favored this idea), by the end of the nineteenth century there was overwhelming evidence that light was a wave phenomenon. As has been known since around 1803 when the English scientist Thomas Young demonstrated it in a series of landmark experiments, light shows the phenomenon of interference, characteristic of waves. Young also showed experimentally that different wavelengths of light correspond to the different colors that we perceive.

The colors observed in soap bubbles are an example of interference with light, produced when the light waves are reflected from the front and back of the thin

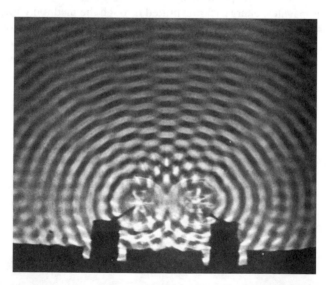

FIGURE 7.8   Interference of two wave motions. The waves are produced by two vibrating rods in a water tank. The circular waves combine to produce a pattern of alternating regions of moving and still water. (From *PSSC Physics Seventh Edition*, by Haber-Schaim, Dodge, Gardner, and Shore. With permission from Kendall/Hunt Publishing Company, Dubuque, IA, 1991.)

film of the bubble and meet. When white light, which consists of light waves of all the colors or frequencies, shines on the bubble, certain frequencies interfere destructively. The colors corresponding to these frequencies disappear from the reflected light. Since the reflected light lacks several frequencies, it appears colored.

Although by the turn of the present century it seemed very clear that light was a wave, this idea was going to be seriously challenged, as we are about to learn.

In 1900, the German physicist Max Planck was attempting to explain the relationship between the temperature of a hot object and the color of the radiation emitted. This problem had been in the minds of physicists for some time because the theory did not fit the experimental data for radiation emitted in the ultraviolet. The assumption had always been that the light was emitted from the hot body in a continuous manner. After many failed attempts, Planck reluctantly incorporated into his theory the assumption that this radiation was being emitted not continuously but in discrete amounts. His equation now matched the experimental data perfectly. According to Planck's new (and, as it turned out, revolutionary) assumption, the energy emitted by a hot object could only have certain discrete values, as if it were carried out in little bundles or packets. Planck called these bundles of energy *quanta*. The discreteness of the energy emitted was expressed through a constant h, now called Planck's constant, which has a value of 6.63×10^{-34} J s.

Planck's discovery was an "act of desperation," as he would call it later, without physical basis and done for the purpose of explaining that particular experimental fact. He would attempt in vain to find an explanation for his radical assumption within classical physics for the rest of his life.

In 1905, the young Albert Einstein, unable to obtain a post as professor of physics because of his rebellious character, was working as a clerk at a patent office in Berne. The job, he said later, was not very demanding and left him plenty of time to work on physics. One of the problems that he considered was the way in which electrons were ejected from certain metals when light was shining on them. Although the phenomenon had been known since 1887, physicists were puzzled by the fact that the kinetic energies of the released electrons did not seem to be at all related to the brightness of the light shining on the metal. Energetic electrons were released from metals even under extremely weak light.

Einstein realized that the puzzle arose from the assumption that light was a wave and that a gentle wave (a weak light beam) could not propel an electron at great speeds. Building on Planck's discovery, Einstein proposed that light does not only behave as a wave, as the interference phenomena showed, but also as a bundle of energy packets that he called light quanta or *photons*. A weak beam of light of a certain color is composed not of weaker bundles of energy but of *fewer* bundles of the same strength as those of an intense beam of light of the same color. Each one of the (few) quanta that make up the weak light beam is as capable of releasing a fast electron as the quanta that make up the stronger beam.

Einstein generalized Planck's theory. He assumed that the quanta of energy that Planck had introduced were also characteristic of light rather than a special property related only to a single phenomenon. Light is not simply *emitted* in bundles of energy; light *is* made up of these bundles.

THE BOHR MODEL OF THE ATOM

We are now ready to understand the basic ideas behind Niels Bohr's model of the atom. Bohr's model was similar to Rutherford's planetary model, but with some important differences. Bohr postulated that the electrons moved around the nucleus in certain specific orbits, which he called *stationary orbits*. Since electrons are negatively charged, they are attracted to the positive nucleus. This attractive electric force provides the centripetal acceleration of the electron as it orbits the nucleus. In these orbits the energy of the electrons (the sum of kinetic and electric potential energy) remained constant; thus the electrons did not radiate energy. Only when the electrons moved from one stationary orbit to another orbit closer to the nucleus, decreasing their potential energy, did they radiate energy. For an electron to jump to a higher orbit it had to absorb energy. Therefore, when the electrons moved from one stationary orbit, or *state*, to another, they absorbed or emitted energy.

Bohr used Planck's formula to compute the allowed stationary orbits of electrons in orbit; he called this procedure his "quantization condition." He postulated that only certain values of the angular momentum L of the orbiting electron were allowed. The lowest value of the angular momentum was simply Planck's constant h divided by 2π. Higher values of L were multiples of this minimum value, namely

$$L = \frac{nh}{2\pi}$$

where n was a positive integer. Thus, the angular momentum of the electron was "quantized"; that is, it took only certain discrete values. For each allowed value of L, a corresponding value of the energy of the electron was calculated. In Bohr's theory, each allowed orbit corresponds to an *energy level*: the higher the orbit, the higher the energy level. Since each one of the electrons in an atom resides in one of the allowed orbits where it has a definite energy, the total energy of the atom must have a specific value. Thus, the atom's total energy is quantized. An analogy using a system from mechanics could help us visualize this concept. A marble on a staircase also has quantized values of its potential energy (Figure 7.9). It can only increase or decrease its energy in fixed amounts which are determined by the height of each step. A marble on a ramp, on the other hand, can have any intermediate value of the potential energy as it rolls up or down the ramp.

The quantization or discreteness of the energy levels in an atom guarantees that the light absorbed or emitted by an atom when one of its electrons undergoes a transition between orbits must be discrete also. Bohr postulated that when an electron jumped from one energy level to another, a photon was emitted or absorbed. The energy of the emitted or absorbed photon must equal the energy difference between the two energy levels. Figure 7.10 shows an *energy level diagram* with several possible transitions. Since each energy level corresponds to an allowed orbit with a specific value of the angular momentum, we use the positive integer n to indicate the different energy levels. The energy level with the lowest level, the *ground state*, has the value $n = 1$; higher energy levels have values of n equal to 2, 3, and so on. The number n is called the *principal quantum number*.

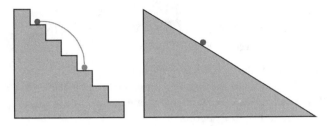

FIGURE 7.9 A marble on a staircase has quantized values of its potential energy since it can only reside on a particular step and cannot float in between steps. In contrast, a marble on a ramp can acquire any intermediate value of the potential energy as it rolls from top to bottom of the ramp.

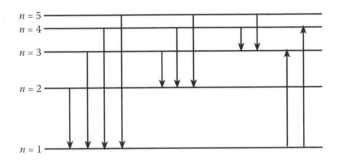

FIGURE 7.10 Simplified energy level diagram for an atom, showing several possible transitions.

Bohr calculated the allowed transitions between energy levels for the hydrogen atom and found that they agreed exactly with the experimental measurements. Soon, however, it became clear that his theory did not work well for other, more complicated atoms. There were other problems, too. No explanation was given for the mechanism of absorption and emission of energy when an electron jumped orbits, or for the assumption that the electron did not radiate when in a stationary orbit. Nevertheless, Bohr's theory was the first attempt at explaining the structure of the atom and, as such, it opened the road for a more complete and successful theory, quantum mechanics. We shall study the fundamental concepts of this theory in Part 7.

COMPOUNDS

Although Bohr's atomic theory does not work for atoms other than hydrogen (and therefore is not completely satisfactory), it is still used to calculate the atomic properties of hydrogen and as the starting point for approximate calculations for other atoms. The Bohr model has also been used to obtain an estimate of the way in which

THE FRONTIERS OF PHYSICS: FILMING AN ELECTRON IN MOTION

Using extremely short flashes of light, physicists at Lund University in Sweden have recently been able to photograph an electron as it leaves an atom. The short flashes are extremely short pulses of intense laser light called attosecond pulses (an attosecond is 10^{-18} second).

Although attosecond pulses had been used before, no one had been able to make a movie with the images, since the individual photographs are too faint. However, the Lund University scientists took several pictures of the same moment and superimposed them to enhance each photograph. To do that, they used a stroboscope technique to isolate each time interval. "A stroboscope enables us to 'freeze' a periodic movement, like capturing a hummingbird flapping its wings," said Johan Mauritsson, a physics professor at Lund. "You then take several pictures when the wings are in the same position, such as the top, and the picture will turn out clear, despite the rapid motion."

atoms interact with each other. More sophisticated calculations based on quantum mechanics have then been undertaken to obtain a more accurate result.

Atoms of one particular element bond to other atoms in very specific ways, which are governed by the properties of the electrons of those atoms and their arrangement around their nuclei. Two or more atoms can combine to form a *molecular compound* or an *ionic compound*. The molecule is the unit of the molecular compound. The combination of two hydrogen atoms (H) and one of oxygen (O) form the water molecule (H_2O). Two atoms of carbon (C), three of hydrogen, and one of chlorine (Cl) form the vinyl chloride molecule, a carcinogenic gas which is manufactured in great quantities because it can be used to form polyvinyl chloride (PVC), a very useful plastics material. Twenty carbon atoms, 28 hydrogen atoms, and one oxygen atom form the retinal molecule, the forerunner of the rhodopsin and iodopsin molecules which absorb the light that reaches the rods and cones, the photoreceptor cells in the retina of your eye.

We can think of a molecule as a stable configuration of several nuclei and electrons. As such, this view is an extension of the concept of the atom. The atoms that form the molecule lose their identities and the molecule becomes the new building block of matter. The bonding of the atoms that form a molecule is due to the sharing of electrons by the atoms. This type of bonding is called a *covalent* bond.

Ionic compounds, on the other hand, are formed by the bonding between ions. Recall that atoms are electrically neutral and are composed of a positive nucleus at the center and negative electrons around this nucleus. Since electrons occupy the outermost parts of the atom, they can be removed from or added to atoms to form *ions*. A positive ion is an atom lacking one or more electrons, whereas a negative ion is formed when one or more electrons are added to a neutral atom.

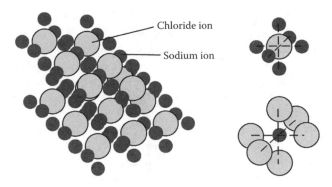

FIGURE 7.11 Graphic representation of the NaCl ionic compound. Each Na^+ ion is surrounded by 6 Cl^- ions and each Cl^- ion is surrounded by 6 Na^+ ions. Since no particular Na^+ ion is attached exclusively to a particular Cl^- ion, there is no unique NaCl molecule.

The bonding in an ionic compound is due to the electrical attraction between oppositely charged ions. The ionic bonding between a sodium ion (Na^+) and a chloride ion (Cl^-) in table salt is an example. In this case, if we remove an electron from Na to form the positive ion of sodium (Na^+) and attach that electron to the neutral chlorine atom to form a Cl^- ion, the two ions, one positive and the other one negative, attract each other. Moving this electron from one atom to the other requires a certain amount of energy. In this particular case, the total energy required is 1.3 eV. However, the energy of attraction between the two ions is greater than 1.3 eV. It is more energetically favorable for the ions to attract each other than to bring the electron back and form neutral atoms again.

Although sodium atoms bind to chlorine atoms in the way just described, molecules are not formed. Each sodium ion is surrounded by six chloride ions and each chloride ion is surrounded by six sodium ions. Since there is no exclusive union between any two given ions, there is no unique NaCl molecule (Figure 7.11).

PHYSICS IN OUR WORLD: WINEMAKING

You may have noticed that wine labels say "Contains Sulfites." Wines actually contain sulfur dioxide, a molecule formed by the combination of one atom of sulfur and two of oxygen. Although the amount of sulfur dioxide present naturally in wines is no more than about 10–20 parts per million, a 1988 law requires winemakers to include the statement on their labels if their wines contain more than 10 parts per million. This law is intended to protect some people who suffer from asthma and who might be sensitive to it (about 5% of asthmatics are sensitive to sulfites).

The sulfur dioxide molecule plays an important role in the winemaking process. Wine is made from fermented grapes. After the ripe grapes are crushed to release their juice, yeasts (single-cell organisms that exist on the grape skins) convert the sugar in the juice into alcohol, a process known as fermentation. When all the sugar has been converted into alcohol, the grape juice has become wine. The enormous variety of wines that are produced in the world comes not only from the different types of grape but also from the kind of container used to ferment the wine (oak or stainless steel), the temperature inside this container and even its size, and how long the wine is stored afterwards.

During the fermentation process, sulfur dioxide is produced naturally in the grape juice. Since it is produced only in very small quantities, winemakers also add it to their wines. Sulfur dioxide prevents the growth of bacteria that would sour the wine into vinegar. Sulfur dioxide also prevents the growth of wild yeasts that would continue the fermentation of sweet wines after they are bottled. Finally, sulfur dioxide is an antioxidant; that is, it prevents oxygen from combining with the wine.

8 The Heart of the Atom: The Nucleus

RAW MATERIAL: PROTONS AND NEUTRONS

Werner Heisenberg, a 29-year-old professor of physics at the University of Leipzig, told Bohr in 1931 that he had given up concerning himself with fundamental questions, "which are too difficult for me." Heisenberg had in mind his failure to explain the physics of the atomic nucleus with the quantum mechanics that he had invented only five years earlier. In the Christmas issue of the *Berliner Tageblatt*, he wrote that progress on fundamental questions such as the quantum mechanics of the nucleus would have to wait for new discoveries about that small piece of matter at the heart of the atom. "Whether indeed the year 1932 will lead us to such knowledge is quite doubtful."

Heisenberg was mistaken. Six months later he came up with the first quantum mechanics of the nucleus, a theory that became the basis for the nuclear physics of today. Rutherford's scattering experiment had given physicists the first clues about the size, charge, and mass of the nucleus of the atom: The nucleus is about 10^{-15} meters across, is positively charged, and contains most of the mass of the atom. The atom itself, on the other hand, is about 10,000 times larger and is electrically neutral. If the atom were expanded to the size of an auditorium, the nucleus would be represented by a pea at the center.

As soon as the existence of the nucleus was demonstrated, physicists started thinking about its composition. Rutherford had identified the basic positive charge, which had been observed as early as 1886 by Eugen Goldstein in Germany, as nuclei of hydrogen. He proposed the name proton for this particle. It was found that the proton was 1836 times more massive than the electron, but contained a positive electric charge of the same magnitude as the negative electric charge of the electron. For the lightest element, hydrogen, things were relatively simple: its nucleus was a single proton with an electron orbiting around it, making the atom neutral. Furthermore, the ratio of the mass of the hydrogen nucleus to that of the orbiting electron was consistent with the ratio of the mass of the proton to the mass of the electron.

Things were not working well for the other atoms. Although it was known that the helium atom had two electrons and thus two protons in its nucleus, its mass was not simply twice the mass of the hydrogen atom. Thus, the nucleus of the helium atom, which Rutherford had identified as being an alpha particle, and which was known to have a positive electric charge equivalent to the negative charge of two electrons, had to contain something more than just two protons.

In 1920 Rutherford predicted the existence of a particle with no electric charge but possessing the same mass as that of the proton. William D. Harkings in the United

States and Orme Masson in Australia also predicted its existence at about the same time. Harkings proposed the name *neutron* for this new subatomic particle. Twelve years later, James Chadwick, a former collaborator of Rutherford's, announced the discovery of neutrons. Chadwick received the 1935 Nobel Prize in physics for this discovery.

After the discovery of the neutron was announced, Heisenberg realized that this neutral particle provided the key to the physics of the nucleus. In a now classic three-part paper entitled "On the composition of atomic nuclei," Heisenberg proposed that atomic nuclei were composed of two types of particle, protons and neutrons. Subsequent experiments verified that this scheme was indeed correct. These two nuclear particles are collectively known as *nucleons*.

PIONEERS OF PHYSICS: HEISENBERG'S FAILING GRADE

Perhaps the most important early twentieth-century physicist after Einstein and Bohr, Werner Heisenberg rose meteorically to the top of his profession. He published his first scientific paper at the age of 20, invented quantum mechanics at the age of 24, attained a full professorship in physics at the age of 25, and received the Nobel Prize at the age of 32. Yet, this brilliant scientist almost failed to pass his final oral doctoral examination.

Due to of his talent and his successes in theoretical physics, Heisenberg completed his dissertation research in a very short time. In Germany at the time, students entering the University could pursue studies leading toward a diploma—equivalent to an American Masters degree—or toward a doctorate. The latter required being admitted by a professor into a research program with no specific course requirements. Heisenberg decided on the doctorate and took courses in mathematics and theoretical physics but neglected laboratory courses. Professor Wilhelm Wien was in charge of the physics laboratory and was also on Heisenberg's final exam committee. When Heisenberg could not explain in detail the operation of an electric battery during the final oral nor answer several questions about microscopes and telescopes, Wien decided to give him a failing grade for the dissertation. Fortunately for Heisenberg, the other professors on the committee gave him better grades and he squeezed through with a pass.

THE COMPOSITION OF THE NUCLEUS

The present view of the nucleus is essentially the one proposed by Heisenberg; that is, the nucleus is composed of protons and neutrons. Protons have an electric charge of 1.6×10^{-19} C, where the C stands for *coulomb*, the SI unit of electric charge. The mass of one proton is 1.673×10^{-27} kg, slightly smaller than the mass of a neutron, which is 1.675×10^{-27} kg. Neutrons, as we said earlier, have no electric charge.

Since protons and electrons have equal and opposite charges and the atom is electrically neutral, the number of protons in the nucleus must be equal to the number of orbiting electrons in the atom. Oxygen, for example, has eight protons in its nucleus and eight orbiting electrons.

The number of protons in the atom determines the *atomic number*, Z. The atomic number of oxygen is thus 8. The total number of nucleons—that is, the total number of neutrons and protons—is called the atomic mass number A, since it gives the approximate total mass of the nucleus. The atom of oxygen, with 8 protons and 8 neutrons, has an atomic mass number of 16. The number of neutrons is given by the neutron number N. Thus

$$A = Z + N$$

The different nuclei are designated by the same symbols used to name the corresponding chemical elements. So we would use H for the nucleus of hydrogen, O for oxygen, and so on. The different types of nuclei are referred to as *nuclides*. A complete description of a nuclide is given by specifying its symbol together with the atomic and mass numbers, in the form

$$_Z^A E$$

where E is the chemical symbol for the element. The neutron number need not be specified since $N = A - Z$. For example, the nucleus of oxygen with eight protons and eight neutrons is written as $_8^{16}O$ because its atomic number is 8 and its mass number is 16. The nucleus of helium, with two protons and two neutrons, is written as $_2^4 He$.

The number of protons in the nucleus or the atomic number determines the element. The nucleus of carbon, for example, contains 6 protons. The number of neutrons can vary, however, and we might find carbon nuclei containing different number of nucleons, such as $_6^{12}C$, $_6^{13}C$, and $_6^{14}C$, with 6, 7, and 8 neutrons, respectively. Nuclei with the same number of protons but a different number of neutrons are called *isotopes*. The nucleus of the most common isotope of hydrogen, as we have seen, consists of a single proton. Another less common isotope of hydrogen, *deuterium* or heavy hydrogen, has a nucleus with one proton and one neutron (Figure 8.1). Water made from deuterium is called *heavy* water. While chemically indistinguishable from regular water, heavy water has different physical properties. For example, it boils at a slightly higher temperature than regular water.

The isotope $_6^{12}C$ of carbon is used to define a new unit of mass for atomic and nuclear calculations. Since the mass of the proton and the mass of the neutron are slightly different and much larger than the mass of the electron, this new unit of mass was defined to take this into account. The *atomic mass unit* (amu) is defined as one-twelfth of the mass of the isotope $_6^{12}C$. Since the mass of the atom is used,

Hydrogen nucleus   Deuterium nucleus

FIGURE 8.1   The nucleus of the most common isotope of hydrogen consists of one proton, represented by a small sphere with a + (for its positive electric charge). Deuterium, another hydrogen isotope, contains one proton and one neutron.

this definition includes the mass of the 6 electrons in this isotope. As we saw in Chapter 1, the conversion factor between atomic mass units and kilograms is

$$1 \text{ amu} = 1.660566 \times 10^{-27} \text{ kg.}$$

The following table lists the masses of the proton, neutron, and electron in amu and kg.

Particle	Mass (amu)	Mass (kg)
proton	1.007276	1.6726×10^{-27}
neutron	1.008665	1.6750×10^{-27}
electron	5.486×10^{-4}	9.1094×10^{-31}

THE GLUE THAT KEEPS THE NUCLEUS TOGETHER

There is an apparent problem with what we have just said about the composition of the nucleus. As we saw in Chapter 3, positive charges repel each other (Figure 8.2). Since the nucleus contains only positively charged protons and uncharged neutrons, then the protons should repel each other and the nucleus should fly apart. In fact, the electrical repulsion between two protons in the nucleus is about one hundred billion times greater than the electrical attraction between the negative electrons and the positive nucleus. Furthermore, the neutrons do not help. They have no charge and there is no reason why they should remain together. Yet the nucleus does not fly apart. What keeps it together? There has to be an attractive force, stronger than the electrical repulsion between two protons when they are close together, that acts not only on the charged protons but also on the uncharged neutrons and keeps them together. This new force is called the *strong nuclear force* (Figure 8.3). At a distance of 10^{-15} m the attractive strong nuclear force between two protons is about one hundred times greater than their electrical repulsion.

The nuclear force is *charge independent*; that is, it acts with the same strength on neutrons as it does on protons (Figure 8.4). It is also a *short range* force; that is, it becomes active at very short distances, over a range of 10^{-14}–10^{-15} meter. If the

FIGURE 8.2   Electrostatic repulsion between two protons.

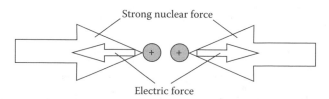

FIGURE 8.3   The attractive nuclear force between two protons separated by a distance of 2×10^{-15} m is about one hundred times greater than the repulsive electric force.

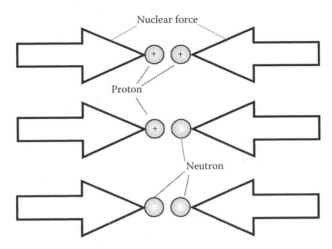

FIGURE 8.4 The nuclear force acts with the same strength on protons or neutrons. We say that the nuclear force is charge independent.

nucleons are separated farther, the force approaches zero. Due to the short range nature of the nuclear force, a nucleon in the nucleus interacts only with its nearest neighbors. In a nucleus with more than about 30 nucleons, the interaction of a nucleon with its immediate neighbors is not affected by the total number of nucleons present in the nucleus. The nuclear force thus becomes *saturated*.

The electrical force, on the other hand, has an infinite range, so a proton feels the electrical repulsion of all the other protons in the nucleus. If a particular nucleus has too many protons, the electrical repulsion on any one proton from all the other protons present may exceed the nuclear attraction thereby making the nucleus unstable. This type of nucleus is called *radioactive* and is the same kind of nucleus as was used by Rutherford in his experiments.

Since the nuclear force binds the nucleons together in the nucleus, work must be done to separate them. This is analogous to the case of two magnets sticking together with their opposite poles facing; work must be done to separate them (Figure 8.5). Conversely, when you bring the magnets together with the opposite poles facing, they pull your hands closer; the magnetic force does work on you. The energy required to move your hands comes from the magnetic potential energy stored in the two magnets. Similarly, when nucleons are assembled together to form a nucleus, *energy is released*. Thus, the total energy of the nucleus (bound system) is less than the total energy of its separated nucleons. This energy difference is called the *binding energy* and is equal to the work needed to separate the nucleus into its component nucleons.

From Einstein's special theory of relativity, it is known that if the energy of the free nucleons is more than the total energy of the nucleus, then the mass of the nucleons must be greater than the mass of the nucleus formed from the same nucleons. As we shall see in Chapter 19, the difference in mass is converted into energy when the nucleons are brought together, according to Einstein's formula $E = mc^2$, where c is the speed of light, equal to 2.997925×10^8 m/s.

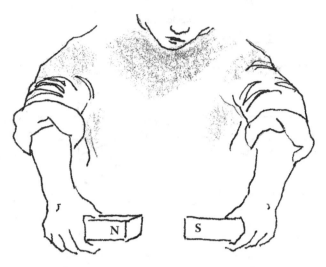

FIGURE 8.5 You must do work to separate two magnets with their opposite poles facing. If you stop pulling them apart, the magnets will move closer together, pulling your hands along and doing work on you. The energy required to move your hands is stored in the two separated magnets.

Einstein's formula allows us to find the energy equivalent in MeV of any given mass. The energy equivalent of 1 amu can be calculated as follows (recall that 1 amu is equal to 1.660566×10^{-27} kg):

$$E = mc^2 = (1.660566 \times 10^{-27} \text{ Kg}) \times 2.997925 \times 10^8 \text{ m/s})^2$$
$$= 14.9244 \times 10^{-11} \text{ kg m}^2/\text{s}^2.$$

Since we are using SI units throughout, the units of energy should come out in joules (as we saw in Chapter 4, 1 kg m^2/s^2 = 1 J). Remembering that 1 eV = 1.602×10^{-19} J, or 1MeV = 10^6 eV = 1.602×10^{-13} J, we have

$$E = mc^2 = (14.9244 \times 10^{-11} \text{J}) \times \frac{1\text{MeV}}{1.602 \times 10^{-13}\text{J}} = 931.5\text{MeV}.$$

Figure 8.6 illustrates the fact that the mass of the nucleus and consequently the mass of the atom ($^4_2$He in the figure) must be smaller than the total mass of the component particles (two protons, two neutrons, and two electrons). The missing mass or *mass defect* is the source of the energy that keeps the atom together.

To get an idea of how strongly the nucleons are bound in the nucleus it is useful to calculate the *binding energy per nucleon*, obtained by dividing the total binding energy by the mass number A. For $^4_2$He , the total binding energy is 28.3 MeV. Since the mass number (the total number of nucleons) is 4, the binding energy per nucleon is 7.1 MeV. If we calculate the binding energy per nucleon for different nuclei and

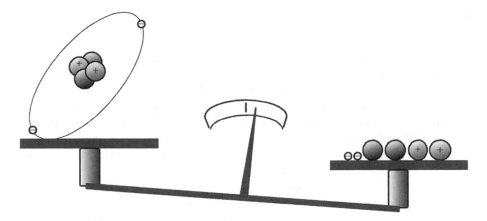

FIGURE 8.6   The $^4_2$He atom has less mass than its component particles.

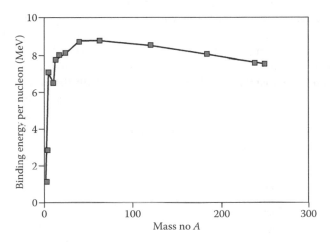

FIGURE 8.7   Binding energy per nucleon versus atomic mass number. Nuclei near the middle of the periodic table (mass number between 50 and 60) are more stable than elements at either end of the table, since the binding energies per nucleon for these middle elements are at maximum. They are, therefore, more tightly bound.

make a graph of the values obtained against mass number, we find that it increases with mass number, from 1.1 MeV for deuterium (an isotope of hydrogen) to about 8.0 MeV for oxygen-16. For nuclei with mass number greater than about 16, the binding energy per nucleon remains nearly the same, increasing slightly from $A = 18$ to $A = 50$ where it peaks at 8.7 MeV. Beyond $A = 50$ it begins to decline to about 7.5 MeV for $A = 250$ (Figure 8.7). The peak at around $A = 50$ (the elements near iron in the middle of the periodic table), means that more energy is required to remove a nucleon from these elements; they are, consequently, more stable.

SIZE AND SHAPE OF THE NUCLEUS

Most nuclei have a nearly spherical shape with a diameter of about 10^{-15} m. Nuclear sizes were originally determined by means of scattering experiments similar to Rutherford's; fast alpha particles were shot at a nucleus and detected at the other side. As in Rutherford's experiments, observing the distribution of the alpha particles after they interacted with the nucleus yielded information about the size of the target nucleus.

Additional information about the nucleus was obtained when fast electrons replaced the alpha particles as projectiles. Electrons offer the advantage that they do not feel the nuclear force and therefore are only affected by the electrical force of attraction to the protons in the nucleus.

Neutrons have also been used to measure nuclear dimensions. Since neutrons have no electric charge, they feel only the nuclear force.

All these experiments have yielded information about the dimensions of the nucleus by measuring different parameters. From the experimental evidence obtained we know that the volume of the nucleus is proportional to the mass number A. Since the mass number is the total number of protons and neutrons in the nucleus, this result means that the nucleons are packed in more or less the same way in all nuclei. We can understand this concept better with an example. Suppose two children are given several ping-pong balls and some modeling clay and told to build a large sphere using the clay as glue (Figure 8.8). If one child uses up 30 balls to make her sphere and the second child uses 60 balls to make a sphere that occupies twice the volume, we can definitely say that the two children packed the balls the same distance apart; that is, they used more or less the same amount of clay to stick each ball to the others. If, however, the second child ended up with a sphere occupying a volume of only one and a half times, we conclude that he packed his ping-pong balls closer together. Nucleons in nuclei are equally packed.

FIGURE 8.8   The two children build spheres with ping-pong balls stuck together with modeling clay. The girl uses up 30 balls to build up her sphere while the boy uses 60. If they use the same amount of clay to stick each ball to the others, the balls will be equally packed. In this case, the 60-ball sphere will have twice the volume as the 30-ball one. Nucleons in nuclei are also equally packed and their volumes are proportional to the number of nucleons in the nucleus (its mass number).

NUCLEAR ENERGY LEVELS

Neutrons and protons in the nucleus, like electrons in the atom, must obey the rules of quantum mechanics. These rules specify that nucleons (protons and neutrons) can move only in certain allowed orbits. These orbits are specified by four labels or *quantum numbers.*

Similarly to the case of the electrons orbiting in the atom, the quantum number *n* determines the energy of the orbiting nucleon. This number can only take whole-number values; that is, 1, 2, 3, 4, etc. The higher the number, the higher is the value of the energy. For most purposes, we can say that a higher value of the energy means that the nucleon travels at a higher speed in its orbit within the nucleus.

The second quantum number, the angular momentum quantum number, determines the shape of the orbit. Higher values of the angular momentum represent more circular orbits whereas lower values represent more elliptical ones. The third and

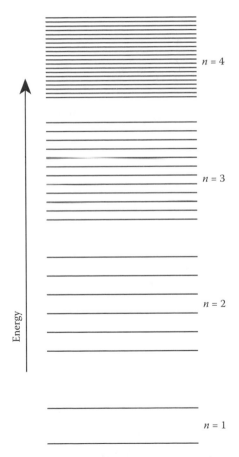

FIGURE 8.9 Nuclear energy level diagram. The nucleon energies are distributed in groups or shells, separated by gaps. (Note that nucleons with higher energies are not necessarily orbiting farther away: the diagram refers only to the energy values.)

fourth quantum numbers determine the direction in which the nucleon orbits and the orientation of this orbit in space.

Using these quantum numbers, the possible energies of the allowed orbits can be determined. As shown in Figure 8.9, these energies are distributed in groups or *shells*. As is the case in most cases governed by the laws of quantum mechanics, we must be careful not to assign too literal a meaning to these shells. All nucleon orbits have nearly the same average radius. The shells refer only to the energy values, not to the orbital sizes.

9 Fluids

STATES OF MATTER

The atoms that make up all matter in the universe arrange themselves in different ways to form rocks, water, galaxies, trees, or people. Different atoms interact in different ways to form a substance. When the interacting atoms of a substance are not moving about too much, they occupy more or less fixed positions and arrange themselves in a geometric pattern which minimizes the interaction energy. This pattern is repeated throughout the substance. When this happens the substance is called a *crystal* (Figure 9.1).

The atoms of a substance can also arrange themselves in other energy configurations which are not ordered and do not show a repetitive pattern, even when the atoms are not moving very much. In this case, the substance is noncrystalline or *amorphous*. Crystals and amorphous materials are *solids*. The forces that bind the atoms together in a solid are strong enough for the solid to maintain its shape.

When the binding forces are weaker, the atoms or molecules do not occupy fixed positions and move at random. These substances are called *fluids*. Liquids and gases are fluids. In a *liquid* the binding forces between the molecules are strong enough to make the liquid stay together but weak enough to allow it to flow. The molecules of a *gas*, on the other hand, are in chaotic random motion and interact only during very short times. Since the molecules of a gas are not bound together, the gas expands to fill any container.

DENSITY

If you have ever held a small bottle filled with mercury, you probably remember the strange impression of holding such a heavy liquid. Actually, mercury is not heavier than water or any other liquid. You can certainly have a container filled with water that weighs as much as your small bottle with mercury (Figure 9.2). The only difference is that the bottle with water has to be much larger and contain much more water. Mercury is *denser* than water. If you take the same volumes of water and mercury, the mercury weighs much more; 13.6 times more, to be exact. This means that the mass of mercury is 13.6 times greater than the corresponding mass of water that occupies the same volume. If you want to have the same mass of both mercury and water, the amount of water required will occupy a volume 13.6 times as large as the volume of mercury.

Thus if two substances have the same volume, the substance with the greater mass will have a greater *density*. That is, density is directly proportional to mass. Conversely, if you have two different substances with the same mass, the denser

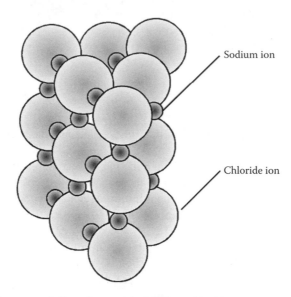

FIGURE 9.1   A representation of a crystal of sodium chloride.

substance occupies a smaller volume. That is, density is inversely proportional to volume. The density of a substance is the *mass per unit volume* of the substance, or

$$\text{density} = \frac{\text{mass}}{\text{volume}}$$

The Greek letter ρ (rho) is generally used to represent density. If we call the mass m and the volume V, we can write this expression in symbols as

$$\rho = \frac{m}{V}$$

Since the gram was originally defined as the mass of 1 cubic centimeter of water, the density of water is equal to 1 gram/cm³ or 1000 kg/m³. 1000 kg is a metric tonne. The densities of other common substances are shown in Table 9.1.

PRESSURE

Fluids can exert forces on other bodies. Since a fluid is in contact with the surface of the container, the force that the fluid exerts on the container is not localized at any one particular point and instead distributes itself over the entire surface. Consider, for example, a glass of water. Each molecule of water moves, on average, at speeds of over a thousand kilometers per hour and collides several billion times a second with other molecules and with the sides of the glass. Each collision with the surface of the glass exerts a tiny force. The sum of the billions of collisions that take place

FIGURE 9.2 The bottle of water in the man's left hand has a volume 13.6 times as large as the small bottle of mercury of the same weight in his right hand.

TABLE 9.1
Densities of Some Common Substances

Substance	Density (kg/m³)	Substance	Density (kg/m³)
hydrogen*	0.090	human blood	1,060
helium*	0.1758	sea water	1,300
air*	1.29	bone	1,700–2,000
oxygen*	1.4	iron	7,800
wood (oak)	600–900	copper	8,930
steam (100°C)	600	silver	10,500
oil	900	lead	11,300
ice	920	mercury	13,600
water (4°C)	1,000	gold	19,300

*At sea level atmospheric pressure and at 0°C.

(a)   (b)

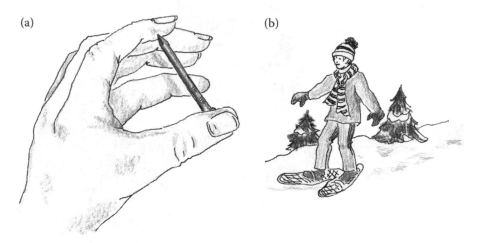

FIGURE 9.3   (a) Since the same force is applied to a smaller area, the sharp end of the nail will push into your skin much more readily than the head. (b) By increasing the area in contact with the snow, the pressure is reduced and the boy does not sink in the snow.

each second, each exerting a small force on a very small area of the glass, results in the total force that the water exerts on the surface of the glass. We can see that it is more convenient to consider the *force per unit area* of the surface, which we call the *pressure*, *P*:

$$\text{pressure} = \frac{\text{force}}{\text{area}} \quad \text{or} \quad p = \frac{F}{A}$$

If we stop for a moment and consider a couple of examples, we might be able to visualize the concept of pressure better. If you hold a nail between your thumb and index finger, the sharp tip of the nail will push into your skin much more readily than the blunt head (Figure 9.3a). You feel that the pressure of the tip is greater, because the force is applied to a much smaller area. The smaller the area to which the force is applied, the greater is the pressure. Conversely, the pressure is smaller if the area increases. This is the principle behind snowshoes (Figure 9.3b). By increasing the surface area in contact with the snow, the pressure that the person's body exerts is decreased. When somebody falls through thin ice in a frozen lake, rescue personnel approach the scene crawling, to minimize the pressure that their bodies exert on the ice.

The SI unit of pressure is *newton per square meter* (N/m^2), which is called the *pascal* (Pa) in honor of the French scientist Blaise Pascal (1623–1662), who did pioneering work with fluids.

ATMOSPHERIC PRESSURE

Galileo in *Two New Sciences* (1638) wrote that a lift-type pump cannot pump water from a well deeper than about 10 meters. These pumps were used widely at the time to obtain drinking water but nobody knew exactly why they worked or why there was a limit to the depth of the well.

It was a student of Galileo, Evangelista Torricelli, who solved the problem. By studying the results of his own experiments and those of his contemporaries Guericke, Pascal, and Boyle, Torricelli was able to explain that a lift pump works because of the pressure of the air. By pumping, the pressure at the top of the pipe is reduced. The pressure of the air on the water surface below pushes the water up the pipe.

The air exerts pressure because the blanket of air surrounding the earth has weight. The force exerted on a 1 square meter area on the surface of the earth at sea level by the column of air above is 101,300 N. The pressure of the atmosphere at sea level is thus 101.3×10^3 N/m² or 101.3 kPa. This is normal atmospheric pressure, called 1 *atmosphere* (atm). Thus,

$$1 \text{ atm} = 101.3 \text{ kPa.}$$

A column of water of about 10 meters high with a cross section of 1 m² also weighs 101.3 kN. Consequently, this column of water exerts the same pressure as the air above the surface of the earth (Figure 9.4). Therefore, if a good lift pump can reduce the pressure inside the pipe to almost zero, the atmospheric pressure can force water up to some 10 meters. It will not be able to raise water up to 11 meters, for example, since this requires a pressure that is larger than the atmospheric pressure at sea level.

For the same reason, if you invert a glass of water into a bowl with water, the water does not flow out of the glass into the bowl. The atmosphere pushes down on the water in the bowl with a force larger than the force exerted by the water in the glass (Figure 9.5).

Torricelli performed his experiments on air pressure not with water but with mercury. Mercury, having a density of 13,600 kg/m³ or 13.6 times the density of water, requires a column that is 13.6 times *smaller* than water to match the atmospheric pressure. A glass tube filled with mercury inverted into a pool of mercury is supported by the atmospheric pressure on the pool (Figure 9.6). At sea level, the air pressure supports a column of mercury 760 mm high. At higher altitudes, the air pressure is less, since there is less air above, and the mercury column that can be supported at these higher altitudes is less than 760 mm. Since it can be used to measure atmospheric pressure, Torricelli's invention is called a *barometer*, from the Greek *baros*, meaning weight.

The pressure of air that supports a column of mercury 1 mm high is called a *torr*, in honor of Torricelli. At sea level the atmospheric pressure has an average value of 101.3 kPa or 760 mm of mercury or 760 torr. This pressure is also called 1 *standard atmosphere* (1 atm).

Knowing the pressure of the air, we can calculate the mass of air above us. Assume that our bodies, when standing up, have a cross-sectional area of about 0.70 m². The weight of the air above is $w = F = PA = (101.3 \times 10^3$ N/m²$) \times (0.70$ m²$)$ or 71×10^3 N. Now, this weight is *mg*, and the value of the acceleration due to gravity, *g*, is still 9.8 m/s² at altitudes of a few tens of kilometers.

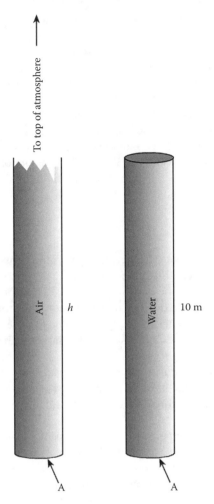

FIGURE 9.4   A column of water about 10 meters high exerts the same pressure as a column of air above the surface of the earth.

Therefore,

$$m = \frac{w}{g} = \frac{71 \times 10^3 \, \text{N}}{9.8 \, \text{m/s}^2} = 7,200 \text{kg}.$$

Certainly, we cannot support a mass of 7200 kg over our shoulders. Why are we not crushed by this enormous weight? The reason is that the pressure of the air acts in all directions, not just downward. Every square centimeter of our bodies is subject to this pressure. The air pushes down and up and sideways on us. And since there is air in our lungs, it pushes from the inside, too. All these forces balance out and we are not crushed by the atmospheric pressure.

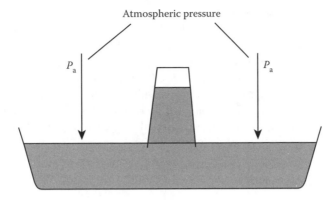

FIGURE 9.5 The atmospheric pressure on the water in the bowl is larger than the pressure of the water in the glass.

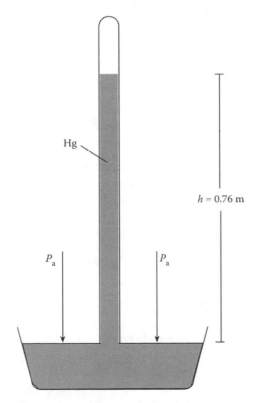

FIGURE 9.6 Torricelli's barometer consists of a glass tube on a pool of mercury. The air above the mercury level in the tube has been evacuated (although mercury vapor is present). The air pressure at sea level is the same as the pressure of a column of mercury 760 mm high.

The air inside airliner cabins is kept at a pressure no less than the pressure of the air at an altitude of 2500 m above sea level. In the United States, aircraft that have accumulated more than 55,000 landings are barred from flying above 7500 m until they pass a stringent inspection. At an altitude of 8000 m, if the fuselage structure has been weakened by cracks, the difference between the inside and outside pressure can cause a rupture. In April 1988 an older airplane with more than 90,000 landings tore open while flying over Hawaii at an altitude of 8000 m. Much of the upper part of the fuselage behind the cockpit was ripped away, causing the death of a flight attendant who was swept out of the cabin.

PRESSURE IN A LIQUID

When you dive into the deep part of a swimming pool, you feel the pressure in your ears. This pressure increases as you descend to greater depths in a lake or the ocean. The pressure buildup is due to the weight of the water above you.

In Figure 9.7 the woman is swimming at a depth h (1 meter, for example). Consider the column of water pushing against a patch of skin on her back, 10 cm on its side. The column of water has a volume $V = Ah$, so its mass is $m = \rho V = \rho(Ah)$. The pressure of the column of water on the patch A is

$$p = \frac{F}{A} = \frac{mg}{A} = \frac{(\rho Ah)g}{A}$$

or

$$P = \rho gh.$$

Notice that the size of the patch is not important since the value of the area A does not appear in our expression for pressure. The pressure on the swimmer is proportional to the depth at which she swims. This is because the density of the liquid is constant no matter what the depth. Liquids are *incompressible*, since the electrons of the atoms and molecules, already very close to each other, resist getting any closer. Since the water molecules move at random in all possible directions, they collide with the swimmer's body from all sides, exerting pressure equally in all directions.

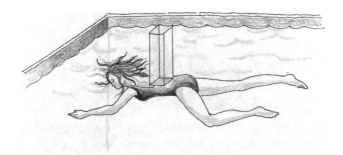

FIGURE 9.7   A swimmer at a depth h feels a pressure due to the water equal to ρgh.

FIGURE 9.8 An external pressure applied to a liquid is transmitted undiminished to all points in the liquid. This is called *Pascal's principle*.

The pressure (due to the liquid) on the woman, swimming 1 meter below the surface, would be $P = (1000 \text{ kg})(9.8 \text{ m/s}^2) \, (1 \text{ m}) = 9.8 \text{ kPa}$. The pressure on a fish, swimming in fresh water 1 meter below the surface would also be 9.8 kPa.

Since there usually is atmospheric pressure acting on the surface of a liquid, the total pressure is the sum of the atmospheric pressure, P_a, and the pressure of the liquid:

$$P_T = P_a + \rho g h.$$

The fact that the pressure in a liquid depends only on the distance below the surface means that the pressure does not depend on the shape of the container.

If we apply some external pressure to a liquid, by means of a piston, for example, as shown in Figure 9.8, *the pressure is transmitted undiminished to all points of the liquid*. This was discovered by the French scientist and philosopher Blaise Pascal in 1651 and is known as *Pascal's principle*.

In Figure 9.8, the pressure exerted by the force F on the lid at the left side of the container is transmitted to all the points in the liquid and to the walls of the container. The pressure is transmitted also to the piston on the right, and if we assume that this piston can move without friction, it will rise as the piston on the left is pushed down.

Pascal's principle is valid for any fluid in equilibrium. (For gases, it must be modified to take into account the change in volume when the pressure changes.) For liquids, Pascal's principle is a consequence of the incompressibility of the liquids. The molecules of a liquid are nearly as close to each other as they are in a solid, so there is not much space left between them. Any pressure applied at some point in a liquid is transmitted to every point of the liquid as each molecule feels the pressure of the molecule behind. Since, as was mentioned earlier, the molecules move at random in all directions, the collisions take place in all possible directions, which results in the pressure being transmitted *equally* in all directions. The swimmer feels the pressure in her eardrums no matter what the orientation of her head is in the water.

The hydraulic lift illustrated in Figure 9.9 is an application of Pascal's principle. A force f applied to the piston of area a results in an external pressure $P = f/a$. This pressure is transmitted to all the points in the liquid in all directions; in particular to the platform holding the car. If the section of the platform in contact with the liquid

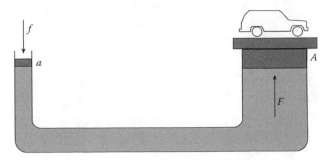

FIGURE 9.9   Hydraulic lift. The pressure $P = f/a$ exerted by the piston of area a is transmitted to all points in the liquid. The pressure $P = F/A$ on the larger area A, must be the same as the pressure on the smaller area on the left. To maintain the same ratio, the force F acting on the larger platform must be larger. This larger force can balance a large weight (like that of a car) resting on the platform.

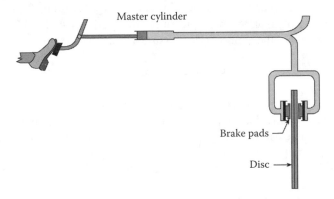

FIGURE 9.10   Brake system of a car.

has an area A, this pressure will be also equal to the force lifting the car F, divided by the area A. Since this area is much larger, the force lifting the car has to be larger to keep the same ratio. The hydraulic lift acts as a force magnifier, so that a small force f on one side produces a large force F on the other that is able to lift the car. To compensate, the piston moves much farther than the platform, a distance A/a as great.

The brake system of a car, illustrated schematically in Figure 9.10, is another application of Pascal's principle. As you step on the brake pedal, the increase in pressure is transmitted through the brake fluid, which causes movable pistons to push the brake shoes or pads against the drums or discs. Friction between these surfaces slows the car.

BUOYANCY

In the third century BC, the Syracusan king Hieron II asked his relative Archimedes, who happened to be the greatest scientist of the time, to determine whether a gold crown the king had commissioned had more than the allowed mixture of silver in it, and to do this without destroying the crown.

Archimedes was famous for the invention of ingenious mechanical devices. He is supposed to have constructed war machines that for three years held the invading Romans at bay in their siege of the city. Cranes mounted on the cliffs above the sea would hold heavy stones that pounded on the soldiers that ventured toward the walls. Similar cranes would be used to lift and overturn the landing ships. There is a story that he constructed a large mirror to reflect the sun's heat on the Roman ships and set them on fire.

Much of this was undoubtedly exaggerated, especially since we know about it only indirectly, from the writings of the later Greeks. Archimedes himself set no value on these devices, regarded them as beneath the dignity of pure science and published only his scientific work.

The importance of his scientific work is probably best exemplified by his method for the calculation of the number π. He calculated the perimeters of polygons inscribed inside and outside a circle and obtained bounds for π that ranged from 3(10/71) to 3(1/7). His method was unsurpassed until Newton invented integral calculus, eighteen hundred years later. It has been asserted that Archimedes' method truly anticipated the integral calculus.

After the king's request, Archimedes was perplexed as to how to go about determining the gold content of the crown without destroying it. As he stepped into his bath one day he noticed that the water overflowed and it occurred to him, or so the story goes, that the amount of water that overflowed must have the same volume as the part of his body under the water. By placing an amount of gold of the same weight as the crown under water, the correct volume of a gold crown could be determined. If the crown had any silver, it would displace more water than the same weight of gold, because silver is less dense than gold. He was allegedly so excited by this discovery that he ran through the streets of Syracuse to the palace without any clothes on shouting "eureka, eureka" ("I've got it! I've got it!"). Archimedes had proved that the crown had more than the required amount of silver in it and King Hieron had the cheating goldsmith executed.

From this incident, Archimedes worked out his principle of buoyancy, known today as Archimedes' principle, which can be stated as follows:

Archimedes' principle: An object partially or completely submerged in a fluid is buoyed up by a force equal to the weight of the fluid displaced by the object.

We all have experienced buoyancy. It is relatively easy to lift a rock underwater. It requires a much greater effort to lift the same rock on dry land. If you have ever attempted to teach someone to swim, you probably noticed how easy it was to hold that person up while in the water. In all these cases, the water exerts an upward force on the object. The source of this upward force is the difference in pressure between the top and the bottom of the submerged object. Since pressure increases with depth, the pressure at the top of the object is smaller than the pressure at the bottom. This pressure difference results in an upward force on the submerged object.

To illustrate Archimedes' principle, let's consider a small aluminum cube immersed in water, as shown in Figure 9.11. Imagine now that, somehow, the solid cube is removed and the cavity left behind is filled with more fluid. Clearly, this added fluid, undistinguishable from the fluid surrounding it, is in equilibrium and

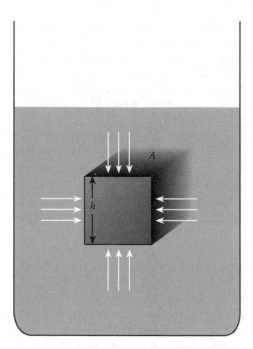

FIGURE 9.11   The buoyant force acting on the submerged cube results from the difference in pressure at the top and at the bottom. This buoyant force is equal to the weight of the fluid displaced by the cube and does not depend on the density of the cube. This is Archimedes' principle.

experiences pressures and forces that depend on the distance below the surface. Since the entire fluid remains in equilibrium, the extra fluid must be subject to the same upward net force as the solid cube it replaced. Moreover, this buoyant force must be just sufficient to balance the weight of the extra fluid. We can see that if we now replace the added fluid with a lead cube of the same dimensions as the original aluminum cube, this new heavier cube will be subject to the same buoyant force as the first cube and as the fluid itself. The force that buoys up the cubes is equal to the force that buoys up the fluid that each cube displaces, and this force is equal to the weight of the displaced fluid.

Clearly, for a given weight, the larger the amount of water it displaces when submerged in water the greater the buoyant force would be. This is the reason why ships—made of steel, which is much denser than water—can float. If we were to take the amount of steel used in building a ship and reshape it into a big solid cube, it would sink when immersed in water (see Figure 9.12). The amount of water displaced by the solid cube of steel is not large enough to provide a buoyant force that could balance the weight of the cube. When this steel is in the shape of a ship, that is, a large hollow bowl, the same weight of steel now displaces a larger volume of water which has a larger weight. By Archimedes' principle, this larger weight of the water displaced produces a larger buoyant force on the ship, enough to keep it afloat.

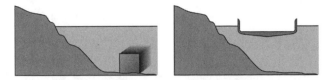

FIGURE 9.12 A big mass of steel in the shape of a cube sinks because the volume of water it displaces is not large enough. The same mass, in the shape of a ship, displaces more water, and the ship floats.

SURFACE TENSION AND CAPILLARITY

Why is a liquid drop spherical? How can a spider walk on the surface of a pond? Why is it possible to carefully place a razor blade on the surface of the water in a glass and have it float? These phenomena result from a property of liquids called *surface tension*. To understand how this works, let's consider a water molecule in the interior of a glass of water (Figure 9.13). This water molecule is surrounded in all directions by other water molecules, so that all the attractive forces from these molecules cancel out. A molecule on the surface of the water, however, has other water molecules only to the sides and below it exerting attractive forces. Therefore, this molecule is pulled in toward the interior of the liquid by these unbalanced forces. The result is that the surface of the liquid becomes compressed, and this acts to make the surface area of the liquid as small as possible.

We can now answer the questions posed above. A liquid drop is spherical because a sphere has the smallest surface area for a given volume. When you place a razor blade or a needle carefully on the surface of the water, the molecules of water on the surface are depressed slightly and the neighboring molecules exert upward restoring forces on them, supporting the razor blade. The surface of the liquid acts as a stretched elastic membrane; this is why it is possible for spiders and small insects to walk on water.

Surface tension can also help us understand the action of a detergent in cleaning. The surface tension of water is about 0.07 N/m. As shown in Table 9.2, water has a large surface tension. When you add detergent to water, the surface tension is reduced. The water then penetrates more easily into the fabric, dissolving dirt.

We can think of surface tension as the energy per unit area of the surface. In equilibrium the surface of a liquid would have the lowest possible energy, which is why liquids minimize their surface area. The units of surface tension are thus units of energy divided by units of area, or J/m^2. Since 1 J = 1 Nm, we can see that these units can be written as $Nm/m^2 = N/m$; that is, force divided by length. These are the units for surface tension used in Table 9.2. Thus, surface tension can also be defined in terms of force per unit length rather than energy per unit area. Experimentally, surface tension is determined by measuring the force required to pull a thin wire in the form of a ring out of the surface of the liquid until the surface stretches to the point of breaking (Figure 9.14).

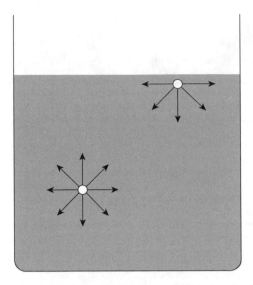

FIGURE 9.13   A molecule in the interior of the water feels attractive forces from all directions, whereas a molecule on the surface feels forces that point to the interior of the liquid, since there are no molecules above it. These molecules on the surface resist being pulled apart so that the surface of the liquid acts as an elastic membrane, able to support small objects.

TABLE 9.2
Surface Tension of Some Substances in Contact with
Air at Room Temperature

Substance	Surface Tension (N/m)
Benzene	0.0289
Mercury	0.465
Water	0.0728
Soapy water	0.025

If you look closely at the surface of the water in a glass, you would notice that near the glass, the surface curves up and the water seems to travel up the surface of the glass a small amount (Figure 9.15). This is what we call a *meniscus*. The forces of attraction acting between the molecules of a liquid are called *cohesive* forces. In addition to these forces, the molecules also experience *adhesive* forces that tend to bind the molecules to the walls of the container. In the case of water and glass, the adhesive forces are *greater* than the cohesive forces between the water molecules so that the molecules of water close to the glass experience a larger force toward the molecules of glass than toward the other water molecules and they are pulled up a small distance away from the surface of the water. The cohesive forces in mercury, on the other hand, are greater than the adhesive forces between mercury and glass, and a mercury atom near the surface of the glass experiences a larger force

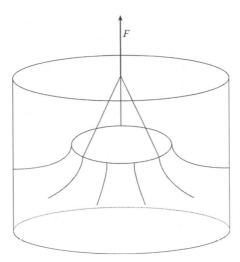

FIGURE 9.14 Measuring the force required to stretch the surface of the liquid to the point of breaking allows experimenters to determine the surface tension of the liquid.

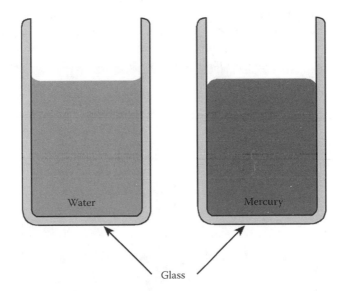

FIGURE 9.15 The curved surface observed on the surface of a liquid in a container is what we call a *meniscus*. The cohesive forces in water are smaller than the adhesive forces between water and glass. This produces a *positive* meniscus. The cohesive forces for mercury, on the other hand, are larger than the adhesive forces between mercury and glass, and this produces a *negative* meniscus.

toward the other mercury atoms, with the result that its surface curves downward, away from the glass. A meniscus that curves upward is called *positive* and one that curves downward, like the surface of mercury in a glass container, is called *negative*.

If we insert a narrow tube in a bowl with water (Figure 9.16), the water will rise in the tube to a higher level than the water in the bowl. This phenomenon is known as *capillarity*, a word that means hairlike. The water in the capillary tube will rise until the adhesive forces pulling the liquid up are balanced by the weight of the liquid in the tube. Therefore, the narrower the tube, the higher the water will rise. If we dip the edge of a cotton towel in water, the water will rise through the fabric fibers because the adhesive forces between cotton and water are larger than the cohesive forces in water. On the other hand, water will not spread through the fabric of a wool sweater with the edge dipped in water, because the adhesive forces between water and wool are smaller than the cohesive forces in water. Unlike cotton, wool fibers are covered with tiny scales. When wool is in contact with water, the fiber's scales rub against each other, pulling the fibers together. The scaly surface of the fibers tends to repel liquids. For this reason, wool fabrics are better for wet weather and cotton is a better material for towels. Cotton is also more comfortable to wear in hot humid weather because it absorbs excess perspiration through capillary action.

Capillarity is also responsible for dampness in basements, as water travels through the narrow cavities in the cinder blocks used to build the basement walls. To prevent this, the exterior walls must be coated with some waterproofing compound. Due to capillarity, soil can hold rainwater in the narrow spaces between tightly packed soil particles, and plants growing in this soil can make use of this water in photosynthesis, the process by which green plants convert water and carbon dioxide into oxygen and various organic compounds in the presence of light.

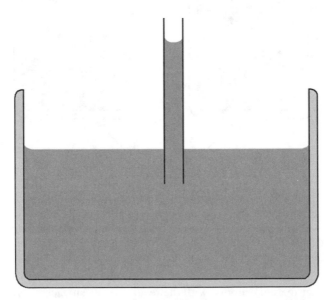

FIGURE 9.16   A liquid rises in a narrow tube because of a phenomenon known as capillarity.

FLUIDS IN MOTION

When you put a nozzle on your garden hose, the water comes out at a higher speed and reaches farther. The explanation for this phenomenon is due to an eighteenth-century Swiss mathematician named Daniel Bernoulli, although he did not set out to explain the behavior of water in a garden hose. *Bernoulli's principle* explains also why an airplane flies and why you can throw a curve with a baseball.

Suppose that you connect two different hoses and turn the water on. Let's assume that the two hoses have different diameters and that the connection between them allows for a smooth transition from one diameter to the other, as shown in Figure 9.17. We have indicated the direction of the flow of water with lines called *streamlines*. Since liquids are incompressible, the amount of fluid passing through area A on the left during a certain interval of time t is exactly the same as the amount of fluid passing through area a on the right during the same time t, and this means that the two volumes, V_L and V_R are the same. Since the areas are different, the volume of water passing through A moves through a length l of hose which is smaller than the length L along which the volume of water through area a must move. Therefore, the velocity of the water on the right, moving along the longer path L, must be greater than that on the left, which moves along the shorter path l, in order for the same amount of water to pass through both the narrow and the wide hoses in the same time. We can now understand why a nozzle on a garden hose shoots the water farther. The nozzle has a cross-sectional area that is smaller than the cross-sectional area of the hose, and so the speed of the water through the nozzle is greater.

If the velocity of the water is greater on the right, through the narrow section, than on the left, through the wide section of hose, work has to be done on the fluid to account for the increase in kinetic energy from left to right. This work is due to the forces exerted by the fluid on the cross-sectional areas in both sections. From the work-energy theorem, the net work is equal to the change in kinetic energy. Since the kinetic energy increases as the water flows from left to right, the work on the left, narrow section must be greater than that on the right, wide section. This implies that the pressure on the fluid at the left, wide section is greater than the pressure on the fluid on the right, narrow section.

Bernoulli's Principle tells us that the pressure is greater when the velocity of the fluid is less, and, conversely, that the pressure is less when the velocity of the fluid is greater.

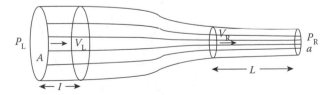

FIGURE 9.17 Two hoses of different diameters connected in such a way as to have a smooth transition from one diameter to the other.

Armed with this beautiful discovery, we can understand what keeps an airplane in the air. Figure 9.18 shows a cross section of an airplane wing. The top surface of the wing is curved and the lower surface is flat so that the air rushing by the upper surface has a longer distance to cover than the air passing by the lower surface. Therefore the velocity of the air above the wing is *greater* than the velocity of the air below the wing. Bernoulli's principle tells us that the pressure is greatest where the velocity is least, so that the pressure on the wing from below is *greater* than the pressure on the wing from above. The pressure difference provides a net upward force on the wing.

It is not very difficult for you to experience Bernoulli's principle first hand. If you pick up a sheet of paper and blow across the upper surface, as shown in Figure 9.19,

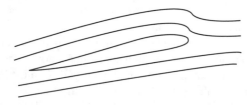

FIGURE 9.18   Cross section of an airplane wing. The air passing the upper surface has a longer distance to cover, so it must move at a higher speed than the air passing the lower surface. The pressure on the underside is greater, where the velocity is smaller, thus providing a net upward force on the wing.

FIGURE 9.19   The pressure from the bottom, where the air is still, is larger than the pressure from above, where the air is moving as you blow. This net pressure upwards lifts the paper.

the velocity of the air above will be greater than the velocity of the air below, which will be near zero. The larger pressure from below pushes the paper up.

PHYSICS IN OUR WORLD: CURVE BALLS

To throw a curve ball, a baseball pitcher gives a spin to the ball. The spinning ball drags some of the air around in the direction of the spin. The speed of air on one side of the ball becomes greater than on the other side. According to Bernoulli's principle, the side of the ball where the air moves faster has a lower pressure than the other side. The difference in pressure causes the ball to curve (Figure 9.20).

We can also understand the behavior of the curve ball by studying the way the air flows relative to the ball. When a ball is thrown through the air at high speed, the air behind the ball moves in an irregular, turbulent way. The streamlines curve symmetrically around the ball. When the ball is spinning, the air is dragged around, causing the region of turbulence to move slightly in the direction of the spin. The asymmetry of the region of turbulence produces a crowding of the streamlines on one side. The pressure is then lower where the streamlines are crowded. This pressure difference produces a net force that deflects the ball.

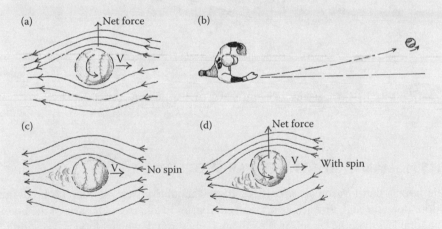

FIGURE 9.20   The physics of a curve ball.

(a) As seen from above, the air moves faster on one side (where the streamlines are closer together) and the pressure is lower there. The ball moves in the direction of the net force. (b) The ball spinning counterclockwise as seen from above curves to the left of the pitcher. (c) A nonspinning ball has symmetric streamlines and the area of turbulence is behind it. (d) In a spinning ball, the area of turbulence is dragged to one side. The ball curves in the direction of the net force exerted by the air.

BASTARD WING

An airplane flying at very low speeds generates little lift and can stall, since the difference in pressures above and below the wing comes from the difference in wind velocity above and below. Birds, on the other hand, usually manage to fly at such low speeds. Their secret lies in a special structure called a bastard wing or alula. This structure consists of a few feathers attached to the first digit or thumb of the wing (Figure 9.21). When birds slow down at takeoff or when landing, they tilt the wing and use the alula to separate its feathers from the rest of the wing, creating a slot that prevents the air from breaking away from the upper surface and distributes more air flow to the top of the wing. The improved difference between the air velocities above and below the wing increases the pressure difference enough to balance the bird's weight.

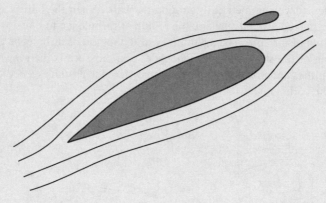

FIGURE 9.21   Cross section of a bird's wing showing the alula or bastard wing.

THE HUMAN CARDIOVASCULAR SYSTEM

The human cardiovascular system is a closed tubular system in which a fluid, the blood, flows through arteries, veins, and capillaries to and from all parts of the body. The total length of the vessels in the human body through which blood is transported reaches many miles.

The heart is the muscular pump that propels the blood into and out of the vessels. An average adult human heart is about $13 \times 9 \times 6$ cm and weighs about 300 g. During a lifetime of 70 years, by beating some three billion times, the heart pumps nearly 250 million liters of blood, which would fill a football stadium almost knee deep. The heart consists of four chambers—two *atria*, left and right; and two *ventricles*, left and right (Figure 9.22). The atria are the receiving chambers for blood from the body which is pumped into the ventricles; and the ventricles pump blood into the lungs and to the rest of the body. Four valves control the flow of blood in the heart: the tricuspid valve, the mitral valve, the pulmonic valve, and the aortic valve.

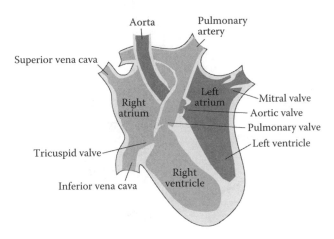

FIGURE 9.22 The human heart consists of four chambers—left and right atria and left and right ventricles.

Blood coming from the head through a large vein or *superior vena cava* and from the arms, liver, body muscles, kidney, and legs through a second large vein, the *inferior vena cava*, enters the right atrium, located in the right upper part of the heart. A third vein, the *coronary sinus*, drains blood from the heart itself into the right atrium. All this blood, which has been depleted of oxygen, is pumped into the right ventricle through the tricuspid valve. As the right ventricle contracts, blood is carried into the pulmonary artery to the lungs where the blood is oxygenated. The pulmonary valve prevents this blood from reentering the heart.

The slightly smaller left atrium receives oxygenated blood from the lungs through four pulmonary veins and pumps it into the thick-walled left ventricle through the mitral valve. At the same time that the right ventricle pumps blood into the pulmonary artery, the left ventricle contracts and forces blood out through the aorta into the arteries of the body.

The system that assures the pumping of the blood in and out of the heart at the right times is controlled by a special structure called the *sino-auricular node*, which acts as the pacemaker of the heart. The heart beats some 70 to 80 times per minute, contracting and expanding. During the period of relaxation of the heart or *diastole*, the pressure is typically 80 mmHg (mercury), while at the peak of its cycle, during the contraction period or *systole*, the pressure is about 120 mmHg.

This blood pressure can be measured with a *sphygmomanometer*, a device that uses a U-tube of mercury to read the pressure during the relaxation and contraction periods of the heart (Figure 9.23). As we saw earlier, pressure in a liquid increases with depth. When a person is lying down, the pressure throughout the major arteries of the body does not change more than 2 mmHg, this small change being due to some resistance to the flow of blood in the arteries. Standing up, however, the difference in pressure between the heart and the foot is about 100 mmHg. Since the upper arm is at about the same level as the heart, the pressure there is close to the pressure at the heart. For this reason, the sphygmomanometer measures the pressure with

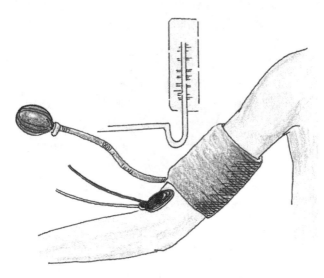

FIGURE 9.23   A sphygmomanometer used to measure blood pressure.

an inflatable cuff wrapped around the upper arm. A column of mercury or a spring scale is connected by rubber tubing to the cuff and a stethoscope is used to listen for noises in the artery. The cuff is inflated until the flow of blood through the brachial artery is stopped by the pressure in the cuff. Air is gradually allowed to escape from the cuff until the pressure in the cuff matches the peak pressure in the heart (systolic). At this point, blood begins to flow through the artery only during the systolic pressure part of the cycle. Since the brachial artery is partially opened, the area is small and the equation of continuity tells us that the velocity should be high. This makes the flow of blood noisy and easy to pick up with the stethoscope. As the cuff continues deflating, the pressure in the sack becomes lower than the systolic pressure but still higher than the diastolic pressure. Blood still does not flow during the diastolic pressure part of the cycle. When the pressure in the cuff matches the diastolic pressure, the artery remains open during the entire cycle. By reading the values of the cuff pressure at the two points, when the pulse is first heard and when the sound of blood through the artery is continuous, the systolic and diastolic pressures can be measured. Blood pressure can change with activity or exercise. The normal systolic pressure of a healthy adult should be less than 120 mmHg and the diastolic less than 80 mmHg.

Part IV

Thermodynamics

10 Heat and Temperature

HEAT AS A FORM OF ENERGY

When you push a book that is lying on a polished table, part of the work that you do on the book is converted into kinetic energy and the book begins to move with a certain velocity (Figure 10.1). After you release it, however, the book does not continue moving with the same velocity; it slows down and eventually stops. What happens to the initial kinetic energy of the book? Since there is no change in the potential energy, as the book remains on the table, the mechanical energy is not conserved. As we pointed out in Chapter 5, friction has taken away some of the energy. Although it might be difficult to verify without sophisticated equipment, the book and the table are warmer than before. From our experience, however, we know that friction causes objects to become warmer. When the weather is cold, for example, we rub our hands together to warm them up; the wheels of a motorcycle become hot during braking due to rubbing with the brake pads; and the tires get warmer due to friction (Figure 10.2).

Heat, then, seems to be a form of energy. This was not known until the middle of the nineteenth century. Before then, heat was thought to be a fluid that was transmitted from hot bodies to cold bodies. This *caloric fluid* (as the French chemist Lavoisier called it) was believed to be a conserved quantity, since it had been observed that when equal parts of hot water and cold water were mixed, the result was warm water at a temperature that was exactly the mean of the initial temperatures. The caloric fluid was thought to have been transmitted from the hot water to the cold water without any losses.

The problem with the idea of the caloric was that it could not explain how heat was produced by friction. The correct interpretation of heat began with the experiments of an American-born scientist who took up British citizenship, held a German title of nobility, and led a strange and colorful life. Benjamin Thompson, Count Rumford, was born in Woburn, Massachusetts, in 1753, a couple of miles from the birthplace—some 50 years before—of his namesake Benjamin Franklin.

While Count Rumford was supervising the boring of a cannon in Munich in 1798 he noticed that the iron became hot enough to glow and had to be cooled with water. Rumford decided to study this phenomenon and performed many experiments in which he immersed the hot metal in water to measure the rate at which the temperature rose. He was able to determine that the amount of heat generated by friction while the boring tool drilled the iron was large enough to melt the metal had it not

FIGURE 10.1   If we push a book lying on a table, the book slows down and stops. What happens to the kinetic energy of the book?

FIGURE 10.2   Friction between the wheels and the brake pads makes the wheels hot. The tires also get warm after the motorcycle has been driven due to friction with the pavement.

been cooled. He realized that the amount of heat that could be generated was not only not constant, as the caloric theory proposed, but unlimited as long as work was done. In his report to the Royal Society entitled "An Inquiry Concerning the Source of Heat which is Excited by Friction," Rumford wrote:

> And, in reasoning on this subject, we must not forget to consider that most remarkable circumstance, that the source of the Heat generated by friction, in these experiments, appeared evidently to be inexhaustible … [I]t appears to me to be extremely difficult, if not quite impossible, to form any distinct idea of anything capable of being excited

and communicated in the manner the Heat was excited and communicated in these experiments, except it be motion.

Rumford's brilliant—and correct—conclusion was that the motion of the drill was being converted into heat and that heat was a form of motion. His calculation of the ratio of heat to work was very close to the accepted values today.

In spite of Rumford's insights, the caloric theory continued as the leading theory for some forty years. Starting in the late 1830s, James Prescott Joule repeated some of Rumford's experiments many times, enhancing and refining them. In one of his experiments, he measured the heat generated by an electric current and compared it with the mechanical energy required to run the simple electric generator that produced the current. In another experiment he measured the heat generated when water was forced through pipes and compared it with the work required to maintain the flow of water through the pipes. He also designed experiments that compared the work required to compress a gas contained in a bottle, which was in turn immersed in water, with the amount of heat gained by the water.

Joule's most famous experiment, however, involved the design of an apparatus in which a brass paddle wheel immersed in water was turned by the descension of weights suspended by string from pulleys (Figure 10.3). The potential energy lost by the weights as they fell was compared with the heat gained by the water.

In his 1890 report to the Royal Society entitled "On the Mechanical Equivalent of Heat," Joule concluded from this experiment "[t]hat the quantity of heat produced by the friction of bodies, whether solid or liquid, is always proportional to the quantity of [energy] expended." Heat, then, as Joule was able to establish, is a form of energy, contrary to the caloric theory.

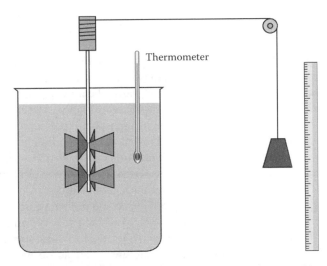

FIGURE 10.3   Joule's apparatus to measure the mechanical equivalent of heat.

PIONEERS OF PHYSICS: COUNT RUMFORD

The son of a farmer, Benjamin Thompson started as an apprentice to a storekeeper. At the age of 19 he married a wealthy and older widow from the town of Rumford (the present Concord), New Hampshire. When the Revolutionary War broke out, Thompson's sympathies were with the English Crown and he served the King by spying on his countrymen.

When the British left Boston, Thompson was forced to flee to London, leaving his wife and daughter behind. After the war ended, he remained in England and was knighted by King George III, but was later accused of being a spy for the French and of accepting bribes. In 1783 the king allowed him to enter the Bavarian civil service and Thompson left for Germany where he worked in the court of Elector Karl Theodor of Bavaria as an administrator. He served the Elector well and was rewarded with the title of Count of the Holy Roman Empire in 1790. Thompson chose the name of his wife's hometown for his title and became Count von Rumford.

MEASURING TEMPERATURE

Temperature is a familiar concept to us as a measure of the hotness or coldness of an object and for this reason we have used the concept in the previous section. The device with which we measure temperature, the *thermometer*, is also familiar to us.

A thermometer measures variations of some physical property that changes with temperature, such as the volume of a liquid or a gas. The idea of representing hot and cold with numbers dates from antiquity. In the second century AD the Greek physician Galen proposed a temperature scale based on ice and boiling water. Arab and Latin physicians actually developed a numeric scale from 0 to 4 to represent coldness and hotness but lacked instruments to measure and relied only on their senses. In the seventeenth century, Galileo invented a thermometer that consisted of a glass tube filled with air connected to a thinner tube marked with divisions and immersed in a vessel of colored water. When it was cold, the air in the tube contracted and the colored water rose in the glass tube. When it was warm, the air expanded and the water moved down in the thin tube. Galileo's thermometer was not very accurate, however, because it did not take into account changes in atmospheric pressure.

In 1657, some of Galileo's disciples improved upon this design and developed thermometers in sealed tubes to avoid changes due to the atmospheric pressure. These thermometers were based on the expansion and contraction of alcohol with changes in temperature and became popular throughout Europe. Alcohol, however, boiled at a low temperature, so high temperatures could not be measured.

In 1714 the German–Dutch physicist Gabriel Fahrenheit, who had emigrated to Amsterdam after his parents died in Germany, decided to substitute mercury for alcohol and for the first time extreme temperatures, well above the boiling point of water and below the freezing point of water, could be measured. Thirteen years before, Newton had suggested that the temperature scale should be set at zero for the freezing point of water and at 12 degrees for the temperature of the human body. Fahrenheit, however, did not want to use negative numbers for the temperatures of the cold winter days and added salt to the water to obtain the lowest freezing point that he was capable of in his laboratory and called it the zero temperature of his scale. Instead of dividing the difference between this freezing point and the temperature of the human body in twelve equal parts as Newton had suggested, he decided to divide this interval into eight times twelve, or ninety-six, to obtain a larger number of divisions in his scale. In 1724, Fahrenheit adjusted this scale so that the boiling point of water would come out to be exactly 180 degrees above the freezing point of water alone, which was 32 degrees in his scale. With this correction, the human body temperature was found to be 98.48 and the boiling point of water, 212°.

The Swedish astronomer Anders Celsius proposed in 1742 a new scale in which the freezing point of water was 100° and the boiling point of water at sea level, 0°. The following year, he reversed his scale, making the freezing point 0° and the boiling point 100°. Since there were 100 degrees between the freezing and boiling temperatures for water, the scale was known for many years as the *centigrade* or "hundred step" scale. Today, we call it the Celsius scale.

On the Fahrenheit scale there are 180 degrees between the freezing and boiling points of water whereas in the Celsius scale there are 100. Thus, each Celsius degree is larger than each Fahrenheit degree by a factor of 180/100 or 9/5. Since the freezing point of water is 32° in the Fahrenheit scale and 0° in the Celsius scale, we can convert any Celsius temperature to its corresponding value in Fahrenheit by multiplying the Celsius temperature by the factor 9/5 and adding 32:

$$T_F = \frac{9}{5} T_C + 32.$$

The Fahrenheit to Celsius conversion can be obtained from this expression by simply solving for T_C:

$$T_C = \frac{5}{9} (T_F - 32).$$

In 1848 the English physicist William Thomson, later Lord Kelvin, devised what is now known as the Kelvin scale, Kelvin was able to show that if a gas was cooled down to 273° below 0°C, the random motion of its molecules would be minimum. Therefore, there could be no temperature below this value. He then proposed a scale that would start at this absolute zero and would increase in intervals that were the same as the Celsius intervals.

The Kelvin scale differs from the Celsius scale only in the choice of zero temperature. To convert from Celsius degrees to *kelvins*, as the intervals in the absolute scale are known, we add 273.15 (a more precise value). Using T for the temperature in kelvins we have

$$T = T_C + 273.15.$$

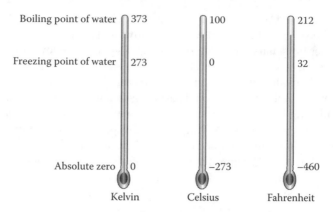

Boiling point of water 373   100   212

Freezing point of water 273   0   32

Absolute zero 0   −273   −460

Kelvin   Celsius   Fahrenheit

FIGURE 10.4   A comparison of the Kelvin, Celsius and Fahrenheit temperature scales.

In Figure 10.4 we compare the three different scales of temperature that we have considered here. Notice that the intervals between any two temperatures are the same only in the Kelvin and Celsius scales.

TEMPERATURE AND HEAT

What exactly does temperature measure? When we measure the temperature of a body, we are actually obtaining information about the average kinetic energy of the atoms and molecules which make up that body. When the body is warm, there is more molecular motion, and temperature is the measure of the kinetic energy of this motion. Temperature is directly related to the average kinetic energy of the atoms and molecules of a body. At higher temperatures, the kinetic energy is greater and this infers that the average molecular speed is greater.

As Joule and Rumford established, heat is a form of energy. What is then the difference between heat and temperature? To answer this question, we need to consider the behavior of the atoms and molecules of a substance. These, as we have seen, are in constant motion. In gases and liquids, they move about at random. In a gas composed of single atoms such as helium, in addition to this kinetic energy due to random motion, the atoms possess kinetic energy due to the spinning of the atoms themselves. However, this *rotational* kinetic energy is very small compared with the *translational* kinetic energy.

In gases like nitrogen and oxygen, composed of *diatomic* molecules, that is, molecules formed by two atoms bound together (Figure 10.5), the molecules can also rotate. The moment of inertia of these diatomic molecules is much larger, however, since the masses of the individual atoms in each molecule are farther apart from the axis of rotation, and the kinetic energy due to this rotation is greater. The two nitrogen atoms in the nitrogen molecule can also vibrate back and forth, and this vibrational motion also contributes to the total kinetic energy of the molecule.

In solids, the situation is more complicated since the molecules, ions, or atoms that make the solid are not free to drift around. These particles have no random translational kinetic energy. There are, however, vibrational kinetic and potential energies, as these particles vibrate back and forth around more or less fixed positions.

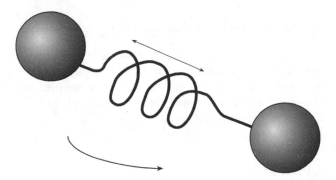

FIGURE 10.5   An oxygen molecule consists of two oxygen atoms bound together by spring-like forces. The molecule can rotate around an axis perpendicular to the line joining the atoms. The atoms can also vibrate back and forth.

The sum of *all* the random energies of the atoms and molecules in a substance constitutes the *thermal energy* of that substance. The thermal energy *transferred* from a warm object to a cooler object as a result of the temperature difference between the two is what we call *heat energy*. Heat, then, is the *flow* of thermal energy, and for this reason, it would not make sense to speak of the amount of heat that a substance has. We can, however, speak of the amount of thermal energy in a substance and the amount of heat *transferred* from one substance to another.

On the other hand, the temperature of a gas or a liquid depends only on the random translational kinetic energy per molecule. (In solids, since there is no translational kinetic energy, the temperature depends on the vibrational energy). The temperature of a substance, then, does not depend on the *number* of atoms and molecules in the substance nor on the *kinds* of atoms that form the substance. If you fill a bowl with water from a swimming pool, the number of water molecules in the pool is, of course, much greater than in the bowl, and the total sum of the translational, vibrational, and rotational kinetic energies of the water molecules in the pool is much greater than the corresponding energies of the water in the bowl. Therefore, the thermal energy of the water in the swimming pool is much greater than the thermal energy of the water in the bowl. However, the temperatures of the water in the swimming pool and in the bowl are the same, since we filled the bowl with water from the pool and temperature is the measure of the average kinetic energy per molecule.

Since heat is a form of energy, the SI unit is the joule. During the times of the caloric theory, a unit of heat called the *calorie* (cal) was introduced. The calorie, still in use today, was defined as the amount of heat required to raise the temperature of 1 gram of water by 1°C. We define the calorie today in terms of joules:

$$1 \text{ cal} = 4.186 \text{ J}.$$

(The nutritionists' calorie is actually 1000 calories or a kilocalorie, as the unit is defined in physics. Due to its widespread use regarding the energy value of foods, to avoid confusion it is sometimes distinguished by writing it with a capital letter C as *Calorie*, Cal.)

PHYSICS IN OUR WORLD: THERMOGRAPHY

As we shall see in Chapter 23, all objects emit radiation. This radiation can be used to determine temperature. At low to intermediate temperatures most of the radiation is in the infrared region. When the temperature increases, the intensity of the radiation emitted increases considerably. An infrared camera can detect this radiation, producing electronic signals that can be displayed on a television monitor or photographed on special film. The image produced is called a *thermogram*, and the technique *thermography* (see Figure 10.6).

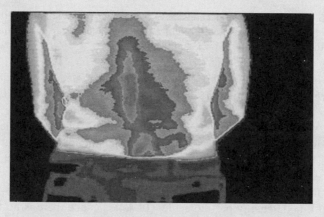

FIGURE 10.6   **(See color insert following page 268.)** Thermograph showing pinched nerves, fractured vertebrae, muscle strain, and dislocations. (Courtesy of Teletherm Infrared, Dunedin, FL, Ashwin Systems International, www.thermology.com.)

The intensity of the infrared radiation emitted by the human body depends on several factors, including state of health. Certain regions of the body have a larger blood supply than others and therefore emit more radiation. However bone tumors, bone infections such as osteomyelitis, arthritic conditions, and muscle and tendon diseases resulting from inflammation emit more radiation than normal. On the other hand, areas where blood flow is reduced emit less radiation than normal, permitting the detection of blockages of the arteries and disorders where blood circulation is below normal, such as atherosclerosis. Thermography has shown itself to be a very important technique for the detection and diagnosis of these and other disorders. A typical thermogram is shown in the color plate.

Thermography can also help in the detection of breast cancer. The x-ray process called mammography remains the recommended procedure for early detection of breast cancer, but when combined with thermography and other techniques such as ultrasound screening, the rate of cancer detection increases substantially.

HEAT CAPACITY

What happens when we heat a substance? When we heat up some water to prepare ourselves a cup of instant coffee, where does the thermal energy transferred to the water—the heat—go? When we place a TV dinner in the oven and heat it, why is the skin of the potato warm but the inside usually very hot? Some foods seem to store more thermal energy than others when placed together in the oven.

The capacities for storage of thermal energy for different substances are different. When heat flows into a substance, this thermal energy goes into the different forms of kinetic energy; translational, rotational, or vibrational. If the substance is a monatomic gas like helium, most of the energy goes into translational kinetic energy. For diatomic gases like nitrogen and oxygen, the energy is shared among the different forms of kinetic energy. If the substance is a liquid or a solid, the interactions between the molecules give rise to other ways for this thermal energy to go. Only the increase in the translational kinetic energy, however, gives rise to an increase in temperature. Adding the same amount of heat to two different substances, such as water for your coffee and soup for your dinner, produces different increases in the temperatures of the two substances, as the amount of heat that goes into translational kinetic energy is different for the two substances.

It takes longer to warm up water for two cups of coffee than for just one cup. The reason is, of course, that heating two cups of water requires more heat than heating just one. A cup of water has a volume of 250 ml and a mass of 250 g. Since one calorie of heat is required to raise the temperature of water 1°C, it takes 250 calories for every degree Celsius that we want to raise the temperature of one cup of water, and 500 calories for two cups. The greater the mass, the greater the amount of heat required to raise the temperature of a substance.

Thus the heat required to warm up a substance is proportional both to the mass of the substance and to the change in temperature. The heat required to increase the temperature of a mass m of a substance an amount ΔT is called the *heat capacity C* of the substance. A more useful quantity, since it has the same value regardless of the mass of the substance, is the *specific heat capacity c*, which is the heat capacity *per unit mass* of a substance. The specific heat capacity, then, is *the heat required to raise the temperature of a unit mass of a substance by one degree.* If Q is the amount of heat required, m the mass of the substance, and ΔT the change in temperature, the specific heat capacity is

$$c = \frac{Q}{m\Delta T}.$$

The SI unit of specific heat capacity is joule per kilogram kelvin (J/kg K), but kcal/kg °C is also commonly used. The specific heat capacities of several substances are listed in Table 10.1. Notice that the specific heat capacity of water is exactly 1.000 kcal/kg °C. This is because the calorie was defined as the heat required to raise the temperature of 1 g of water by 1°C. To raise the temperature of one kilogram of water by 1°C requires one kilocalorie.

TABLE 10.1
Specific Heat Capacities of Some Substances

Substance	Specific Heat (kcal/kg°C)	J/kg K	Substance	Specific Heat (kcal/kg°C)	J/kg K
Water	1.000	4184	Glass	0.15	630
Human body	0.83	3470	Steel	0.107	490
Ice	0.50	2090	Copper	0.093	390
Steam (100°C)	0.46	1925	Mercury	0.033	138
Wood	0.42	1750	Gold	0.032	134
Aluminum	0.21	880	Lead	0.031	130

HEAT OF FUSION AND HEAT OF VAPORIZATION

When heat is added to a substance, its temperature usually increases. There are, however, some special situations where the temperature stays constant as heat is added. If you add heat to an ice cube in a closed container at 0°C and monitor its temperature, you would notice that the mixture of ice and water resulting from the melting of the ice remains at 0°C until all the ice has melted. It seems as if the water-ice mixture is able to absorb the heat without any change in temperature. The ice, however, is melting; that is, it is changing its phase from solid to liquid and this requires energy.

A similar phenomenon takes place when water changes into steam at 100°C. If you start with tap water at 20°C and add heat by placing the container with water on a stove, for example, the temperature of the water will steadily increase until it reaches 100°C. The temperature will then remain at 100°C until all the water has turned into steam. The temperature increase will resume after that.

The change of a liquid into solid is called *fusion*, and the change of a liquid into its gas phase is called *vaporization*. The amount of heat required for a substance to change phase depends on the type of substance, its mass, and whether the phase change is from solid to liquid or from liquid to gas. For water, 80 kcal of heat are needed to melt one kilogram of ice. Ice, like most solids, is a crystal and, as we saw at the beginning of Chapter 9, the molecules in a crystal occupy fixed positions and are held in those positions by strong intermolecular forces. When the ice melts, work must be done against these forces, which means that energy must be supplied to the solid. As more heat energy is supplied, the water molecules are able to break loose from their neighbors and the crystal structure is broken. The molecules that formed the crystal and occupied fixed positions while in the solid phase are now free to move about; the water is now in the liquid phase (Figure 10.7). Since any other substance would have a different molecular structure and different binding forces between its molecules, the energy required for the molecules of this substance to break loose from their neighbors is different.

The heat energy that is supplied to the solid goes into increasing the potential energy of the molecules of the solid. Since temperature is a measure of the random

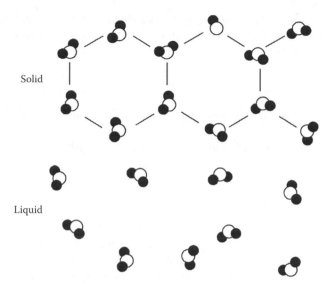

FIGURE 10.7   The heat that is added to the ice goes into increasing the potential energy of the water molecules in the crystal structure. When these molecules gain enough energy, they break loose from their neighbors and are free to move. This is the liquid phase of water.

kinetic energies of the molecules, the temperature remains fixed while the solid melts.

Although the molecules of a liquid are free of the strong binding forces that keep them in more or less fixed positions while in the solid phase, they are still bound by weaker intermolecular forces that allow them to move around. These weaker bonds must also be broken when the liquid vaporizes. In the gaseous phase, however, the molecules are separated from each other by considerable distances and this separation requires energy. In the transition from solid to liquid, the molecules remain more or less at the same distances and it is only the breaking of the crystal structure that requires energy. The heat required to vaporize a substance in the liquid phase is therefore larger than that required to melt it. In the case of water, for example, the heat needed to vaporize one kilogram of water is 540 kcal, almost seven times the heat required to melt one kilogram of ice.

When a gas condenses into its liquid phase, energy is released. This amount of energy is the same amount that was required to vaporize the liquid. Water, then, releases 540 kcal for every kilogram of steam that condenses. Similarly, when 1 kilogram of liquid water solidifies, 80 kcal of energy are released. The heat *absorbed* or *released* by one kilogram of a substance during a phase transition is called the *latent heat, L*. If the transition is from solid to liquid or vice versa, this latent heat is called latent heat of fusion, L_f. If the transition involves the liquid and gas phases of a substance, it is called the *latent heat of vaporization, L_v*. Thus, for water,

$$L_f = 80 \text{ kcal/kg} = 335 \text{ kJ/kg} \qquad L_v = 540 \text{ kcal/kg} = 2259 \text{ kJ/kg}.$$

Since 80 kcal of heat are needed to melt 1 kg of ice, 160 kcal would be needed to melt 2 kg, and 800 kcal to melt 10 kg. If we have m kg of ice, we would need 80 m kcal of heat to melt them:

$$\text{Heat to melt a mass } m: \quad Q = mL_f.$$

Similarly, to vaporize a mass m of a liquid requires an amount of heat $Q = mL_v$.

$$\text{Heat to vaporize a mass } m: \quad Q = mL_v.$$

EVAPORATION AND BOILING

When a glass of water is left overnight on the kitchen counter, for example, we notice that the level of the water in the glass drops a certain amount. Some of the water has evaporated. This happens because some of the molecules of water have escaped and became gaseous molecules. As we have seen, the molecules of a liquid are in continuous random motion and possess random kinetic energies. Some molecules, then, would move with greater velocities than the average and if they happen to be near the surface of the liquid, moving in an upward direction, they might have enough energy to escape. This escape is what we call *evaporation*. Since the molecules that escape are, on average, the ones with greater kinetic energies, the average value of the kinetic energy of the molecules that remain in the liquid is reduced. Since temperature is the measure of the average kinetic energy per molecule of the liquid, the temperature of the liquid drops as the liquid evaporates (Figure 10.8).

Evaporation, then, is a cooling process. This explains why you feel cool as soon as you come out of a swimming pool; as the water on your skin evaporates, the remaining water, which is in contact with your skin, is left at a lower temperature. You may also have noticed that when a nurse rubs alcohol on your skin, it feels cool.

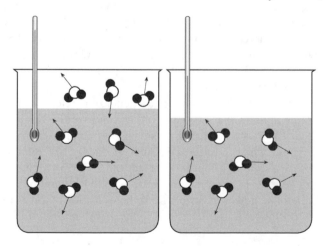

FIGURE 10.8   As the fastest molecules leave the surface of the water, the temperature of the remaining liquid drops.

This is because alcohol evaporates very rapidly. Another liquid that evaporates fast is gasoline, and if you ever accidentally spilled some of it on your hand while pumping gas at a self-service gas station, you might have noticed that it also leaves your skin cooler. And the reason why a fan offers some relief on a hot summer day is because it blows air around your body, speeding the evaporation of perspiration.

When the temperature of the liquid is high enough, close to the boiling point, evaporation is increased substantially, since a greater number of molecules possess enough energy to escape from the liquid. At these high temperatures, evaporation can also take place *inside* the liquid, away from the surface, where bubbles of gas form. When these bubbles of gas begin to appear in the interior of the liquid, boiling starts.

For the bubbles of gas to form, the molecules inside the bubble must have kinetic energies large enough for the pressure from inside the bubble to match the pressure from the liquid on the bubble. The pressure that the liquid exerts on the bubble is the atmospheric pressure plus the pressure due to the liquid at that depth. In most cases the pressure due to the liquid is very small compared to the atmospheric pressure and we can say that boiling begins when the pressure of the gas is equal to the atmospheric pressure (Figure 10.9). Since the gas in the bubbles is less dense than the liquid surrounding them, they are buoyed to the surface of the liquid.

As we have seen, water boils at 100°C. This, however, is only true when the atmospheric pressure is 1 atm. At different atmospheric pressures, the boiling point of water—or of any other liquid—changes. Since boiling begins when the pressure of the gas in the bubbles that form in the interior of the liquid is equal to the atmospheric pressure, the molecules of the liquid require a greater kinetic energy to be able to form bubbles that can sustain an increased atmospheric pressure. Water, then, boils at a higher temperature when the atmospheric pressure increases, and, conversely, boils at a lower temperature when the atmospheric pressure decreases. If you go camping in the mountains, water boils at a temperature lower than 100°C; since the rate at which foods cook in boiling water depends on the temperature of the

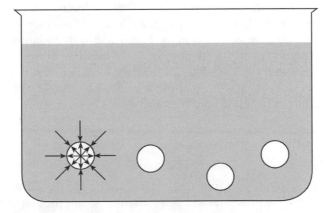

FIGURE 10.9   When the temperature of the liquid is high enough (100°C for water), evaporation takes place even inside the liquid and bubbles begin to appear. When the pressure of the gas in the bubbles equals the atmospheric pressure, boiling begins.

water, it takes longer to boil an egg. In Cocoa Beach, Florida, water boils at 100°C and it takes about 10 minutes after the water reaches the boiling point to hard-boil an egg. On the highest mountain in the Eastern United States, Mt. Mitchell, North Carolina, elevation 6600 feet, water boils at 93°C; it takes almost twice as long to hard-boil an egg there.

A pressure cooker confines the steam under the sealed lid until the pressure is nearly 2 atmospheres (Figure 10.10). At this pressure, water boils at 120°C and foods cook faster at this higher temperature.

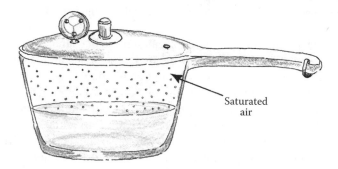

Saturated
air

FIGURE 10.10   A pressure cooker confines the steam until the pressure under the sealed lid reaches 2 atm. Foods cook faster because of the higher temperatures.

PHYSICS IN OUR WORLD: INSTANT ICE CREAM

The traditional method for making ice cream involves churning a mixture of milk, eggs, sugar, and flavorings while it is chilled slowly. The churning process prevents the formation of large crystals, thus producing a smooth texture, and also traps air inside the mixture, which accounts for the fluffiness and lightness of good ice cream.

Physicist Peter Barham of the University of Bristol in the United Kingdom has developed a technique to make ice cream in less than 20 seconds. Instead of chilling the mixture slowly, as in the traditional method, Barham chills it instantaneously. He pours liquid nitrogen on the mixture. Since nitrogen liquefies at temperatures below −196°C, the mixture cools so fast that very few crystals develop. When the liquid nitrogen comes into contact with the much warmer mixture, it begins to boil, creating bubbles in the mixture.

HUMIDITY

Due to the evaporation of water from lakes, ponds, rivers, and the sea, the air contains some water vapor. If we place a lid on a pan with water so that the evaporated gaseous water molecules cannot escape (Figure 10.11), an equilibrium state will eventually be reached where the number of molecules leaving the surface of the water equals the number of molecules that return to it after bouncing off the lid and walls of the pan. When

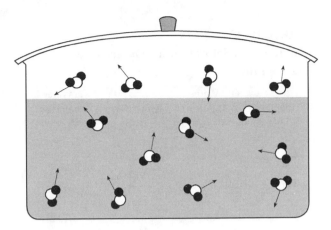

FIGURE 10.11 The number of molecules leaving the surface of the water is equal to the number of molecules reentering the liquid. The air under the lid is said to be saturated.

this happens, the air above the water becomes *saturated* with vapor, in other words, the air holds the maximum amount of water vapor that it can at that temperature.

The amount of water vapor that the air can hold depends on the temperature of both the water and the air. As we discussed in the previous section, an increase in the water temperature results in greater evaporation, and this results in more water vapor in the air. If the air temperature increases, the average kinetic energy of the air molecules increases, and this means that more collisions among the different molecules present in air take place. The water molecules in air collide with greater speeds and have less likelihood of sticking to each other, forming small water droplets that can rain down on the water surface. Thus, the amount of water vapor required for saturation increases with temperature.

Humidity is a measure of the amount of water vapor present in the air at any given time. *Absolute humidity, AH,* is the actual amount of water vapor that the air contains; that is, the total mass of water vapor in the air per unit volume, generally given in g/m³. Weather forecasters, however, give humidity in percentages; this is actually a *relative humidity, RH,* which is the ratio of the mass of water vapor per unit volume of air (the absolute humidity) to the mass per unit volume of water vapor required to saturate the air. In Table 10.2, we list the values of this *humidity at saturation, HS,* for different temperatures. Therefore, if we know the absolute humidity at any given time, the relative humidity can be calculated as follows:

$$RH = \frac{AH}{HS}.$$

On a nice spring day, the temperature might reach 20°C during the day. If the relative humidity is 50% (or 0.50), and the absolute humidity remains more or less the same throughout the evening, the air will become saturated just before the temperature drops to 8°C. At this temperature, drops begin to condense from the vapor. If the temperature is above the freezing point of water, 0°C, dew forms; if it is below 0°C, frost forms. We call the temperature at which the air starts to saturate the *dew point.*

TABLE 10.2

Humidity at Saturation for Different Air Temperatures

Temperature (°C)	Water Vapor/m³ Air (g/m³)	Temperature (°C)	Water Vapor/m³ Air (g/m³)
−8	2.74	16	13.50
−4	3.66	20	17.12
0	4.84	24	21.54
4	6.33	28	26.93
8	8.21	32	33.45
12	10.57	36	41.82

THERMAL EXPANSION

Most objects expand when the temperature is increased and contract when the temperature is decreased. As we saw in an earlier section, some thermometers are based on this phenomenon. Concrete highways and bridges must be built with gaps to take into account the expansion when temperatures rise. Large bridges may have expansion joints as wide as 40–60 cm to allow for expansion.

Temperature, as we have seen, is a measure of the average kinetic energy of the molecules of a body. As the temperature of a body increases, the average kinetic energy of its molecules also increases. This increase in kinetic energy means that the molecules move through larger distances, requiring more room to do so. For this reason, a substance expands when its temperature increases.

A 10-m steel beam used in the construction of a bridge reduces in length to 9.996 m when the temperature drops to −10°C. When the temperature rises to 35°C during a hot summer day its length increases to 10.001 m. Although the increase in length amounts to only 5 mm for the 10 m beam, a large bridge with a total length of 1000 m would need 100 of these beams and the total increase in length between winter and summer would be 50 cm.

We can see in the above example that the increase in length of the steel beam due to the increase in temperature is proportional to the length of the beam. A 10 m beam increases only 5 mm when the temperature changes from −10°C to 35°C, whereas a 1000 m beam (or one hundred 10 m beams placed one after the other) would increase its length by 50 cm. This thermal expansion is also proportional to the temperature change. Finally, if the beams were made of aluminum (not practical for bridges) the thermal expansion would be twice as large. Thus, the increase in length depends on the original length of the beam, the temperature change, and the kind of material used in its construction.

We call L_0 the initial length of the beam when the temperature is −10°C and ΔL its increase in length as the temperature rises to 35°C in the summer. This increase in length ΔL is proportional to the initial length L_0 (10 m in our example), the change

TABLE 10.3
Coefficients of Thermal Expansion of Some Substances

Substance	Coefficient (1/°C)	Substance	Coefficient (1/°C)
Rubber	80×10^6	Copper	17×10^6
Ice	51×10^6	Iron (Steel)	12×10^6
Lead	30×10^6	Concrete	10×10^6
Aluminum	24×10^6	Glass (ordinary)	9×10^6
Silver	20×10^6	Glass (Pyrex)	3.2×10^6
Brass	19×10^6	Carbon (Graphite)	8×10^6
		Carbon (Diamond)	1.2×10^6

(a)   (b)

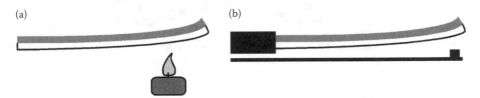

FIGURE 10.12 (a) A bimetallic strip made with brass (shown as shaded) and aluminum would bend upward so that the aluminum is on the outside since it expands more. (b) If this bimetallic strip is used as a simple thermostat, it can control a heater, an oven, or any other device that needs to operate when the temperature falls below a certain value.

in temperature ΔT, and a property that depends on the material used (steel in our example) called the *coefficient of linear expansion*, α (the Greek letter alpha). We can express this as

$$\Delta L = \alpha L_0 \Delta T.$$

In Table 10.3 we list the coefficients of linear expansion for some common substances. This coefficient, which has units of inverse degrees Celsius (1/°C), measures the fractional change in length with a given change in temperature for a substance.

The fact that different substances expand differently with a given temperature change, as illustrated by the different values of the expansion coefficients, has been used in the construction of thermostats. A thermostat is an electrical switch that is activated by changes in temperature, and is used to control electrical appliances that depend on temperature changes, like air conditioners, heaters, and toasters. A simple thermostat can be constructed with two strips of two different metals bonded together as illustrated in Figure 10.12a. If one of the strips of metal is brass and the other aluminum, when the temperature increases, the aluminum strip expands more than the brass strip, since aluminum has a larger thermal coefficient. Since the two strips are joined together, the result is a bending of the two strips so that the aluminum side is on the outside and the brass side is on the inside.

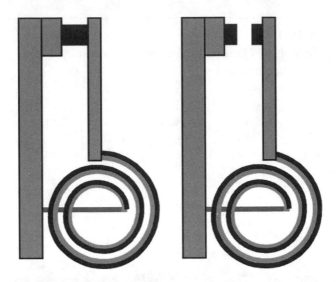

FIGURE 10.13   A bimetallic strip shaped like a coil is a common type of thermostat. As the temperature changes, the coil opens or closes the circuit.

As shown in Figure 10.12b, this simple thermostat can be used to control the operation of any appliance that works when the temperature falls below a certain value, like a heater or an oven. If we reverse the strip so that the aluminum is toward the top part of the figure, the strip would bend downward when the temperature increases above a certain value. In this case, it can be used to control an air conditioner, for example.

A common type of thermostat uses a bimetallic strip shaped like a coil (Figure 10.13). With changes in temperature, the different rate of expansion of the two metals in the strip subjects the coil to stresses that open or close the contact.

THE UNUSUAL EXPANSION OF WATER

Water is the important exception to the rule that substances expand when the temperature increases. Although this also applies to water in its three phases at most temperatures, between 0°C and 4°C water actually *contracts* when the temperature increases. This is due to the structure of the water molecule. Each oxygen atom in a water molecule can bind itself to two additional hydrogen atoms from other water molecules and each one of its two hydrogen atoms can in turn bind itself to one oxygen atom in an adjacent water molecule. This means that each water molecule can participate in four bonds with other water molecules, as shown in Figure 10.14a. In liquid water, about 80% of these bonds are complete, whereas in ice all water molecules are bonded to other water molecules to form the crystal structure of the solid (Figure 10.14b).

The crystals in ice have what is described as an *open structure* because of the large open, unoccupied spaces between the molecules. Most other solids may have up to 12 molecules joined together as nearest neighbors and their structures are not

(a)   (b)

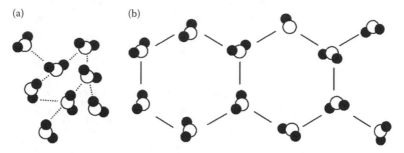

FIGURE 10.14 (a) The four bonds of liquid water: each oxygen atom can bind itself to two hydrogen atoms in neighboring molecules and each one of the two hydrogen atoms in a water molecule can bind to oxygen atoms from two nearby molecules. (b) The open structure of ice.

FIGURE 10.15 The lower density ice floats, insulating the water beneath and slowing further freezing. Aquatic life then survives in many lakes and ponds.

open like that of ice. The structure of liquid water is less open than that of ice because the number of bonds in the liquid decreases with increasing temperature, and this makes the water molecules at temperatures near 0°C more tightly packed than they are in ice. As more bonds are broken when the temperature of the water increases from 0°C to 4°C, the molecules are able to occupy more of these open spaces, so that the density of water *increases,* reaching its maximum value of 1000 kg/m³ at 4°C. At temperatures higher than 4°C, the larger kinetic energies of the water molecules require that more room be available for the increased motion of the molecules, and water begins to expand with increases in temperatures, as all other substances do.

Life as we know it on the earth exists in part because of these two unusual phenomena; the lower density of ice as compared with water, and the contraction of water as the temperature increases from 0°C to 4°C. When the temperature drops in the winter, the surface water of lakes and ponds cools down to 4°C and becomes denser. The warmer, less dense water beneath is buoyed up to the surface, where it gets chilled to 4°C, becoming in turn denser. The cool descending water brings oxygen down with it. This process continues, keeping the water deep below the surface at 4°C, until the atmospheric temperature drops further and ice forms on the surface of the water. Since this lower density ice floats, lakes and ponds freeze from the top down, and this layer of ice insulates the water below, slowing further freezing. Fish and other forms of aquatic life survive the winter in the slightly warmer and oxygenated water below the ice (Figure 10.15).

11 The Laws of Thermodynamics

THE FOUR LAWS OF THERMODYNAMICS

Why does time seem to flow in only one direction? Can the flow of time be reversed? The directionality of time is still a puzzle because all the laws of physics except one are applicable if time were to be reversed. As we shall see in this chapter, the second law of thermodynamics is the exception. The flow of time seems to arise from the second law.

There are four laws of thermodynamics: the second law was discovered first; the first was the second; the third was the third, but it probably is not a law of thermodynamics after all; and the zeroth law was an afterthought. We shall occupy ourselves in this chapter with the study of these laws.

THE IDEAL GAS LAW

The study of thermodynamics is intimately connected with the study of the behavior of gases. The reason is that gases, being much simpler, are better understood than liquids and solids. An *ideal gas* is any gas in which the cohesive forces between molecules are negligible and the collisions between molecules are perfectly elastic; that is, both momentum and kinetic energy are conserved. Many real gases behave as ideal gases at temperatures well above their boiling points and at low pressures.

The English scientist Robert Boyle, the 14th child of the Earl of Cork, was an infant prodigy. At the age of eight he spoke Greek and Latin and at 14 traveled to Italy to study the works of Galileo. He returned to England in 1645 to find his father dead and himself wealthy. In 1654, he became a member of the "Invisible College," which later became the Royal Society, where he met Newton, Halley, and Hooke.

In 1662, while experimenting with gases, he was able to show that if a fixed amount of a gas was kept at a constant temperature, the pressure and the volume of the gas follow a simple mathematical relationship. Boyle discovered that gases were compressible, so that when the pressure was increased (as when the piston in Figure 11.1a is pushed down by the extra sand) the volume of the gas decreased. Boyle further found that if the container was placed on a hotplate kept at a constant temperature, the increase in pressure was matched by the decrease in volume. This meant that if the pressure was doubled, the volume halved; if the pressure was tripled, the volume decreased to exactly one third. Boyle expressed this relationship between pressure and volume as

$$PV = \text{constant} \qquad \text{[at constant temperature]}$$

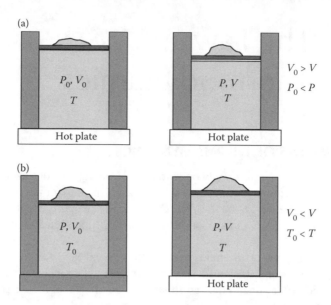

FIGURE 11.1 (a) A container filled with a gas kept at the same temperature by placing it on a temperature-controlled hot plate. The pressure is changed by adding sand on the movable piston. When the pressure is increased, the volume decreases. (b) A fixed amount of sand on top of the frictionless lid maintains the gas in the container at a constant pressure. Increasing the temperature of the gas increases its volume in a linear way.

This expression is known today as *Boyle's law.* Several years after Boyle's experiments, it was found that the constant in Boyle's law was the same for all gases.

Throughout the eighteenth century, many scientists investigated the expansion of gases when heat was added, but their results lacked consistency and no conclusion regarding the dependence among volume, pressure, and temperature was reached. In 1804, the French chemist Joseph Louis Gay-Lussac was able to show that if the pressure of the gas was kept constant (as illustrated in Figure 11.1b with the constant weight of the sand on the frictionless lid), then the change in volume was proportional to the change in temperature. He investigated this relationship between temperature and volume with air, hydrogen, oxygen, nitrogen, nitrous oxide, ammonia, carbon dioxide, hydrogen chloride, and sulfur dioxide, and found that it held consistently.

Since the volume and the temperature of a gas at constant pressure are directly proportional, a plot of volume versus temperature for different gases should give us straight lines for each gas. Gay-Lussac found that if these lines were extrapolated, they all cross the temperature axis at exactly the same point (Figure 11.2a). This point is the absolute zero of temperature, 0 K or −273.16°C. This was the basis for the introduction of the absolute or Kelvin scale of temperature.

The absolute zero is the minimum temperature attainable because at this temperature the volume of the gas would be zero, as we can see in the graph of Figure 11.2b. A plot of volume versus absolute temperature for an ideal gas yields

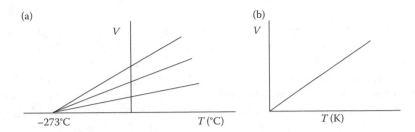

FIGURE 11.2   (a) A plot of volume versus temperature for different gases yields straight lines. When extrapolated, these lines intersect the temperature axis at −273°C. (b) A plot of volume versus absolute temperature for an ideal gas is a straight line through the origin.

a straight line that passes through the origin, as shown in Figure 11.2b, since T in Kelvin is zero for $V = 0$. Therefore,

$$V = \text{constant} \times T \qquad \text{[at constant } P\text{]}.$$

This is *Gay-Lussac's law*, also known as *Charles' law*, because the French physicist Jacques Alexandre Charles had independently made the same discovery a few years earlier but had failed to publish it.

We can combine Boyle's law and Gay-Lussac's law into one single expression,

$$PV = \text{constant} \times T$$

which is known as the *ideal gas law*.

We can extract a third relationship from the ideal gas law. When the volume of the gas remains constant, the pressure is proportional to the temperature,

$$P = \text{constant} \times T \qquad \text{[at constant } V\text{]}.$$

We should keep in mind that the temperature T in all the gas laws is in kelvins.

PHYSICS IN OUR WORLD: AUTOMOBILE ENGINES

The gasoline engine used in automobiles is a heat engine which generates the input heat from the combustion of gasoline inside the engine. For this reason, gasoline engines are called internal combustion engines.

An automobile's gasoline engine consists of the cylinder head, the cylinder block, and the crankcase. The cylinder head has two sets of valves, intake and exhaust. When the intake valves are opened, a mixture of air and gasoline enters the cylinders. When the exhaust valves are open, the burned gases are expelled from the cylinders. The valves are opened and closed by the *camshaft*, a system of cams on a rotating shaft, while the moving pistons turn a shaft, the *crankshaft*, to which they are connected.

The camshaft and the crankshaft are interconnected by a drive belt or chain so that as the pistons move, turning the crankshaft, the camshaft is also turned, opening and closing the valves.

Most automobiles have a four stroke cycle engine. In the *intake stroke* the downward motion of the piston draws fuel into the cylinder. The volume of the cylinder increases from a minimum volume V_{min} to a maximum volume $V_{max} = rV_{min}$, where r is the *compression ratio*. For modern automobiles, the compression ratio is about 8. In the *compression stroke* the intake valve closes as the piston reaches the end of the downstroke, and the piston compresses the air–fuel mixture to V_{min}. In the *power stroke*, an electric spark from the spark plug ignites the gases, increasing their temperature and pressure. The heated gases expand back to V_{max}, pushing the piston and doing work on the crankshaft. Finally, in the *exhaust stroke*, the exhaust valve opens, and the piston moves upward, pushing the burned gases out of the cylinder. The cylinder is now ready for the next cycle.

THE ZEROTH LAW OF THERMODYNAMICS

Although we have not labeled them as such, we have already studied the zeroth and first laws of thermodynamics in the previous chapters. The zeroth law deals with bodies in thermal equilibrium. If we place two objects that are at different temperatures in contact with each other and wait a sufficient length of time, the two objects will reach the same temperature—they are in thermal equilibrium with each other.

How do we check that the two objects have reached thermal equilibrium? We need to use a third object, a thermometer for example, to verify that the two objects are in thermal equilibrium. When we use a thermometer to measure the temperature of an object, we bring it into contact with the object and wait some time before reading the temperature. What we have done is to wait until the object and the thermometer reach thermal equilibrium; that is, until the thermometer is at the same temperature as the object. If we now use the same thermometer to measure the temperature of a second object, and after waiting a long enough time to make sure that the thermometer has reached thermal equilibrium with this second body, the reading of the thermometer is the same as when it was in thermal equilibrium with the first object, we can say that the two objects are in thermal equilibrium with each other. In other words,

If two objects are each in thermal equilibrium with a third object [the thermometer], they are in thermal equilibrium with each other.

This seemingly obvious statement is the zeroth law of thermodynamics.

THE FIRST LAW OF THERMODYNAMICS

The first law of thermodynamics is a generalization of the principle of conservation of energy to include thermal energy. In Chapter 10, we used the term *thermal energy* to describe the sum of all the random kinetic energies of the atoms and molecules in a substance. As we learned earlier in the book, atoms in a system have binding energies, the nuclei of these atoms also have binding energies, and the molecules have energy in the chemical bonds. These energies stored in the molecules, atoms, and nuclei are different forms of *potential energy*. The *total energy* of a system—a gas, for example—includes all forms of energy, both thermal and potential, and is called the *internal energy*, U, of the system.

If a system is isolated, that is, if it does not exchange energy with its surroundings, the total energy must remain constant. This is the principle of conservation of energy, familiar to us from Chapters 4 and 5. It is also the first law of thermodynamics; we can state it very precisely as follows:

> In an isolated system, the total internal energy remains constant, although it can change from one kind to another.

A system can interact with its surroundings in two ways: First, when it *does work* or *work is done on it*, and second, when *heat is exchanged* with the surroundings. To understand how this comes about, consider the gas in the well-insulated container of Figure 11.3. The airtight frictionless lid can slide up or down and we have added enough sand on top of it to balance the pressure exerted by the gas.

We now remove the insulation from the bottom of the container and place it on a stove so that an amount of heat Q is added to the gas while at the same time we add enough sand so that the lid remains in place (Figure 11.4a). This means that the gas does not do any work on the surroundings and all the heat added goes to increase the internal energy of the gas. That is,

$$Q = \Delta U.$$

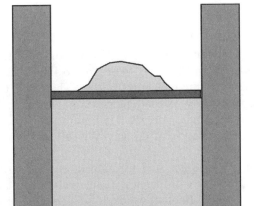

FIGURE 11.3 A gas in a well-insulated container. The pressure of the lid, sand, and the atmospheric pressure balance the pressure of the gas.

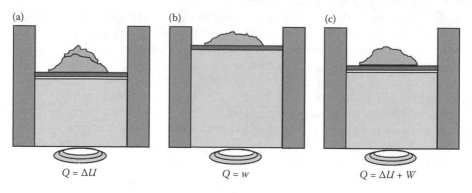

FIGURE 11.4 (a) Adding enough sand to the lid keeps the volume of the gas constant. In this case, since no work is done by or to the gas, the heat added goes into increasing the internal energy of the system. (b) Removing sand from the lid lets the gas expand. The heat added goes into doing work. (c) If the amount of sand is not altered in any way, the heat added increases the internal energy of the gas and the gas does work on the surroundings. This is the first law of thermodynamics.

In this process, the volume of the gas remains constant, and we call it an *isochoric* (equal volume) process. When the mixture of air and gasoline ignites in a car's engine, the volume of the mixture remains unchanged and the heat from the ignition increases the internal energy. This process is close to an isochoric process.

Suppose that instead of adding sand to the lid to keep the gas at a constant volume, we slowly remove enough sand so that the gas is allowed to expand (Figure 11.4b). In this case *all* the heat added goes into doing work on the surroundings, and the internal energy of the gas remains the same; that is,

$$Q = W.$$

Recall that the temperature of a gas depends only on the random kinetic energy per molecule. In an ideal gas there are no intermolecular interactions, so the potential energy of the molecules is zero. Therefore, the internal energy of an ideal gas depends only on the temperature of the gas. Since the internal energy of the gas in the situation depicted in Figure 11.4b does not change, there is no change in the temperature of the gas. We call this process an *isothermal* process.

We can see that if we do not alter the amount of sand on the lid and simply add heat to the gas, the gas will increase its internal energy *and* do work on the surroundings (Figure 11.4c). In other words, if an amount Q of heat flows into the gas resulting in an increase in its internal energy ΔU, and the gas does an amount of work on the surroundings, then

$$Q = \Delta U + W.$$

What this expression tells us is that

$$\begin{bmatrix} \text{heat added} \\ \text{to system} \end{bmatrix} = \begin{bmatrix} \text{increase in internal} \\ \text{energy of system} \end{bmatrix} + \begin{bmatrix} \text{work done} \\ \text{by system} \end{bmatrix}.$$

This is the principle of conservation of energy, which we call here the first law of thermodynamics. In this case, the external pressure that the atmosphere, the sand and the lid exert on the gas is constant. This pressure must be matched by the expanding gas; thus, the gas exerts a constant pressure on the surroundings. This is an *isobaric* (equal pressure) process.

The first law of thermodynamics shows that if a system undergoes a volume and pressure change, the internal energy of the system changes by an amount that is given by $Q - W$. In Chapter 5 we learned that the work done in moving an object in a gravitational field does not depend on the path through which the object is taken between the initial and final points. The gravitational force is a conservative force. In thermodynamics, when the system changes from some initial state to some final state, the quantity $Q - W$ is found experimentally to depend only on the initial and final coordinates and not on the path taken. This quantity $Q - W$ is the change in the internal energy of the system,

$$\Delta U = Q - W.$$

THE SECOND LAW OF THERMODYNAMICS

The second law of thermodynamics was made famous several years ago by the English novelist and physicist C. P. Snow in his well known essay "The Two Cultures," where he suggested that some understanding of it should be expected of every educated person. His choice was a fortunate one because the second law is one of the most important laws in all of science. This, however, does not make it difficult to understand.

The German physicist Rudolf Gottlieb, known today by the name of Clausius, stated it more than a hundred years ago as follows: "*Heat does not pass spontaneously from cold to hot.*" This statement of the second law is known as the *Clausius statement*.

It would not violate the principle of conservation of energy if heat were to pass spontaneously from cold to hot. During a warm summer day a puddle could spontaneously release heat to the surroundings, cooling down and solidifying into ice, without violating the conservation of energy principle. It would not violate the principle of conservation of energy either if water waves were to converge on a stone at the bottom of a pond and propel it out of the water into the hands of a child standing nearby. These events are never observed. Many times we have seen blocks of ice melting into puddles and children throwing stones into ponds, causing ripples that spread away from the stone but never the reverse processes (Figure 11.5).

FIGURE 11.5 A child throwing a stone into a pond is a common sight during warm days. The reverse situation, in which water waves converge onto a stone lying at the bottom of a pond and propel it out of the water into the child's hands, is never seen.

The fact that these events are always observed taking place in one direction and not in the other is related to the *direction of time*. People grow older, not younger; stars are formed from rotating clouds of hydrogen gas, begin their thermonuclear processes out of which heavier elements are produced and energy is released, expand and contract after millions of years, and finally explode as supernovas or become black holes; the reverse sequence of events is never observed (Figure 11.6).

There is a trend in nature toward a greater degree of disorder. It is not just your room that is hard to keep organized; the entire universe keeps getting more disorganized. In 1865, Clausius introduced the term *entropy*, from a Greek word that means *transformation*, as a measure of the disorder of a system. (The conventional symbol for entropy is *S*.) In terms of entropy, the second law can be stated as follows:

> The entropy of the universe never decreases; all natural changes take place in the direction of increasing entropy.

There is only one way that all the pieces of a jigsaw puzzle can be organized to make a picture but many incorrect ways in which they can be put together. It is, therefore, extremely unlikely that by throwing the pieces together they will fall in the correct order. When you accidentally drop a stack of papers and hurriedly pick them up, they become out of order. There is one correct order for the papers and many incorrect ones. There are only a few ways of placing the things in your room that are esthetically pleasing to you and very many ways that are not. Your room is therefore more likely to be in an esthetically displeasing order.

When you open a soft drink bottle, some of the carbon dioxide that had been dissolved under pressure in the liquid and is mixed with air above the liquid leaves the bottle and diffuses into the atmosphere. The reverse of this process, where the diffused gas collects itself and enters the bottle so that you can replace the cap and restore the previous order, is never observed. The propane gas in a tank of a recreational vehicle diffuses into the air after you open the valve. The original situation, with the propane gas in the tank separated from the air outside, is more ordered than the latter, where the propane is mixed with the air. Entropy increases in each one of these normal processes.

FIGURE 11.6   An exploding supernova discovered and photographed in January 1987 in the Magellanic Cloud, a satellite galaxy of the Milky Way. These gigantic explosions represent the end of the life cycle for certain stars. They also represent the beginning, as carbon, oxygen, silicon, iron, and other heavy elements that were produced in the old star are spewed out in the explosion. New stars are born out of the matter of the explosion mixed with the surrounding gas in the galaxies. Our sun and planet Earth contain the ashes of early supernova explosions. (Courtesy of NASA, Space Telescope Science Institute, Hubble Heritage Team.)

We should note that nothing drives the molecules of a gas into a state of greater disorder; there is no special force behind this phenomenon. The diffusion of a gas is purposeless and is only the result of the random motion of its molecules.

When we let a gas expand, as when we let the air out of a tire, its temperature decreases. As the gas expands, it does work on the surroundings and this means a reduction in the kinetic energy of the gas. The air leaking out of a tire does work by pushing the outside air in the vicinity of the valve through molecular collisions. The energy of the gas jostles out into the environment.

This diffusion of energy out into the environment explains Clausius' statement of the second law. The tungsten atoms in the filament of a light bulb are vibrating rapidly when the light is on. The vibrating atoms near the surface of the filament collide with nearby air molecules, which in turn collide with other air molecules farther out. The energy of the filament is thus diffused into the cooler environment. Heat passes from the hot filament to the colder air and never the other way around.

When we add heat to a substance its molecular motion increases, and this results in more disorder. Adding heat to a substance, then, increases its entropy and the more heat we add, the greater the increase. A given amount of heat that is added to a substance is more effective in producing disorder if the substance is cold than if it is already hot. A misplaced item stands out more in a well organized room; you would hardly notice the same misplaced item in a messy room. We can see that the change in entropy of a substance is directly proportional to the amount of heat added and inversely proportional to the temperature. If a system absorbs an amount of heat Q at an absolute temperature T, the change in entropy is

$$\Delta S = \frac{Q}{T}.$$

The units of entropy are joules per kelvin, J/K.

(Courtesy of Sidney Harris.)

THE FRONTIERS OF PHYSICS: ENTROPY THAT ORGANIZES?

By mixing tiny polystyrene spheres of two different sizes, physicists have been able to turn the tables on entropy and use it to line up the spheres in an organized pattern. Arjun G. Yodh of the University of Pennsylvania and his research group placed a drop of salt water containing a mixture of these spheres between two microscope slides. They used many 0.08 micrometer polystyrene spheres and mixed in a few 0.46 micrometer ones and etched a small groove in the slide.

Due to their random motion, the small spheres in the salt water collided with themselves and with the fewer larger spheres. The collisions produced a state of maximum entropy, in which the small spheres spread out in disordered state. The random motion of the larger spheres also tended toward an increase in entropy. However, the much larger number of smaller spheres obstructed the motion of the larger spheres. Larger spheres reaching the edge of the groove, for example, would not be allowed to enter it; the jostling motion of the smaller spheres pushed them back. Even when the experimenters placed the large spheres in the groove, they were driven out by the smaller particles. Many of the larger spheres get trapped at the edge of the straight groove, forming a row.

The scientists measured the cumulative force keeping the larger particles from entering the channel to be 40 femtonewtons (10^{-15} N). This was the first measurement of the "entropic interaction."

THE THIRD LAW OF THERMODYNAMICS

Absolute zero is the lowest temperature possible. The German physical chemist Walther Hermann Nernst (1864–1941) proposed in 1907 that the absolute zero temperature cannot be reached. In an experiment, the temperature of a system can in principle be reduced from the previous temperature obtained, even if by a very small amount, as has been done recently, achieving temperatures of the order of 10^{-7} K (one ten-millionth of a kelvin). What Nernst proposed was that although it is possible to get closer and closer to absolute zero, to actually get there requires an infinite number of steps. This, of course, makes the absolute zero temperature unattainable. This statement is known as the third law of thermodynamics.

> Third Law of Thermodynamics: It is impossible to reach the absolute zero temperature in a finite number of steps of a process.

Nernst was awarded the 1920 Nobel Prize in Chemistry for this discovery.

Some scientists think that this statement is actually an extension of the second law and not a separate law. Whether a separate law or only an extension of the second law, the statement that absolute zero cannot be reached is universally accepted.

ENTROPY AND THE ORIGIN OF THE UNIVERSE

The fact that all natural processes take place in the direction of increasing entropy means, as we have seen, that the entropy of the entire universe is constantly increasing. Nature's tendency toward greater disorder diminishes the amount of energy available to do work. Think, for example, of a bouncing ball. With each bounce, the mechanical energy of the ball decreases and the ball reaches a lower height every time, until, finally, it lies still on the ground. While the ball is falling, although there is still random molecular motion, all its molecules also possess the same translational motion toward the ground, producing an ordered state. When the ball hits the ground there is a slight increase in the random motion of the molecules of both ground and ball as a result of the collision, and this increases the thermal energy of the ball (and the ground). There is, therefore, less mechanical energy for the second bounce. When the ball loses all of its mechanical energy to thermal energy through the repeated collisions and bounces no more, the ordered translational motion of its molecules has been lost to a random disordered motion that has produced an increase in thermal energy. As the ordered state disappears, the entropy of the system increases.

If we think of the bouncing ball as a very simple machine to do work (using it to catapult a small stone, for example, as illustrated in (Figure 11.7), the energy that is available to perform this work decreases after every bounce and disappears altogether when the ball lies still on the ground. The total energy of the system (ball and ground) remains constant; the mechanical energy has been transformed into thermal energy, no longer available to do work.

In all processes that involve exchange of heat, there is an increase in entropy and a decrease in the energy available to do work. The entropy of the universe must eventually reach a maximum value when everything is in a state of perfect disorder and the total energy of the universe is distributed uniformly. This led scientists to predict the "heat death" of the universe when processes are no longer possible.

We now know that the heat death of the universe has already happened. Even though, as we have said, the entropy of the universe is continuously increasing, the total entropy of the universe is essentially constant. The reason is that almost all of the entropy of the universe is in the radiation of photons and not in matter—by a factor of 400 million to 1—and the entropy of the radiation has already reached its peak.

The universe consists of matter in the form of galaxies and radiation in the form of photons. The galaxies are conglomerates of billions of stars: our own galaxy, the Milky Way, is estimated to contain 100 billion stars. From observational evidence, astronomers calculate that the universe contains some 100 billion galaxies each

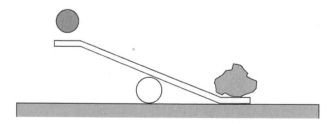

FIGURE 11.7   A bouncing ball can be used to catapult a small stone. The energy available to do this work decreases with every bounce as it dissipates into heat.

separated from the others by a few million light-years. (One light-year is the distance light travels in a year, nearly ten trillion kilometers.) In all these galaxies, stars are constantly forming, evolving, exploding, and collapsing. All these processes increase the total entropy of the universe. However, the increase in entropy due to all the changes that have taken place since the beginning of the universe amounts to only one ten-thousandth of the entropy of the radiation.

When the universe was formed some 13 billion years ago, it was all radiation. It is a triumph of modern physics that we are today able to trace the history of the universe beginning at a hundredth of a decillionth (10^{-35}) of a second after the very moment of creation, an interval of time so small that we are incapable of imagining it. We will be studying some of the details of this wonderful theory in Chapter 26. We can, however, outline the main ideas here, so that we can understand how the total entropy of the universe can be considered to be almost constant.

The Big Bang theory of the origin of the universe was proposed by the Russian–American physicist George Gamow toward the end of the 1940s to explain a strange phenomenon that had been discovered some 20 years earlier by the American astronomer Edwin Hubble. Hubble had undertaken a study of one hundred galaxies with the 100-inch telescope at Mount Wilson and had made a startling discovery. He noticed that all these galaxies were invariably moving *away* from us. For a while it looked as if Ptolemy was right after all; the earth seemed to be at the center of the universe.

Soon Hubble realized that this was not the case. By comparing the velocity of recession with the known distances to the galaxies, he discovered that the more distant the galaxy, the faster it receded. The velocities that he measured were not minor. Galaxies in the constellation Ursa Major, for example, were found to be receding at a rate of 42,000 km/s or one-seventh the speed of light. Hubble showed that the velocities of recession of galaxies were related to their distances by a simple expression, now known as *Hubble's law*. This law states that

$$v = Hd$$

where v is the velocity of recession of a galaxy, d is the distance to the galaxy, and H is *Hubble's constant*. This constant has a value of 15–30 km/s per million light-years. In other words, a galaxy one million light-years from us moves away at a speed of 15 to 30 km/s. This discovery led Hubble to realize that although every one of the distant galaxies that he studied was receding from us, the earth was not at a special place in the universe, but rather that every galaxy was moving away from every other galaxy.

Hubble's discovery means that the universe is expanding. Imagine baking a loaf of raisin bread; as the dough rises, the raisins move apart from each other (Figure 11.8). If we select one particular raisin, we would see that all the other raisins move away from it and the farther a raisin is from the one we selected, the faster it moves away. We did not pick a special raisin; it works the same with any one raisin as they all move away from each other. The expanding dough carries the raisins along so that the distance between them increases. The expanding space carries the galaxies along so that from our own galaxy we see the other galaxies moving away from us.

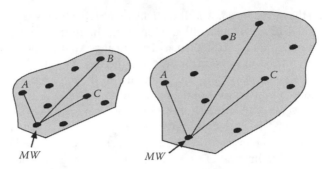

FIGURE 11.8 The raisin bread analogy of the expansion of the universe. As the loaf of bread is baked, the expanding dough carries the raisins away from each other. Raisins A, B, and C move away from raisin MW as the dough rises. MW can be any raisin in the loaf.

Intelligent beings on a planet in a distant galaxy would see our galaxy and all the other galaxies moving away from theirs.

An expanding universe means that at earlier times the galaxies were closer together and this suggests a beginning for everything. This is what motivated George Gamow to propose the Big Bang theory of the origin of the universe.

As we said earlier, the universe consists of matter and radiation. The average density of matter in the universe has been estimated to be about one nuclear particle per cubic meter. The average density of radiation in the universe today is about 400 million photons per cubic meter. Since the entropy of a system is proportional to the number of particles, we can see that the entropy of the universe is mostly in the photons, by a factor of 400 million.

Since almost all the entropy of the universe is in the photons, the increase in entropy that takes place when processes occur in the planetary systems and galaxies of the universe, although numbered in the billions, can add very little to the entropy that already exists in the photons. The heat death of the universe, then, effectively happened shortly after the Big Bang, with the creation of all the photons.

ENTROPY AND THE ARROW OF TIME

The concept of time is intimately linked to the concept of entropy. Our perception of time arises from the accumulation of information in our brains; we remember the past. Events, changes, and processes need to occur so that experiences can be sensed and stored in our memories. As we have seen, changes in the universe take place only when the entropy of the entire universe increases. The accumulation of information in our brains takes place only when these changes take place; that is, when the entropy of the universe increases. Nature has used this increase in entropy to give direction to the arrow of time.

Except for the second law of thermodynamics, all the laws of physics remain unchanged if the direction of the arrow of time is reversed; they are time-reversible. The laws of mechanics, for example, are time-reversible. If we make a movie of an oscillating pendulum, a perfectly elastic bouncing ball, or any other

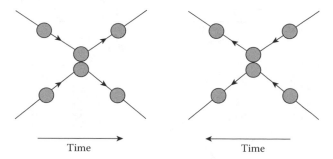

FIGURE 11.9 The collision of two molecules is time-reversible. We could not tell the difference if a video of this collision was to be shown in reverse.

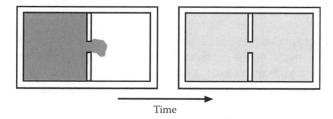

FIGURE 11.10 The diffusion of a gas through an opening in the partition is not a time-reversible process. We could immediately tell if a video of this process were shown in reverse.

purely mechanical process (where no exchange of heat takes place) and then run the movie backwards, it is impossible for us to tell the difference. The pendulum oscillates back and forth and the ball bounces the same as when the movie is run forward.

Since the laws of mechanics and electromagnetism do not depend on the direction of time, the elastic collision of two molecules in a gas appears equally possible if the flow of time were to be reversed, as illustrated in Figure 11.9. However, when we consider all the molecules in the gas, the situation is no longer time-reversible. If the gas is originally confined to the left half of a container by means of a partition, as seen in Figure 11.10, and a small hole is made on the partition, the gas will diffuse until it is uniformly distributed throughout both sides of the container. If a video of this process were to be shown in reverse, we would immediately recognize that this is the wrong direction. Although the motion of any individual molecule of the gas is time-reversible, the behavior of the whole gas is not.

Lord Kelvin recognized this *reversibility paradox* in the 1870s. Kelvin himself and later the German physicist Ludwig Boltzmann (1844–1906) realized that it was the *statistical* nature of the second law that explained the paradox. There are many more disordered arrangements of a system than ordered ones, and for this reason the disordered arrangements are much more likely to occur. Given enough time, even the

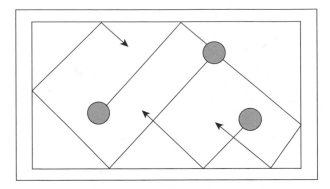

FIGURE 11.11   After a long enough time, the interacting molecules will find themselves occupying the original positions. This is known as Poincaré recurrence. In a real system, it will actually take many times the age of the universe for this to happen.

very few ordered arrangements can occur, although for real situations, enough time means a time longer than the age of the universe.

This last statement, on the recurrence of a mechanical system, was proposed in 1889 by the French scientist Henri Poincaré, and is known as *Poincaré recurrence*. Consider, for example, a gas enclosed in a sealed container (Figure 11.11). If we concentrate our attention on a single molecule, we would find it undergoing collisions with other molecules and with the walls of the container. Suppose, for simplicity, that there are only three molecules in this gas. If we wait a long enough time, we will eventually see the three molecules occupying the same positions that they occupied at the beginning, when we first started observing them. The time it takes for this to happen is known as *Poincaré cycle time*. For an actual macroscopic system, the Poincaré cycle time turns out to be many millions of times the age of the universe. No wonder we never see it happening!

The Russian–Belgian physical chemist Ilya Prigogine, who won the Nobel Prize in 1977 for his work in irreversible thermodynamics, postulates that the Poincaré recurrence does not happen because very small changes made to a system can drastically change its future behavior, preventing it from ever reaching the same initial state. The following example will help us clarify his hypothesis. Suppose somebody traces a large circle on the ground and asks you to follow it as closely as possible on a bicycle, so that, after you complete the circle, you end up at the same position from where you started. If you are an average bicycle rider, you can easily correct small deviations from the circular path, and it would not be very difficult for you to accomplish the task. Suppose, however, that, as you are about to complete your circle, you momentarily lose your balance and deviate from the path, not ending at the same spot. You could continue riding, completing a second loop, and perhaps a third, until you make it to the right spot. If you wait a long enough time and complete enough loops, your final position is indistinguishable from the initial position.

Now, suppose that the game becomes more challenging, and the circle is traced on to the edge of a very tall and narrow circular wall (Figure 11.12). Here you would

FIGURE 11.12 A cyclist riding on the edge of a very tall and narrow circular wall.

have to be extremely cautious because even a tiny deviation from the circle would cause you to fall off the edge, preventing you from ever completing the circle. In this case, a small deviation would completely change the system so that the same initial state could not be reached.

Which one of these opposing views is the correct one? This is a question that has not yet been resolved. We can see that thermodynamics is a part of physics that is still being developed and that it has implications that reach into the heart of our understanding of the world.

Part V

Electricity and Magnetism

12 Electricity

ELECTROMAGNETISM

Amber is a beautiful stone that has been used since prehistoric times to make jewelry and ornamental carved objects. A fossil tree resin from pine and other softwood trees, amber was also of interest to the ancient Greeks, who called it ελεκτρον (*elektron*). In the seventh century BC, Thales of Miletus observed that when amber was rubbed vigorously with cloth, it attracted small bits of straw, feathers, or seeds. Other materials that show this property were discovered in the centuries that followed.

Lodestone is another mineral with unusual properties. Known also as *magnetite* or magnetic iron ore, lodestone is an iron oxide mineral that attracts iron. Known for this property as far back as 500 BC, lodestone turns to a north–south direction when floating in a liquid or suspended from a string. The Roman poet Lucretius advanced a theory about the cause of magnetism in his poem *On the Nature of the Universe.*

A detailed study of these properties only began with the work of William Gilbert (1544–1603) in England. Gilbert received a medical degree from Cambridge and established himself as a physician of renown, becoming president of the College of Physicians and later court physician to Queen Elizabeth I. He became interested in the work on magnets of the French scholar and engineer Petrus Peregrinus de Maricourt and started performing very careful experiments to determine the nature of magnetism. His experiments eventually led him to investigate also the properties of amber and to realizing that its attraction was different from magnetism which involved only iron. In his book *De Magnete*, published in 1600, he not only presented a systematic discussion on magnetism but also gave a discussion on the force of attraction caused by friction in amber. He coined the word *electric* for "bodies that attract in the same way as amber."

Electricity and magnetism developed as two different sciences until the early nineteenth century when the Danish physicist Hans Christian Oersted observed that there was a connection between them. This connection was developed further by, among others, the English scientist Michael Faraday. It was, however, the Scottish physicist James Clerk Maxwell who brought together electricity and magnetism in a complete form by the formulation of his theory of electromagnetism in the form of the four equations that bear his name. We shall return to the fascinating story of the development of electromagnetism in Chapter 14.

ELECTRIC CHARGE

For twenty-four centuries, from Thales to Gilbert, the *attractive* properties of amber and of "bodies that attract in the same way as amber" were known. In 1733,

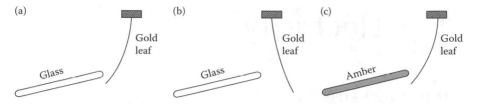

FIGURE 12.1   (a) A glass rod that has been rubbed attracts a gold leaf. (b) When the gold leaf is touched with the glass rod, the rod and the leaf repel each other. (c) A gold leaf charged with the glass rod is attracted by a charged amber rod.

the French chemist Charles François de Cisterney du Fay performed a series of experiments in which he touched a gold foil with a glass rod that had been electrified by rubbing it with silk (Figure 12.1). Before it was touched by the glass, the foil was *attracted* toward the rod but after it was touched, the gold foil was *repelled* away from it. Moreover, contrary to his expectation, the foil was *attracted* toward an amber rod that had been rubbed with wool. A gold foil that was touched first by the electrified amber rod and was repelled by it was then attracted toward the glass rod.

Du Fay supposed that there were two kinds of electrification and he called the type produced by the glass *vitreous* (from the Latin word for glass) and the type obtained with the amber rod *resinous*. Du Fay generalized his findings by stating that *bodies with the same type of electrification repel each other, whereas bodies with different type of electrification attract each other.*

In 1747, Benjamin Franklin, the great statesman, inventor, writer, and the first American physicist, conducted experiments that showed that one type of electrification could be neutralized by the other type. This indicated to him that the two types of electricity were not just different; they were *opposites*. He further proposed that all objects possessed a *normal* amount of electricity some of which was transferred to another body by rubbing. When this transfer of electricity took effect, the first body had a deficiency on its normal amount of electricity, and this could be indicated by a *negative* sign, while the body that received the electricity ended up with an excess, which could be indicated by a *positive* sign. Since there was no difference in the behavior of the two types of electricity, Franklin had no way of knowing which one was positive and which one was negative. He arbitrarily decided that rubbing glass with a silk cloth transferred electricity to the rod and, therefore, was positive, while rubbing amber with wool made it lose electricity, and was negative (Figure 12.2). We now use the term *electric charge* and speak of positive electric charge and negative electric charge.

Franklin had a 50% chance of being right with his convention of signs. He lost. We now know that electrons are the carriers of electric charge when the rods are rubbed and, in Franklin's sign convention, they have negative charge. In the case of the glass rod, when it is rubbed with silk, electrons actually leave the rod and join the silk molecules. Some of the glass molecules near the surface of the rod are left lacking a negatively charged electron and become positive ions; that is, the silicon and oxygen atoms that make up the glass molecules

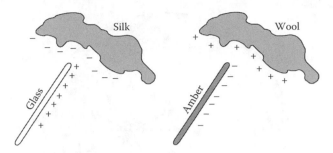

FIGURE 12.2 A glass rod that has been rubbed with silk acquires a positive charge. An amber rod that has been rubbed with wool becomes negatively charged.

have one electron fewer than the corresponding number of protons in their nuclei and the molecule is left with a net positive charge. The silk, on the other hand, acquires extra electrons in the rubbing process and becomes negatively charged. In a typical experiment, about one billion electrons get transferred in the rubbing process. Franklin had assumed that positive charges were transferred. Although we still use Franklin's convention of signs, we need to be aware of the correct interpretation.

COULOMB'S LAW

What is the nature of the attractive force between a positive and a negative electric charge or the repulsive force between two positive or two negative charges? The English chemist Joseph Priestley, in a brilliant insight, provided the answer. Priestley had been asked by his friend Franklin to investigate a phenomenon that Franklin had encountered. About 1775, Franklin noticed that small charged cork balls were not affected when hung from a thread close to the inner surface of a charged metal can, although they were attracted by the can when placed near the outside surface. Priestley correctly realized that a similar situation occurs in mechanics. If an object of mass m were to be placed inside a hollow planet, the net gravitational force acting on the object would be zero (Figure 12.3). Since the gravitational force is an inverse square law—that is, it is proportional to the inverse of the square of the distance—Priestley proposed that the force between two electric charges also varies as the inverse of the square of the distance between the charges.

Charles Augustin Coulomb was born to wealthy parents in 1736 in Angoulême, France. He studied mathematics and science and became a military engineer, serving in the West Indies until 1776, when he returned to Paris. Due to his precarious health and his desire for a quiet life, he retired to the town of Blois where, while the French Revolution started, he dedicated himself to scientific experimentation. In 1777, Coulomb invented a *torsion balance* to measure the force between electrically charged objects. (Shortly thereafter, Cavendish used a similar torsion balance to "weigh the earth;" that is, to determine the value of G.) A small charged sphere was attached to a horizontal, insulating rod which was suspended by a thin silver wire (Figure 12.4). Close to this sphere, he placed a second charged sphere. As the

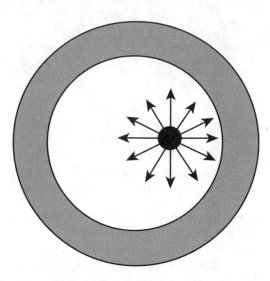

FIGURE 12.3   The net gravitational force acting on the object of mass m inside a shell is the sum of the gravitational forces exerted on the object by all the parts of the shell. All these forces balance out and the net force is zero.

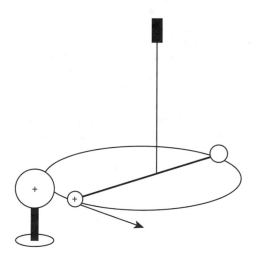

FIGURE 12.4   Coulomb's torsion balance. The electrical force between the suspended sphere and the fixed sphere can be determined from the twisting of the wire.

spheres were attracted to or repelled from each other (depending on whether the charges were opposite or the same), the wire twisted. The twisting angle allowed Coulomb to determine that the force between the spheres for different separations was proportional to the inverse of the square of the separation, as Priestley had proposed. If r is the distance between the centers of the spheres, the electrical force F between them is given by

$$F \propto \frac{1}{r^2}.$$

Coulomb also showed that the electric force also depends on the magnitude of the charges. Although at the time there was no method for measuring the amount of electric charge on an object, Coulomb ingeniously figured out a way of comparing charges. Bringing together two identical spheres, one charged and the other uncharged, he found that the original charge was distributed in equal parts between the two spheres, so that each sphere held one half of the original charge. Bringing other uncharged spheres into contact with one of the charged ones, he could produce fractions of one fourth, one eighth, and so on, of the original charge. This allowed Coulomb to establish that the electric force between two charged objects is also proportional to the product of the magnitudes of the charges.

The results of Coulomb's experiments on the forces exerted by one charged object on another can be summarized in what we now call *Coulomb's law*:

> The force exerted by one charged object on another varies inversely as the square of the distance separating the objects and is proportional to the product of the magnitude of the charges. The force is along the line joining the charges and is attractive if the charges have opposite signs and repulsive if they have the same sign [see Figure 12.5].

If we call q_1 and q_2 the magnitudes of the two charges, and r the distance between their centers, we can state Coulomb's law in a single equation:

$$F = k \frac{q_1 q_2}{r^2}$$

where k is known as *Coulomb's constant*. The SI unit of charge is the coulomb (C) and the value of Coulomb's constant is

$$k = 9 \times 10^9 \ \text{Nm}^2/\text{C}^2.$$

In practical situations charges as large as one coulomb are very seldom encountered. The force that two objects each holding a charge of 1 C would exert on each other

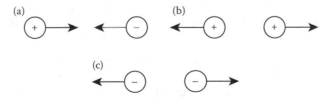

FIGURE 12.5 The electric force exerted by one charge on another is along the line joining the charges. (a) Charges of opposite signs attract each other. (b), (c) Charges of the same sign repel each other.

when they are separated by a distance of 1m is 9×10^9 N, which is about equivalent to the weight of one million tons. Typical charges produced by rubbing small objects are of the order of nanocoulombs (nC) to micro-coulombs (μC).

The most fundamental unit of charge is the charge of one electron or one proton. The fundamental charge e has a value

$$e = 1.602 \times 10^{-19} \text{ C}.$$

THE ELECTRIC FIELD

How does the electric force between two charged objects separated by a distance r propagate from one object to the other? In the nineteenth century the English physicist Michael Faraday introduced the concept of *field* as an intuitive way of looking at the electrical interaction between charges. Although, as we shall see in the final chapters of the book, there are other ways of looking at this interaction, the concept of field is still a very powerful and useful one.

There are situations in everyday life that we can use to illustrate the concept of field. A line of people waiting for tickets for a rock concert is an example. When the tickets run out, the word spreads out very quickly. Some people hear it directly from the person at the ticket window, others are told by the people who heard it first, these people tell others, and still others guess it from the movement of the crowd or from people's disappointed expressions. Knowledge about the lack of tickets spreads out through the crowd without a need for everyone in line to speak directly to the person at the window. The region around the ticket office where the people interested in obtaining tickets are located constitutes a "field."

In physics, field is used to specify a quantity for all points in a particular region of space. The temperature distribution of the water in a pond is an example. The property of the space around the earth where any object experiences its gravitational attraction constitutes the *gravitational field* of the earth. Similarly, the *electric field* describes the property of the space around an electrically charged object. The presence of a charged body at a particular point *distorts* the space around it in such a way that any other charged body placed in this space feels a force that is given by Coulomb's law.

Figure 12.6 shows a positive charge q at some point in space. If we place a small positive test charge q_0 at some other nearby point, the charge q will exert a force on this test charge which points away from the location of q and along the line joining them. The electric field strength **E** is the Coulomb force divided by the magnitude of the test charge q_0,

$$\mathbf{E} = \frac{\mathbf{F}}{q_0}.$$

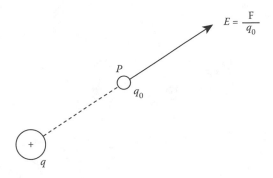

$$E = \frac{F}{q_0}$$

FIGURE 12.6   The electric field strength **E** at some point P near a positive charge q is determined by placing a small positive charge q_0 at point P. The value of **E** is the magnitude of the Coulomb force felt by the test charge divided by the magnitude of the test charge.

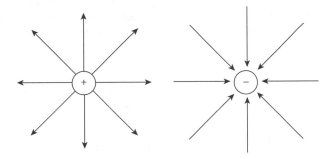

FIGURE 12.7   Electric fields around a positive and around a negative charge.

The electric field strength is a vector quantity since it has magnitude and direction and is proportional to the Coulomb force, which is a vector quantity. The direction of the electric field vector is the direction of the force on a positive test charge. The units of **E** are newton per coulomb, N/C.

By placing the test charge at several points in the vicinity of the charge q, we can map the electric field around this charge. Figure 12.7 shows the electric fields due to a positive and a negative charge. Since the test charge is always positive, the electric field around the negative charge points inward. The electric field line configuration for a pair of charges is a superposition of the field lines for two single charges, as shown in Figure 12.8a. Figure 12.8b shows the electric fields for two unlike charges and for two like charges. Notice also, the field lines in between the two parallel metal plates holding opposite charges (Figure 12.9). In this case, the field is constant throughout the region enclosed by the plates, as evidenced by the parallel, equally spaced field lines. Figure 12.10 is a photograph of two rods carrying equal and opposite charges. The electric fields are made visible by seeds floating in an insulated liquid.

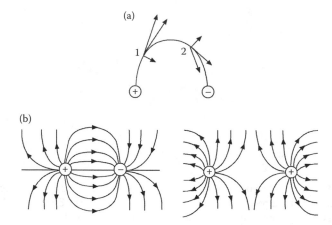

FIGURE 12.8 The electric field due to a pair of charges is a superposition of the fields for two single charges. (a) The field at points 1 and 2 is the resultant of the field due to the positive charge, a vector pointing away from the positive charge, and the field due to the negative charge, a vector pointing toward the negative charge. (b) Electric fields due to two unlike charges (left) and two like charges (right).

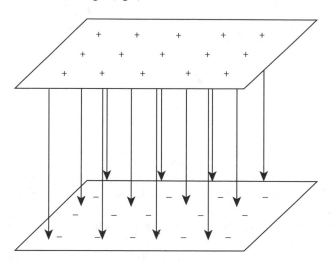

FIGURE 12.9 Electric field due to two oppositely charged metal plates.

THE FRONTIERS OF PHYSICS: ELECTRONIC PAPER

One implementation of electronic paper, the technology behind the portable readers now on the market, uses microcapsules containing clusters of black and white electrostatic-charged particles whose motion is controlled by an applied electric field. Each microcapsule is 40 μm in diameter and is filled with a liquid in which the black and white particles can move. The black particles are given a positive charge and the white particles are charged negatively.

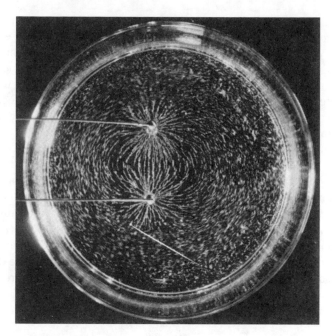

FIGURE 12.10   Electric fields produced by two rods carrying equal and opposite charges. The patterns are made visible by grass seeds floating in an insulating liquid. (From *PSSC Physics Seventh Edition*, by Haber-Schaim, Dodge, Gardner, and Shore. With permission from Kendall/Hunt Publishing Company, Dubuque, IA, 1991.)

The microcapsules are sandwiched between two transparent and conductive electrodes. When an electric field is applied between the electrodes, the black and white particles move to the opposite sides in the microcapsule. When the white particles are on the upper surface, the area covering the microcapsule appears white. Reversing the polarity of the field moves the black particles to the upper surface of the microcapsule, and it appears black.

THE FUNDAMENTAL CHARGE

In 1891, a young American student who had just completed his undergraduate education with a major in Greek at Oberlin College was approached by the school with an interesting proposition. Due to a shortage of scientists in the United States, the school had been unable to hire a qualified physics instructor and asked the recent graduate if he would accept the challenge. Robert A. Millikan not only accepted the offer but fell in love with the subject and took his Masters degree while he taught introductory physics at the school. He went on to receive the first PhD in physics that Columbia University ever awarded. After working as a postdoctoral fellow under Max Planck in Germany he returned to the United States and accepted a position at the University of Chicago in 1910.

In 1897, J. J. Thomson had succeeded in determining the charge-to-mass ratio of the electron, in what is now considered to be one of the landmark experiments in the history of science. After his determination of this ratio, Thomson and his collaborators J. S. Townsend and H. A. Wilson attempted to determine the charge *e* of the electron in a series of experiments in which an ionized gas was bubbled through water to form a cloud. By determining the mass and total electric charge of the cloud and estimating the number of ions in the cloud, they obtained a value for the magnitude of the charge of the electron of about 1.0×10^{-16} C. However, the method used in the estimation of the number of ions in the cloud could not be verified.

In 1909, Millikan began a series of experiments in which charged oil drops were balanced in midair for several hours by a constant electric field obtained with two parallel plates charged with opposite charges (Figure 12.11). An oil drop of mass *m* would be attracted toward the ground with a force $F = mg$, which is the drop's weight. When the drop holding a charge *q* is placed in a region of a constant electric field, an electric force $F_{el} = qE$ in the upward direction is exerted on the drop, which could be made to balance the weight of the drop by applying the appropriate field. That is,

$$qE = mg.$$

By varying the electric field between the plates, the oil drops could be moved up or down. If the charge on a drop changed during the observation, its velocity was observed to change. This allowed Millikan to conclude that *the electric charge*

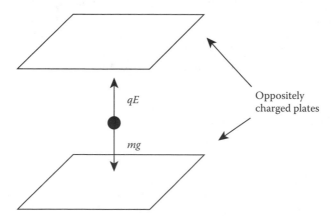

FIGURE 12.11   Millikan's oil drop experiment. The weight of the oil drop, *mg*, is balanced by the electric force acting on the charged drop, *qE*.

occurred always in multiples of an elementary unit which was the magnitude of the charge of the electron e. He obtained a value for the charge of the electron with an accuracy of one part in one thousand. In 1923, Millikan was awarded the Nobel Prize in physics for these experiments.

THE FRONTIERS OF PHYSICS: ELECTROSTATICS ON MARS

Mars, our neighbor in the Solar system, has been the subject of human interest since early times. Today, the possibility that there is water under its surface that could harbor living organisms has made the exploration of Mars a matter of enormous relevance to us. NASA and the European Space Agency have aggressive plans to explore the planet.

Scientists at NASA are attempting to solve the many problems that current and future missions to the planet may face. The Viking Lander missions of the 1970s together with the more recent Mars Exploration Rovers have shown that large areas of the surface of Mars are covered by fine particles in a fairly homogeneous, thin layer. These dusty conditions combined with the frequent dust devils and occasional large dust storms, as well as with the extremely low absolute humidity near the surface, create an environment conducive to electrostatic charge buildup. The surfaces of landers, rovers, or equipment may acquire an electric charge when placed in contact with the dust particles in the soil or suspended in the atmosphere. These dust particles, which can themselves acquire an electric charge as they collide with each other when blown by the wind, get attached to some of these surfaces by electrostatic forces. The result could be clogged filters, inefficient dust covered solar cells or thermal radiators, and obscure viewports.

To assess these problems, NASA developed a dedicated multisensor *electrometer*, an instrument that can measure the electrostatic charge generated when different materials are rubbed (Figure 12.12). This instrument was designed by scientists at NASA's Jet Propulsion Laboratory (JPL) and at the author's laboratory at NASA Kennedy Space Center (KSC) for a future lander mission to Mars. An early version of the instrument was developed for the heel of the lander robot arm. During operation, the electrometer was to be rubbed against the Martian soil. The current design embeds several electrometer sensors along the perimeter of a rover wheel. As the rover traverses the Martian terrain, the electrometer sensors measure the electrostatic charge exchanged between the soil and the materials on the sensors. The electrostatic charge depends strongly on the water content as well as on the mineral content of the soil. Laboratory experiments have shown that this instrument will allow scientists to identify soil composition and moisture content on Mars.

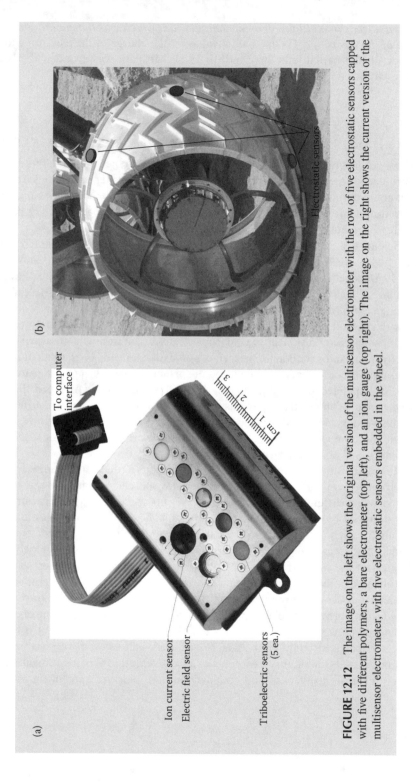

FIGURE 12.12 The image on the left shows the original version of the multisensor electrometer with the row of five different polymers, a bare electrometer (top left), and an ion gauge (top right). The image on the right shows the current version of the multisensor electrometer, with five electrostatic sensors embedded in the wheel.

ELECTRIC POTENTIAL

In our study of energy, we found that the concept of potential energy was very useful, in particular when we used it to describe the behavior of objects in the vicinity of the gravitational pull of the earth. As we learned in Chapter 4, the gravitational force is a conservative force, which means that the work done in moving an object from one point to another in a gravitational field depends only on the initial and final points and not on the path through which the object moves.

An electric charge q_0 placed in the electric field of another charge q feels a force F that is proportional to the product of the magnitudes of the two charges and inversely proportional to the square of their separation r_1. When the charge q_0 is brought in closer to the charge q, at a distance r_2, the only thing that changes in the expression for the force between the two charges is the distance between them. The path taken in moving the charge q_0 from the first position to the second does not matter (Figure 12.13). Like the gravitational force, the electric force is also a conservative force.

If the charges q and q_0 are both of the same kind, we would need to do work to bring q_0 closer to q. When we do that, the total energy of the particle carrying the charge q_0 increases by an amount equal to the work done. This increase in total energy appears as an increase in *electric potential energy*, ΔPE. When the positive charge q_0 is moved *away* from the charge q, there is a *decrease* in electric potential energy, since the two positive charges repel each other. On the other hand, if the two charges are oppositely charged, as when a positive charge q_0 is placed in the electric field of a negative charge q or a negative charge q_0 is placed in the electric field of a positive charge q, moving q_0 closer to q would result in a decrease in electric potential and moving them apart would result in an increase in electric potential, since the two charges attract each other.

The magnitude of the change in electric potential energy depends on the magnitude of the charge q_0. It is convenient, then, to have a quantity that does not depend on this charge and for this reason we define the *electric potential difference V* as the

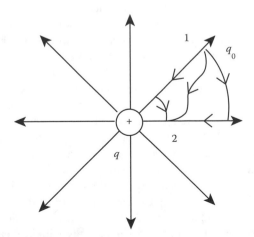

FIGURE 12.13 The work done in moving the charge q_0 from Point 1 to Point 2 is independent of the path taken. We say that the electric force is a conservative force.

change in electric potential energy of a charge q_0 divided by the magnitude of that charge; that is,

$$V = \frac{\Delta PE}{q_0}.$$

The SI unit of potential difference is the volt (V). From the definition, 1 volt equals 1 joule divided by 1 coulomb, or

$$1V = \frac{1J}{1C}.$$

The volt was named in honor of Count Alessandro Volta, a professor of physics at the University of Pavia, Italy, who invented the electric battery. We sometimes refer to potential difference as *voltage*, a term derived from the name of the unit. From the definition of potential difference we can see that the *electron volt, ev,* is a unit of energy. A particle with a charge equal to that of the electron which is moved between two points in an electric field so that their potential difference is 1 volt will change its potential energy by 1 electron volt:

$$\Delta PE = qV = 1 \text{ eV} = 1.6 \times 10^{-19} \text{ C V} = 1.6 \times 10^{-19} \text{ J}.$$

STORING ELECTRICAL ENERGY

As we have seen, we must do work to move an electric charge in an electric field and this changes the potential energy of the charge. Could we *store* this potential energy for later use? In 1746, the Dutch physicist Pieter van Musschenbroeck, who was professor of physics at the University of Leyden, attempted for the first time to do just that. He suspended a metal jar filled with water from insulating silk threads and led a brass wire from a charged cylinder into the water. A student who was assisting Musschenbroeck with the experiment happened to touch the brass wire and became the first person to receive an artificially produced electric shock. Musschenbroeck realized that he had accidentally discovered a way of storing charge. News of the experiment spread rapidly and soon "Leyden jars" were being built and improved upon in many other laboratories. Today, a device that has the capacity of storing electrical energy, like the Leyden jar, is called a *capacitor.*

We can store electrical energy by creating an electric field. One way of creating an electric field is with two uncharged metal plates separated by a distance d. When some small amount of charge is transferred from one plate to the other, a small potential difference appears (Figure 12.14). If we continue the process of transferring charge, we end up with a potential difference V between the plates and an electric field E in this region. It takes work to separate the charges and create the field and this work becomes the potential energy that is stored. This particular device is called a *parallel plate capacitor.*

Capacitors can have different shapes. The amount of charge that can be stored in a capacitor at a given potential depends on its physical characteristics. Volta introduced the expression "electrical capacity," in analogy with heat capacity, to indicate the storage capacity of these devices. Today we call this concept *capacitance.* The smaller the voltage needed to store a given charge, the greater is the capacitance.

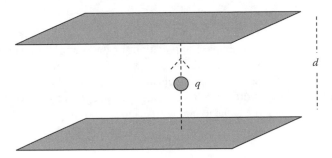

FIGURE 12.14 When a small amount of charge is transferred between two uncharged metal plates, a small potential difference in created. This device is a parallel plate capacitor.

Many capacitors have a nonconducting material between the charged plates. This increases the capacitance. If the insulating material is air, the increase is very small; about 6 parts per 10,000. For other materials the increase can be much greater. Glass, for example, increases the capacitance by a factor of between 5 and 10. The increase in the capacitance is due to a reduction in the potential difference between the plates (since capacitance is inversely proportional to the potential difference). When the plates are charged, the charges in the nonconducting material between the plates reorient themselves, so that the positive charges point toward the negatively charged plate and the negative charges toward the positive plate. This orientation sets up an electric field in the nonconductor that is in the opposite direction to the electric field between the plates. When the field is reduced, the potential difference is also reduced, thus increasing the capacitance.

An animal cell is a living example of a capacitor. The membranes of cells are generally composed of lipid or fat molecules and protein molecules oriented so that the inner part of the membrane containing lipid and cholesterol molecules is sandwiched between layers containing protein molecules. The membrane wall of a cell separates two regions that contain potassium ions (positive) and chloride ions (negative). Thus, a cell is like a very small capacitor where the positive and negative charges are separated by the nonconducting membrane wall. The potential difference across a typical cell is of the order of 100 mV.

THE FRONTIERS OF PHYSICS: STORING SINGLE ELECTRONS

A new device, developed by Mark W. Keller and his collaborators at the National Institute of Standards in Boulder, Colorado, allows these researchers to individually count and store millions of electrons, one by one, into a specially developed capacitor. With their electron pump, as the device is called, individual electrons are transferred into a capacitor for storage. The pump consists of an array of six microscopic bullet-shaped regions of aluminum, separated by small walls of aluminum oxide. Electrons are allowed to pass through the aluminum islands only when an electric current is applied to the islands. Timed electrical pulses push the electrons from island to island until they reach the capacitor.

The researchers have used their pump to count millions of electrons, missing only one in 70 million. The main application for the pump, however, is not to count electrons, but to increase the accuracy in the determination of capacitance. By knowing exactly how many electrons are stored in a capacitor (which gives the total charge stored) and measuring the voltage, Keller and his collaborators can calculate the capacitance with accuracy not possible previously.

PHYSICS IN OUR WORLD: INKJET PRINTERS

An inkjet printer uses a print head that shoots ink at the paper. With one of the leading technologies—the thermal inkjet printer—a resistor inside the print head heats a thin layer of ink which expands into a vapor bubble. The expansion forces the ink through a small nozzle which causes it to break up in droplets a tenth of a millimeter in diameter. The print head shoots about 6000 droplets every second at speeds of about 15 m/s, first through a charging device and then through the charged plates of a capacitor (Figure 12.15). In places where the paper is to be left blank, an electric change is deposited on the droplets. As the charged droplets pass through the electric field that exists inside the capacitor, they are deflected away from the paper, back to the ink reservoir. When the ink is to hit the paper, the charging device is turned off and the ink droplets are left uncharged. The neutral droplets fly undeflected through the capacitor and hit the paper.

Color inkjet printers have four ink reservoirs containing black, cyan, magenta, and yellow ink, the standard colors used in commercial printing. By mixing these four colors, all other colors can be obtained.

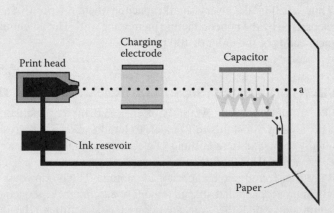

FIGURE 12.15 The printhead of an inkjet printer shoots charged ink droplets that are guided by the electric field of a capacitor.

13 Applied Electricity

CONDUCTORS AND INSULATORS

As we have seen, rubbing a glass or an amber rod with a piece of cloth produces an electric charge on the rod (Figure 13.1a). If the rod is made of metal, no charge develops (Figure 13.1b). However, touching a metal rod with a charged object will cause the metal to become charged (Figure 13.1c). The reason for this behavior is that the atoms in a metal have some electrons that are not tightly bound to their nuclei and are free to move about. Any excess charge readily moves in conductors. In metals, then, electric charges move or flow through the material. We say that metals are good *conductors* of electricity. In glass, amber or other materials like them, on the other hand, electrons are not free to move; they are bound to individual molecules or atoms. Any excess charge placed on them remains (unless they are touched by some other object). These materials are called *insulators*.

Plastics, wood, and rubber are examples of good insulators. Pure water is also an insulator. Tap water, however, contains salts that form ions which can move through the liquid, making it a good conductor.

There are some materials, called *semiconductors*, which are intermediate between conductors and insulators. Modern electronics has developed due to the discovery of the properties of these materials. The electrical conductivity of semiconductors can be enhanced by the addition of traces of other elements with a slightly different electronic structure. As we shall see, these *impurities* provide an additional electron or the lack of one, which results in a negative or positive charge that can move around.

ELECTRIC CURRENT AND BATTERIES

During the second half of the eighteenth century, a flow of electric charge could be produced only by discharging a Leyden jar. In 1800, Alessandro Volta, a professor of physics at the University of Pavia, Italy, discovered that a stack of discs of silver and zinc, interspersed with wet pasteboard and held with an insulating handle, produced a separation of charge and a potential difference between the two metals. When the first and last discs were connected by a conductor, Volta obtained a flow of electric charge or *current* through the conductor with the same properties as the current obtained by discharging a Leyden jar, but with the important difference that the flow was more or less *continuous*. Volta called his device a *battery* (Figure 13.2).

An electric current exists whenever there is a net flow of charge. A potential difference is needed for a flow of charge to exist, in much the same way that a stone would not fall to the ground unless it is at some height above the ground; that is,

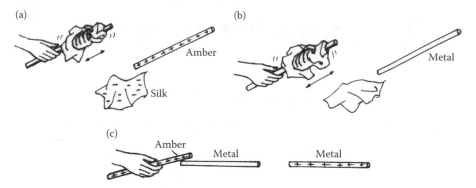

(a)

Amber

Silk

(b)

Metal

(c)

Amber

Metal

Metal

FIGURE 13.1   (a) Rubbing an amber rod with silk produces an electric charge on the rod. (b) No charge is produced on the metal rod. (c) Touching a metal rod with a charged amber rod causes the metal to become charged.

Moistened
leather disc

Zinc

Silver

FIGURE 13.2   Alessandro Volta reported his battery to the Royal Society as, "an assemblage of a number of good conductors of different sorts … 30, 40, 60 pieces or more of copper, or better of silver, each in contact with a piece of tin, or what is much better, of zinc, and an equal number of layers of water … or pieces of cardboard or leather … well soaked." (From © Bettman/CORBIS Seattle, WA. With permission.)

unless there is a gravitational potential difference with respect to the ground. In a television tube, for example, electrons are accelerated due to a potential difference. The electron beam that strikes the phosphor-coated screen and produces the image constitutes an electric current. An electric current can exist in a conductor provided a potential difference exists between two points in the conductor. When the ends of a metal wire, for example, are connected to a battery, an electric field is created inside the wire which acts on the electrons that are free to move in the metal, moving them, thus producing an electric current. If during a time t an amount of charge q flows past a particular point in a conductor, the electric current i is given as

$$\text{electric current} = \frac{\text{charge that flows}}{\text{time}} \quad \text{or} \quad i = \frac{q}{t}.$$

Electric current is the rate at which charge flows in a conductor. The unit of current is the ampere (A), named in honor of the French physicist André Marie Ampère, and is equal to one coulomb per second. It is a fundamental SI unit.

Maintaining an electric current requires the maintenance of a potential difference between two points. Volta's battery did just that; it provided electrical energy so that a potential difference could be maintained between the two metals (called *electrodes*).

Before a good understanding of the atomic nature of matter was achieved, it was believed that the charge carriers that moved in a conductor were the positive charges and the direction of electric current was chosen to be that of the positive charges. Today, we know that it is electrons that move but the convention of the direction of current has not been changed (Figure 13.3). We can still apply this convention if we understand that the view of the negatively charged electrons moving in one direction with respect to the stationary positive ions is equivalent to the positive ions moving in the *opposite* direction with respect to the negative electrons.

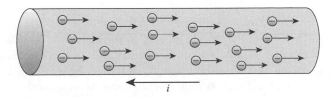

FIGURE 13.3 The conventional direction of the flow of current in a conductor is opposite to the direction of motion of the electrons, the carriers of the current.

PHYSICS IN OUR WORLD: ELECTRIC CARS

The concept of automobiles powered by an electric motor that is run by batteries is not new; the essential battery technology was developed toward the end of the nineteenth century and by 1900 many electric cars were being manufactured. Due to the weight of the large batteries required and the need to recharge them at fairly short intervals, electric cars were heavy and slow to operate. Development of lighter materials and recent advances in battery technology have made feasible electric cars possible. General Motors' EV1, the first electric car intended for the general public, used 26 12-volt lead-acid batteries that took 3 hours to charge and gave the car a range of 70–90 miles. After the EV1 was developed, General Motors, along with the other American automakers, went to court to delay the implementation of new and more stringent emission standards. The delay removed the economic advantage to build the electric car, and the company pulled the plug on its development. About a thousand EV1s were built but by company but none were sold; they were only leased for a limited time.

General Motors' subsequent electric vehicle, the Chevy Volt, is powered by 220 lithium-ion cells that take 8 hours to charge when connected to a 120-V outlet and three hours from a 240-V outlet. The car has a 40-mile range on its all-electric operation. When the 16 kWh lithium-ion battery pack is depleted, the Volt can switch to a small internal combustion generator to power its electric engine. The generator can add 300 miles to the Volt's range of operation.

OHM'S LAW

In a metal, the atoms are arranged in a crystal lattice with a large number of electrons that are free to move around in the metal. These electrons, called *conduction* or *valence* electrons, are in continuous motion like gas molecules, bouncing off the lattice ions in such a way that, in the absence of an electric field, their average velocity is zero. Bumper cars at an amusement park move fairly fast between collisions; they, however, do not go anywhere very much (Figure 13.4). Unlike the bumper cars, electrons do not collide with each other but with the atoms in the lattice. In metals, about one electron per atom is used in the conduction process and the atoms in the lattice are actually positive ions, since they have contributed one of their electrons to this process.

When a potential difference is applied across the metal, an electric field $\mathbf{E}$ appears in the metal and a force of magnitude eE acts on the electrons. A conduction electron is then accelerated to large speeds in the direction of this force. Before long, however, the electron collides with an ion in the lattice, bounces off, and is accelerated again in the same direction only to collide again with another ion. The net result is that conduction electrons move along the wire at very small *drift* velocities, of the order of 1 mm/s, in spite of the large velocities acquired in between collisions (Figure 13.5a). We could illustrate the effect of this applied potential with the bumper cars example if we imagine a surface where the cars move to become tilted

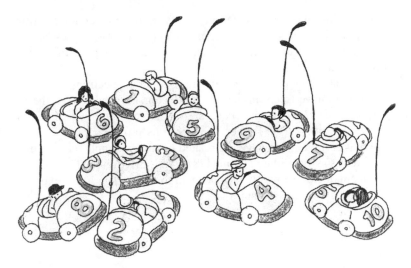

FIGURE 13.4 Bumper cars at an amusement park collide continuously with each other and do not go anywhere.

(a)   (b)

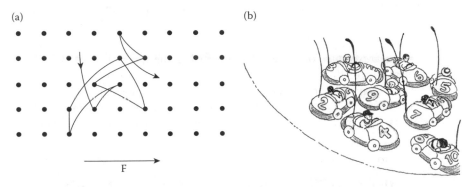

F

FIGURE 13.5 (a) The drift velocity of the conduction electrons in a metal is very small compared with the velocities of the electrons between collisions with the lattice ions. (b) A gravitational potential provided by the inclined surface makes the cars drift toward the lower part.

(Figure 13.5b). Although the cars still collide with each other and move in all directions, the sloped surface makes them move in that direction somewhat more often than in the others. There is a *drift* velocity toward the lower end of the surface. We should be careful with this analogy, however. A single electron does not necessarily move from one end of the conductor to the other. It is energy and momentum that is transmitted through the conductor.

A metal, then, offers some *resistance* to the flow of current through it. When we apply a potential difference V to the two ends of a metal wire, a current i appears in the wire. This current increases as the applied potential difference increases. In our analogy with the bumper cars, raising the slope increases the flow of cars toward the lower part, thus increasing the "current." The magnitude of the current in the wire depends on several factors. One is the type of metal used; different lattice configurations interact in different ways with the conduction electrons. The size of the wire also affects the amount of current. A thin wire presents a greater resistance to the flow of current than a thicker wire, and a longer wire also presents a greater resistance.

We can see that, although we speak of the conduction electrons as free to move, they do not really move freely inside the conductor. For this reason, the presence of the external force eE does not accelerate the electrons in a way that makes the current *increase* continuously. Rather, the electrons promptly reach a steady-state situation so that current and voltage are related by a very simple relationship, as the German physicist Georg Wilhelm Ohm (1787–1854) discovered. The current i flowing through a conductor is directly proportional to the voltage V that exists between the two ends of the conductor, that is

$$i \propto V \quad \text{or} \quad i = \text{constant} \times V.$$

This simple expression, known as *Ohm's law*, is usually written as

$$i = \frac{V}{R} \quad \text{or} \quad V = iR$$

where R is the *resistance* of the conductor and has the units of *ohms* (Ω) (Greek capital omega). From this expression, we can see that 1 Ω equals 1 V/1 A.

Although it bears the name of a *law*, Ohm's law is not a fundamental law of nature like Newton's law of universal gravitation. Rather, it is a result of experimental observations and is valid only for certain materials within a limited temperature range.

THE FRONTIERS OF PHYSICS: ELECTRIC DENTISTS

According to recent statistics, almost 85% of all 17-year-olds have already had several cavities. The reason for its prevalence is the difficulty in detecting cavities early on. Tooth decay starts in the enamel coating of the tooth where foods are fermented by bacteria, producing acids that erode the mineral in the tooth enamel. By the time this demineralization becomes detectable, it may be too late for fluoride treatments to have any healing effect.

A method recently discovered by researchers at the universities of Dundee and St Andrews in Scotland, and the University of Nijmegen in the Netherlands makes use of the change in electrical resistance of the eroded regions in the tooth enamel. These regions are filled with fluids that have smaller resistances than enamel. Measuring the resistance at different places in the tooth reveals the presence of cavities. The researchers used the technique on extracted teeth with perfect accuracy. The next step in the researchers' schedule is to obtain funds to make this promising method available clinically.

SIMPLE ELECTRIC CIRCUITS

The electric circuit in a flashlight is one of the simplest we can study. It consists of one or more batteries, a metal conductor, a switch, and a light bulb (Figure 13.6a). A diagram of this circuit is shown in Figure 13.6b. A hydraulic analog of this circuit appears in Figure 13.6c. Water flowing from the tank (high potential) falls through the pipe where it moves a paddle wheel, thus performing work. Low pressure water is then pumped up to the tank. Water flows in this circuit only when the valve is open. Similarly, in our electric circuit, the battery "pumps" up the bulb terminals to a high potential. Work is performed as the current flows through the narrow filament of the bulb, producing light and heat. Low potential current returns to the battery where it is "pumped" up again. Current flows in the circuit only when the switch is closed.

Again, we must be careful with our analogy. In the hydraulic case, the water molecules actually move around in the pipes. In the electric circuit, the electric field that exists in the conductor acts on the conduction electrons which then move. The ions remain fixed in the lattice.

The light bulb in our simple circuit dissipates energy (as light and heat) by means of its resistance. Electric heaters, toasters, electric stoves, and ovens are other examples of devices that dissipate energy by means of their resistance. As electrical energy dissipates, electric potential drops. There are situations where it is desirable to lower this electric potential or voltage in a circuit. For these cases, a *resistor* is used. A resistor is merely a device that dissipates electrical energy; it

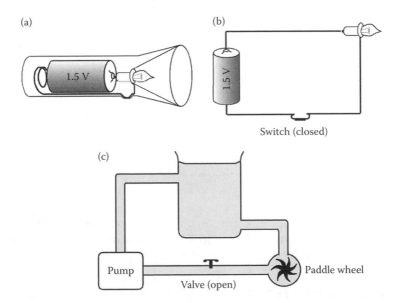

FIGURE 13.6   (a) A simple circuit in a flashlight. (b) Representation of this circuit with a single battery, a conducting wire, a switch, and a light bulb. (c) Hydraulic analog with a pump, pipe, a valve, and a paddle wheel. Water is pumped to a high potential (tank) from where it falls, transforming potential energy into kinetic energy. Some of this energy is used to drive the paddle wheel. This low potential water is returned to the pump where the circuit is completed.

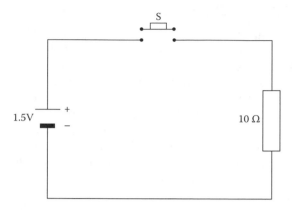

FIGURE 13.7 A schematic diagram of an electric circuit.

can be a strong narrow wire offering a greater resistance to the flow of current than the wires used in the circuit. Usually, however, it is made of ceramic materials with a low conductance.

To simplify the graphic representation of electric circuits, certain standard symbols are used. The internationally accepted symbols are:

	A battery
	A resistor
	Switches
	A capacitor
	A conductor

A schematic diagram of an electric circuit consisting of a 1.5 V battery, a 10 Ω resistor and a switch is shown in Figure 13.7. It is customary to include the values of the different components.

RESISTOR COMBINATIONS

There are two ways of connecting items such as light bulbs into an electrical circuit: series and parallel. In a parallel connected circuit such as household lighting, every lamp is connected separately to the input line, and any one of them can be turned on or off without affecting the others. Series connections have the lamps "strung out," so that if one of the lamps fails they all go out. Christmas tree lights are connected this way. We shall study these two types of circuits briefly.

In a *series* circuit, like the Christmas tree light bulbs, current passes through each one of the light bulbs one after the other. In Figure 13.8 we have a series combination of two light bulbs of resistances R_1 and R_2. The total resistance in the circuit is the sum of the resistances of the two bulbs; that is, $R_s = R_1 + R_2$. This last expression means that if we have two light bulbs with resistances of 5 and 8 Ω connected to a 1.5-V battery, we could replace them with a single light bulb having a resistance of 13 Ω, without changing the amount of current flowing through the circuit.

From the law of conservation of energy, the potential or voltage delivered by the battery in Figure 13.8 must equal the potential drop through the first light bulb plus the potential drop through the second. This is analogous to the situation shown in Figure 13.9. The potential energy gained by the girl as she climbed to the top of the double slide equals the sum of the potential energies lost in the first and the second

(a)   (b)

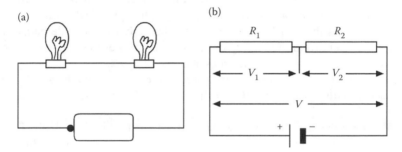

FIGURE 13.8 (a) Two light bulbs connected in series to a battery. (b) Schematic diagram of this circuit. The resistances of the two bulbs are shown as R_1 and R_2, and the potential drops as V_1 and V_2, respectively. The voltage supplied by the battery is shown as V.

FIGURE 13.9 The child sliding down the double slide loses potential energy in the two sections of the slide which is equal to the potential energy gained when she climbed to the top. This is analogous to the potential differences in the circuit of Figure 13.8.

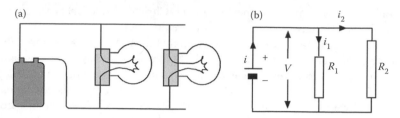

FIGURE 13.10   (a) Two light bulbs connected in parallel. (b) Schematic diagram of this circuit.

slopes. Similarly, the potential increase delivered by the battery should equal the sum of the potential drops through the two light bulbs.

In a *parallel* connection, the current splits into one or more branches. In Figure 13.10, two light bulbs of resistances R_1 and R_2 are connected in parallel, forming two branches. The current i coming from the battery splits into i_1, which flows through R_1, and i_2, flowing through R_2. Since these are the only two possible paths for the current, the current from the battery must equal the sum of the individual currents flowing through each one of the branches. That is, $i = i_1 + i_2$.

We can see that both light bulbs are connected directly to the battery by means of conducting wires. This means that each light bulb draws current from the battery *independently* of the presence of the other light bulb. We could, in principle, connect as many light bulbs as we wanted without affecting the first light bulb. The only limitation is the ability of the battery to provide electrical energy to many light bulbs. In a parallel connection, the total resistance of the circuit *decreases* as more branches are added, since there are more paths for the current to flow.

ELECTRICAL ENERGY AND POWER

An electric current can exist in a region of space that has been evacuated, for example the electron beam in a television tube. In particle accelerators, like SLAC, the two mile electron accelerator at Stanford University, electrons are accelerated along the two mile tube and emerge at one end traveling at speeds close to the speed of light to strike various targets. Scientists study these collisions to gain a better understanding of the submicroscopic world.

When current exists in a conductor, however, the collisions of the electrons with the ions in the crystal lattice transfer energy to these ions, resulting in an increase in the internal energy of the material. As we know, an increase in internal energy means an increase in thermal energy and, therefore, an increase in temperature. The flow of current in a conductor produces heat.

The amount of heat produced by an electric current flowing in a conductor depends on the magnitude of the current and on the resistance of the conductor. The filament of a light bulb, for example, is made of tungsten, a metal that melts at 3387°C. When enough current flows through it, the metal heats up to around 2600°C, radiating energy as heat and visible light. Current flowing through the heating element of a toaster, a metal coil, produces heat at a lower temperature, which is used to toast the slices of bread. When this same amount of current flows through the toaster cord, a copper wire of much greater cross sectional area and therefore much lower resistance, no appreciable heat is produced.

When an electric current i flows through a light bulb, an amount of charge q flows through it in a certain time interval. The rate at which the battery performs work in moving this charge is called *electric power*. Electric power is equal to voltage multiplied by current, or

$$P = Vi.$$

By Ohm's law, $V = iR$; therefore, electric power can be expressed in terms of current and resistance as

$$P = Vi = i^2R.$$

The SI unit of power, remember, is the watt, W, which is joules per second. Since $P = Vi$, then $1\ \text{W} = 1\text{V} \times 1\text{A}$.

James Joule, who developed an experiment to measure the mechanical equivalent of heat, also developed a way of measuring the heat dissipated by an electric current and determined that it was proportional to the square of the current in the conductor. For this reason, the expression $P = i^2R$ is known as *Joule's law*.

SEMICONDUCTORS

Electric current can flow in a vacuum, through a conductor such as a metal wire, and through *semiconductors*. As we said early in the chapter, semiconductors are materials with electrical conductivities that are intermediate between those of conductors and insulators.

The properties of a solid depend not only on its constituent atoms but also on the way these atoms are stacked together. When two identical atoms are brought together, for example, each one of their energy levels split in two, with the separation between the pairs of levels depending on the distance between the atoms. If four atoms are brought closer together, their energy levels split into four. In a solid, there are billions of atoms very close together, and each one of their energy levels splits into billions of very closely packed levels. In one gram of sodium there are about 10^{22} atoms of sodium and each one of the energy levels in sodium splits into 10^{22} levels so close together that they can be considered to be a single *energy band*. The bands from all the different energy levels are separated by a *forbidden energy gap*, in which no electron can exist (Figure 13.11). In good conductors, the outermost occupied energy band, called the valence band, is not completely filled. In sodium, a good conductor, this valence band is only half-filled so that an electron in one gram of sodium has about 5×10^{21} different possible, allowed energy states. In nonconductors, on the other hand, the valence band is completely filled and there are no available levels for an electron to move into. The energy gap between the valence band and the next band, called *conduction band*, is large. When the energy gap is small, the solid is a semiconductor. The conduction band is where the electrons that conduct electricity reside.

Silicon and germanium are the most common semiconductor materials used in electronics. Pure semiconductors, however, are of no great practical importance. In

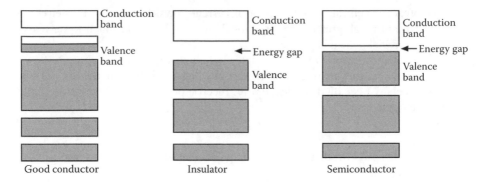

FIGURE 13.11 Energy band structure of a solid. In insulators, the energy gap is large. In semiconductors, it is small.

contrast to a metal, where almost every atom contributes one or two electrons that conduct electricity, in a pure semiconductor, only one atom in a billion contributes an electron. When small amounts of impurities, at the level of one in a million, are added to a semiconductor in a process called *doping*, the conduction properties of semiconductors can be enhanced. Silicon and germanium have four valence electrons which almost fill the valence band. An "impurity" atom with five valence electrons (such as arsenic or phosphorus) contributes an extra electron which does not bond with the surrounding silicon atoms (Figure 13.12a). This additional electron, being loosely bound, can easily jump up to the conduction band when it gains a small amount of energy, thus contributing to the conductivity. Semiconductors doped with these *donor* atoms are called *n-type* semiconductors because the charge carriers are the negatively charged electrons.

When a piece of silicon is doped with impurity atoms with only three valence electrons, like boron or aluminum, there is an electron deficiency which leaves the lattice lacking one bond (Figure 13.12b). This electron deficiency constitutes what has been called a *hole*. In this case, there is a tendency to capture electrons from the valence band, in effect moving a nearby electron to that site and transferring the hole to the location previously occupied by the electron. Since the hole represents an absence of a negative electron and moves in the opposite direction to the electrons, these holes can be considered positively charged. Semiconductors doped in this way are called *p-type* semiconductors because the charge carriers are the positively charged holes.

Transistors, diodes, and other solid state electronic devices can be manufactured by joining different types of semiconductors. The simplest solid state device is a *p–n junction* diode, which allows current to pass in only one direction. When an *n*-type semiconductor and a *p*-type semiconductor are brought together, some electrons from the *n* region drift into the *p* region and some holes from the *p* region drift into the *n* region. This migration of electrons and holes is due to the unequal concentration of charges in the two regions. The electrons that drift into the *p* region move into the holes near the boundary, neutralizing the free charge carriers. However, because each region is electrically neutral (since the electrons and holes come from neutral impurity atoms), this diffusion of charges near the boundary creates a layer

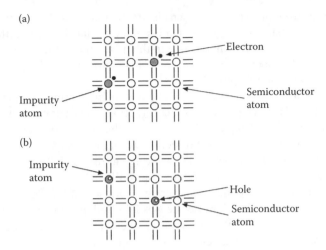

FIGURE 13.12 (a) In an *n-type* semiconductor, the loosely bound extra electron from the impurity atom contributes to the conductivity. (b) In a *p-type* semiconductor, the missing electron leaves a *hole* in the electron structure. When a nearby electron moves in to fill the hole, it leaves a hole where it was, in essence moving the hole in the opposite direction.

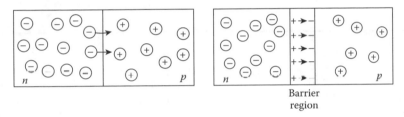

FIGURE 13.13 The migration of electrons in the *p–n* junction creates a barrier region that prevents further diffusion.

of positive charges in the *n* region and a layer of negative charges in the *p* region. A barrier region is then formed where an electric field appears (Figure 13.13). This electric field prevents any further diffusion of charges between the two regions since an electron from the *n* region that makes it into the barrier will be repelled by the layer of electrons in the *p* region; likewise, a hole that moves into the barrier region will be repelled by the layer of holes in the *n* region.

Suppose we connect a battery to a *p–n* junction with the positive terminal connected to the *p* side and the negative to the *n* side, as in Figure 13.14a. Electrons coming from the battery *recombine* with the holes of the barrier layer in the *n* region, neutralizing them. This recombination has the effect of lowering the potential difference that was set up by the electric field in the barrier region. Electrons can now move from the *n* side into the *p* side to complete the circuit, thus allowing the flow of current. The junction is said to be *forward biased*.

When the polarity of the battery is reversed so that the positive terminal is connected to the *n* side and the negative to the *p* side, electrons are pulled from the *n* region and holes from the *p* region. When electrons or holes are drawn from near the barrier region, the charge of opposite sign is enhanced thus increasing the potential difference across the barrier. An increase in this potential further prevents the flow of charge across the junction and no current flows through it. The junction is now connected in *reverse bias* (Figure 13.14b).

A *p–n* junction acts like a discriminating switch, preventing the flow of current in one direction and allowing it in the other direction. Such a device, called a *diode*, is able to *rectify* an alternating current; that is, to change it into direct current since it allows current to pass in only one direction when the voltage is applied.

In 1947 three American physicists working at the Bell Telephone Research Laboratories, John Bardeen, Walter Brattain, and William Shockley, invented the *transistor*. Their first transistor, a point-contact design, consisted of a wedge of semiconductor of about 3 cm on each side. This device was followed in 1951 by the more reliable *p–n–p* transistor with a thin layer of *n*-type semiconductor sandwiched between two thicker layers of *p*-type semiconductor.

The three regions in a *p–n–p* transistor are called the *emitter, base,* and *collector.* Suppose that we connect a *p–n–p* transistor to two batteries, as shown in Figure 13.15, so that the emitter–base *p–n* junction on the left is forward biased while the base–collector junction on the right is reverse biased. The positive terminal connected to the emitter side on the left pushes holes in the *p* region toward the

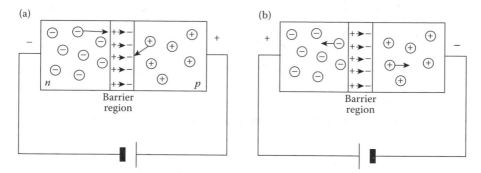

FIGURE 13.14　(a) Current flows through the junction when it is *forward biased*. (b) When the junction is *reverse biased* no current flows through it.

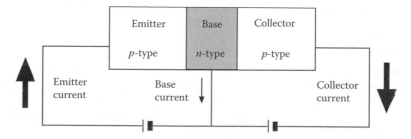

FIGURE 13.15　A *p–n–p* transistor. A small change in the voltage to the base produces a large change in the collector.

emitter–base junction. Since this junction is forward biased, it acts as a diode, allowing the flow of holes from emitter to base. Once in the *n* region, these holes would come under the influence of the second battery, moving into the collector due to the attraction of the negative terminal of this battery.

Not all the holes that enter the base region travel through to the collector. About 2% of them combine with the free electrons in this region, thus producing a small base current. Since the base is only about one micrometer in thickness, most of the holes (the remaining 98%) pass through, forming the collector current. The result of an increase in the base potential is a large increase in collector current and a small increase in base current. If we were to connect the voltage coming out of a compact disc player in series with the battery on the emitter side and a loudspeaker in series with the battery on the collector side, any small change in the input voltage of the CD player would produce a large change in the current flowing through the loudspeaker, thus amplifying the input voltage. A transistor is useful because it can amplify a small signal into a larger one.

SUPERCONDUCTORS

As we have learned, the resistance to the flow of current is caused by electrons being scattered by interactions with the vibrating atoms in the metal lattice. When the temperature decreases, the vibrations decrease and therefore the resistance decreases. However, we would not expect the resistance to drop to zero but rather to reach a minimum constant value at a certain low temperature. Below this temperature, an additional decrease in the vibrations of the atoms in the lattice would not be expected to appreciably affect the flow of current. The presence of defects in the lattice and the fact that the vibrations never stop completely regardless of how close we get to the absolute zero would seem to indicate that the resistance should never reach zero.

In 1911, the Dutch physicist Kamerlingh Onnes discovered that, below a certain critical temperature, the electrical resistance of certain metals vanished completely, the metals becoming *superconductors*. He had recently accomplished the liquefaction of helium and was measuring the electrical resistance of metals at the newly achieved low temperatures (helium boils at 4.2 K at a pressure of one atmosphere) when he discovered that below 4 K the resistance of mercury dropped to zero. Soon he found that other metals showed the same property at these very low temperatures.

Superconductivity seems to contradict what we have learned of the behavior of matter. How can superconductivity be understood? In 1935, the brothers Heinz and Fritz London were carrying out experiments in superconductivity in Oxford when they realized that the observed effects could be explained, at least partially, if the conduction electrons moved as a unit, as if they were linked together like the cars of a train. Although this idea explained why the electrons could move throughout the crystal lattice without being stopped, it also presented problems; electrons are negatively charged and repel each other. It was difficult to understand how they could remain together.

During the early 1950s, the English physicists David Bohm and Herbert Fröhlich advanced a theory that could explain the problem of the electron repulsion. The key is in the ions of the lattice. A metal contributes an average of one electron per atom to the conduction process. These conduction electrons move about freely, leaving behind the positive ions that form the lattice. According to Bohm and Fröhlich,

FIGURE 13.16   When the ions of the lattice are drawn toward the path of the electrons, a positive region is created that attracts other electrons.

when the electrons in a superconductor pass through the lattice ions, the negative charge of the electron attracts the positive ions, drawing them closer together. Since the ions are much more massive than the electrons, they take a little longer to separate back to their original positions, thus creating a slightly greater concentration of positive charge that attracts other electrons which might be following a similar path (Figure 13.16). Eventually, the ions push away from each other, due to their mutual electrical repulsion, moving past the initial position. This gives rise to a vibration, called a *phonon*, in the crystal lattice.

In 1956, Leon N. Cooper, at the University of Illinois, proposed that the interaction between the vibrating lattice and the electrons does not just unify the electric charge but creates *pairs* of electrons that behave as a single particle. In the physics department at the University of Illinois there was at the time a shortage of space and Cooper had to share an office with John Bardeen. John R. Schrieffer was Bardeen's graduate student and the three decided to extend Cooper's work on *Cooper pairs*, as the electron pairs came to be known, to the entire lattice. After spending a great deal of time on the problem, Schrieffer felt he was getting nowhere with it and was thinking of changing his thesis research. At about this time, Bardeen had to travel to Stockholm to receive the Nobel Prize for his work on the invention of the transistor and asked Schrieffer to work on the problem for one more month. In this month Schrieffer realized that the two electrons in a Cooper pair had opposite velocities that added up to a net momentum of zero. This allowed him to express the problem in a more manageable form. His result became the basis for a more complete theory of superconductivity, later known as *BCS theory*. The three received the Nobel Prize in physics in 1972 for this theory, making Bardeen the only person ever to win two Nobel Prizes in the same subject.

According to BCS theory, when the temperature falls below the critical temperature, the electrons act collectively as they interact with the lattice ions, forming Cooper pairs. The energy of these superconducting electrons is *lower* than when they act individually, at higher temperatures. There is an energy gap between the superconducting state and the normal, non-superconducting state. At very low temperatures, below the critical temperature, there is not enough energy available to excite the electrons from the lower superconducting state to the higher non-superconducting state. The conduction electrons therefore remain in this superconducting state. Scientists have been able to maintain steady currents in superconducting rings for several years with no measurable reduction.

Until early 1986, superconductivity required extremely low temperatures obtainable only by cooling the materials with expensive liquid helium. By 1973, however,

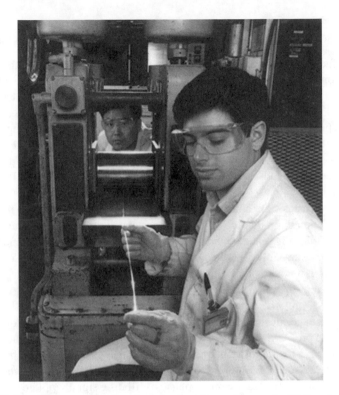

FIGURE 13.17 Superconducting wire developed at Argonne's National Laboratory. (Courtesy of Argonne National Laboratory.)

scientists discovered by accident that certain oxides of some rare earth elements, which are normally insulators, could become superconductors. In January 1986, J. Georg Bednorz and K. Alex Müller, of the IBM Zurich Research Laboratory in Switzerland, found that a compound of barium, lanthanum (a rare earth), copper, and oxygen became superconductor at 35 K. Shortly after the results were published, the American physicist Paul C. W. Chu, of the University of Houston and his collaborator, professor Mau-Kuen Wu of the University of Alabama, began a series of experiments with similar compounds using different rare earth elements, achieving superconductivity at the incredible temperature of 98 K. This means that the new superconductors can be cooled to these levels with liquid nitrogen, which liquefies at 77 K and is inexpensive. Müller and Bednorz won the 1987 Nobel Prize in physics for their discovery.

Currently, researchers have achieved superconductivity at temperatures as high as 138 K. The development of new materials, such as superconducting wires and films (Figure 13.17), has lead to extremely efficient electric motors, generators, and transmission lines, which could lead to new ways of storing energy.

The original BCS theory not only did not predict superconductivity at these higher temperatures but it actually fails to explain it at temperatures above 40 K. Theoreticians are hard at work trying to modify the theory, or to come up with a new one, so that high temperature superconductivity can be understood.

THE FRONTIERS OF PHYSICS: CONTROLLING MOONDUST

The surface of the moon is covered with a fine layer of dust with sizes in the micrometer and submicrometer range. This layer of dust is expected to be electrostatically charged. Since the moon has practically no atmosphere (it has a tenuous atmosphere with an atmospheric pressure in the 10^{-13} kPa range) the full spectrum of the sun's electromagnetic radiation can reach the surface, charging the dust. In addition, the absence of an atmosphere and of a magnetic field allows the high energy electrons and protons in the solar wind to reach the surface completely unimpeded. These energetic particles can also charge the surface dust.

The electrostatically charged lunar surface dust adheres easily to many surfaces, a phenomenon that was directly experienced during the manned Apollo missions. Dust adherence adversely affected seals, mechanisms, cameras, and other equipment. However, due to the short duration of each one of the six Apollo missions that landed on the surface, dust adhesion was more of a nuisance than a problem.

As NASA prepares to return to the moon to build a sustainable long-term human presence, dust adhesion will become a more important problem. To mitigate the problem, the author's laboratory at NASA Kennedy Space Center is developing a technology that can maintain surfaces free of dust by removing it from the surfaces and by preventing its accumulation. In the *dust shield* technology, a low frequency multi-phase oscillating signal applied to conductive electrodes embedded in the instrument's substrate generate an electrodynamic wave that carries along charged dust particles (Figure 13.18). In a three-phase dust shield, three separate oscillating signals are applied to consecutive electrodes to produce a non-uniform electric field that travels across the surface.

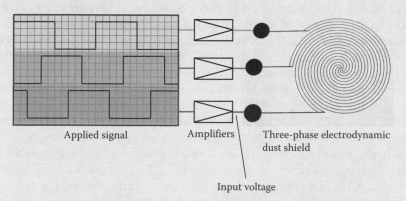

Applied signal   Amplifiers   Three-phase electrodynamic dust shield

Input voltage

FIGURE 13.18 Multi-phase oscillating signal input that generates an electrodynamic wave moving along the surface of the dust shield.

Dust shields with transparent electrodes of indium tin oxide are being developed to protect camera lenses, spectrometers, and other optical instruments (Figure 13.19). Dust shields for solar panels, thermal radiators, and other equipment, as well as flexible dust shields to protect spacesuits are also being developed in the laboratory.

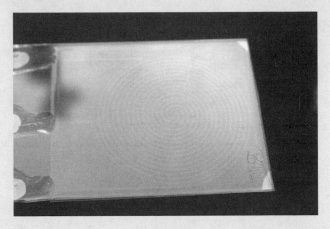

FIGURE 13.19   Transparent dust shield with transparent indium tin oxide electrodes to protect camera lenses and optical instruments. (Photograph by the author, courtesy of NASA.)

14 Electromagnetism

THE DISCOVERY OF MAGNETS

"At this point, I will set out to explain what law of nature causes iron to be attracted by that stone which the Greeks call from its place of origin *magnet*, because it occurs in the territory of the Magnesians." Thus wrote the Roman poet Lucretius in his book *De Natura Rerum* ("On the Nature of the Universe") published in 55 BC. Written in the form of a long poem, his book barely survived, having been lost throughout the Middle Ages. In 1497, a surviving manuscript was discovered and a treasure of world literature was saved for posterity. His poem continues:

> Men look upon this stone as miraculous. They are amazed to see it form a chain of little rings hanging from it. Sometimes you may see as many as five or more in pendent succession swaying in the light puffs of air; one hangs from another, clinging to it underneath, and one derives from another the cohesive force of the stone. Such is the permeative power of this force.
>
> In matters of this sort it is necessary to establish a number of facts before you can offer an explanation of them. This may mean approaching the problem by a very roundabout route. For this reason I beg you to lend me your ears and your mind with particular attentiveness.

Lucretius then proceeds to explain that magnets attract iron because they emanate a stream of atoms that pushes away the air between the magnet and the iron, producing a vacuum that the atoms of iron promptly move to fill. This process is aided by the air behind the iron which pushes it from behind toward the void.

In the thirteenth century the French scholar Petrus de Maricourt, known as Peregrinus ("the Pilgrim"), made what were probably the first experiments that have any bearing on modern ideas of magnetism. Peregrinus, an engineer in the army of Louis IX, started trying to design a motor that would keep a planetarium rotating for some time and thought of using magnets to accomplish the task. In 1269, while in Italy during the siege of Lucera, Peregrinus described in a letter, *Epistola de Magnete*, the results of his investigations with magnets. He explained how he was able to determine the north and south poles of a magnet and his discovery that like poles repel each other while unlike poles attract. In an ingenious experiment, Peregrinus took a natural magnet which he shaped like a sphere and marked the directions taken up by a magnetic needle placed near the surface. All these directions "will run together in two points just as the meridian circles of the world run together in two opposite poles of the world." He interpreted the behavior of the magnetic needle as pointing to the pole of the celestial sphere.

Peregrinus further wrote that magnetic poles could not be isolated, because every time a magnet was split in half, two complete magnets, with north and south poles, were formed. In a second part of this work, Peregrinus presented a detailed work on the magnetic compass and described an improved floating compass encircled by a graduated scale.

The modern treatment of magnetism started in 1600 when William Gilbert published his monumental treatise *De Magnete*. Gilbert explained why the compass needle lines up in a north–south direction: the earth itself is a magnet. He demonstrated his theory with a globular lodestone similar to the one that Peregrinus had used and which he called *terrela*. He laid a magnetic needle at different places on the surface of his spherical magnet and found that the needle acted just like a compass needle, pointing toward the north–south direction of the spherical lodestone.

THE MAGNETIC FIELD

We learned in Chapter 12 that the electric field describes the property of the space around an electrically charged object. Similarly, we can say that the *magnetic field* describes the property of the space around a magnet. In Figure 14.1, we can see the magnetic field lines around a permanent magnet, which resemble the electric field lines around two unlike charges, as seen in Figure 12.10. Since magnetic poles may exist

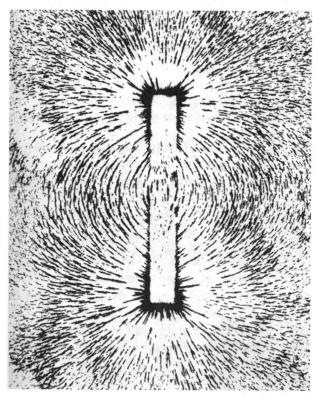

FIGURE 14.1   The magnetic field around a magnet, shown here with iron filings spread on a flat surface around a magnet. (From *PSSC Physics Seventh Edition*, by Haber-Schaim, Dodge, Gardner, and Shore. With permission from Kendall/Hunt Publishing Company, Dubuque, IA, 1991.)

only in pairs, we do not see straight magnetic field lines as was the case with the electric field lines around a single charge shown in Figure 12.7. In 1931, the English physicist P. A. M. Dirac postulated the existence of *magnetic monopoles* to round off the symmetry between electricity and magnetism. A magnetic monopole, if it exists, would be a single north or south pole, flying free. In 1974, Gerard 't Hooft in the Netherlands and Alexander Polykov in Russia independently suggested that some of the new physics theories implied that these particles should exist. On February 14, 1982, Blas Cabrera of Stanford University recorded the passage of what appeared to be a magnetic monopole. However, due to the failure to detect a second monopole in the years since, Cabrera later said that the event he reported might not have been a real one.

The presence of a magnet in a particular place *distorts* the space around it in such a way that any other magnetized body in this region feels a *magnetic force*. As Peregrinus discovered, when two magnets with their north poles facing each other are brought close together, they repel each other. If they are brought closer with their south poles facing, the magnets also repel each other. However, when the north pole of one magnet faces the south pole of the other magnet, the two magnets attract each other (Figure 14.2). The magnetic poles of magnets exert forces on each other. About 200 years ago, the English scientist John Michell found that the force between magnetic poles obeys the inverse-square law. The magnetic force F between them is inversely proportional to the square of the distance r between the two poles:

$$F \propto \frac{1}{r^2}.$$

In 1785, Charles Coulomb experimentally measured the magnetic force between two poles using a torsion balance similar to the one he used to determine the nature of the electrical force between two charges, and confirmed Michell's inverse-square law for magnetic poles. In addition to being proportional to the square of their distances, the magnetic force between two poles is directly proportional to the product of their strengths, or

$$F \propto \frac{q_m q'_m}{r^2}$$

where q_m and q'_m are the strengths of the two interacting magnetic poles.

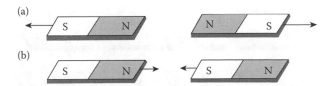

FIGURE 14.2 Forces between two magnets. (a) If the magnets are brought together with like poles facing each other, they repel. (b) If they are brought closer with unlike poles facing, the magnets attract each other.

PHYSICS IN OUR WORLD: MAGNETO-OPTICAL DRIVES

A type of computer storage device common in some countries, the magneto-optical drive, uses light from a laser to heat a tiny spot on a magnetic disk to change its magnetic polarity. These disks, encased in small portable cartridges, come in two sizes, 13 cm and 9 cm in diameter, and are inserted into the drive mechanism through a slot.

The process of writing information to the magneto-optical disk is similar to that of writing to any digital magnetic storage device, such as hard disks, floppy disks, or digital audio cassettes. This process requires setting the polarity of different spots on the disk. A negative polarity is interpreted by the computer as a 0 and a positive polarity as a 1.

Before the polarity is set, a laser beam first heats a small spot on the rapidly spinning disk to a temperature of about 150°C. At this temperature, called the Curie temperature, the atoms in the material of the disk have magnetic dipoles which are oriented at random. When an external magnetic field is present, such as that supplied by a small electromagnet located under the disk, the magnetic dipoles align along the direction of the external field. The total magnetic field of all the atomic dipoles in the spot produces its magnetization. The electromagnet switches polarity to write 0s or 1s (negative and positive polarity).

ELECTRIC CURRENTS AND MAGNETISM

If, as Peregrinus said, magnetic poles cannot be isolated, what is then the meaning of the magnetic pole strength q_m? Although, as we said above, magnetic monopoles could exist, they cannot be obtained by splitting a magnet as splitting a magnet in two produces two complete magnets (Figure 14.3). Therefore, if a magnet cannot be considered to be made up of two separate magnetic monopoles, what is then the source of magnetism? The answer has to do with an accidental discovery that the Danish physicist Hans Christian Oersted (1771–1851) made in 1819. Since many

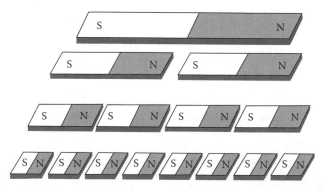

FIGURE 14.3 Cutting a magnet in two produces two smaller but complete magnets. This process can continue down to atomic dimensions without ever isolating a single magnetic pole.

properties of magnetic behavior resemble electrical behavior, scientists had long suspected that there might be a connection between electricity and magnetism and had attempted to measure the effects of electric currents on magnetic compasses. During a physics lecture, Oersted was trying to show that an electric current flowing through a wire lying on a table did not deflect a compass needle. After he placed the compass on the table at various locations near the wire with the needle always pointing north, Oersted picked up the compass and held it *above* the wire. The compass needle twitched and pointed in a direction *perpendicular* to the wire, and when he reversed the current, the needle swung and pointed in the opposite direction, always perpendicular to the wire. Clearly, a force was acting on the compass needle.

Oersted's results were the first ever found in which the force was not in the same direction as the line connecting the sources of the force. It had apparently not occurred to anybody to look for a force that was not parallel to the direction of the flow of current. However, a similar discovery had been reported in the August 3, 1802, issue of the *Gazetta di Trentino* by the Italian jurist Gian Domenico Romagnosi, but had been ignored. Oersted wrote a pamphlet in Latin, as was customary in those days for scientific papers, where he described his discovery, and sent the paper off to many scientific societies. A translation of his paper appeared in 1820 in the *Annals of Philosophy* and before the end of the year, the French scientist Andre-Marie Ampère had extended Oersted's work and had concluded that all magnetism was due to small electric currents. Ampère gave mathematical form to Oersted's discovery. His formulation is known as *Ampère's law*. These discoveries were the first steps toward a complete understanding of the close relationship between electricity and magnetism.

To examine Oersted's discovery more closely, consider a straight segment of wire carrying a current *i* where several small magnets or compasses allow us to observe the direction of the magnetic field in the vicinity of the wire. If we place the magnets around the wire and turn off the current, all the compasses point north; as soon as the current flows through the wire, the magnetic needles of the compasses point in such a way as to form a circle whose center is at the wire and whose plane is perpendicular to the wire (Figure 14.4). When the compasses surrounding the wire are placed farther away from the wire, the magnetic needles again form a circle centered at the wire. According to Ampère's law, the electric current flowing in the wire produces a *circular* magnetic field around the wire.

The magnetic field lines due to a current follow concentric circles that surround the wire. The direction of the field can be found with the *right-hand rule*: if we grasp

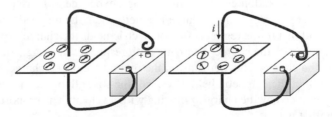

FIGURE 14.4   A wire carrying a current surrounded by small magnets. The magnets form a circle around the wire.

FIGURE 14.5   The right-hand rule to determine the direction of the magnetic field around a current-carrying wire. If the right thumb points in the direction of the current, the curled fingers indicate the direction of the magnetic field.

the wire with the right hand so that the thumb points in the direction of the current, the curled fingers indicate the direction of the magnetic field (Figure 14.5). We can see that the direction of the magnetic field is perpendicular to that of the current.

The simplest application of Ampère's law involves the calculation of the magnetic field due to a current i flowing through a long straight wire. The magnetic field B at a distance r from the wire is given by

$$B = k\frac{i}{r}$$

where k is a constant. The SI unit of magnetic field is the *tesla* (T) which is a relatively large unit. For this reason, the *gauss*, equal to one ten-thousandth of a tesla, is also defined. 1 gauss $= 10^{-4}$ T. In the expression for the magnetic field B, the constant k is equal to 2×10^{-7} T m/A.

A MOVING CHARGE IN A MAGNETIC FIELD

An electric current, as we know, is the rate at which charge flows. A moving electric charge creates a magnetic field, even if it is a single charge instead of a current. This magnetic field exerts a force on a magnet. Therefore, a magnetic field must in turn exert a force on a moving charge. Two electrically charged objects at rest with respect to each other exert a force on each other that is given by Coulomb's law. However, if the charges are moving in relation to each other, the situation gets more complicated, since a moving charge creates a magnetic field which in turn exerts a force on the other charge.

What is the nature of the force exerted by a magnetic field on a moving charge? The answer to this question can be found through experiment. If we place an electric charge at rest in a region near a large magnet, where the magnetic field is fairly uniform, we find that if this region is free of electric fields the charge experiences no net force. If, however, the charge is moving with respect to the magnetic field, a net force acts on the charge. This depends on the magnitude of the charge, its velocity, and the strength of the magnetic field. If we call the strength of the magnetic field B, the force experienced by the charge q moving with a velocity v is proportional to the product qvB; that is,

$$F \propto qvB.$$

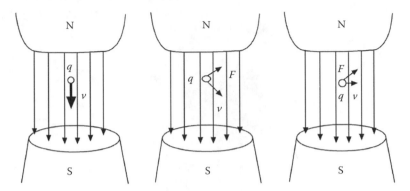

FIGURE 14.6 A charge q moving with a velocity v in a magnetic field of strength B is subject to a force F. This force varies from zero, when the velocity is *parallel* to the field, to a maximum value, when the velocity is *perpendicular* to the direction of the magnetic field.

However, when the particle is moving in a direction parallel to the magnetic field, this force is zero. If the particle moves in a direction that is *perpendicular* to the field, the force is at maximum. Motion along a direction other than parallel or perpendicular produces a force that falls in between this maximum value and zero (Figure 14.6). The force, then, is proportional to the component of the velocity along a direction perpendicular to the magnetic field, or

$$F \propto q \, v_{\text{perp}} \, B.$$

As shown in Figure 14.6, the force is always *at right angles* to both the velocity and the direction of the magnetic field. If the strength of the magnetic field is given in teslas, the expression for the magnetic force on the moving charge can be written as

$$F = q \, v_{\text{perp}} \, B.$$

For our discussions, we will consider only the simpler case where the velocity is at right angles to the direction of the magnetic field. In this case, the force can be written as

$$F = qvB.$$

If a charged particle, such as a proton, enters a region where there is a uniform magnetic field, and the particle moves with a velocity v perpendicular to the direction of the magnetic field, a magnetic force F, perpendicular to both the velocity and the direction of the magnetic field, acts on the particle (Figure 14.7). Since the force is perpendicular to the velocity, it does no work on the particle and its kinetic energy remains constant. Therefore, the particle's speed remains constant. This magnetic force, however, accelerates the particle in a direction that is perpendicular to the velocity, producing a centripetal acceleration that makes the particle move in a circle with a constant speed.

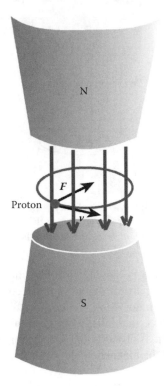

FIGURE 14.7   A proton moving with a velocity v perpendicular to the magnetic field experiences a force F which makes it move in a circle with a constant speed.

In Figure 14.8 an electron leaves a curved track in a liquid hydrogen bubble chamber as it moves through a uniform magnetic field pointing into the page. Although electrons cannot be seen, their tracks are made visible briefly in a bubble chamber as thin lines of boiling liquid. In a hydrogen bubble chamber, liquid hydrogen is heated under pressure to a temperature slightly below its boiling point. When a charged particle moves through the liquid, it ionizes some of the hydrogen molecules. If the pressure is suddenly decreased, the liquid starts boiling preferentially around the ion paths. In Figure 14.8 the radius of curvature decreases because the electron is slowed down by the liquid hydrogen.

PARTICLE ACCELERATORS

The circular motion of a charged particle moving in a direction that is perpendicular to a magnetic field was used in the early 1930s by physicist Ernest Orlando Lawrence of the University of California at Berkeley to accelerate protons to very high energies.

The desire to accelerate particles originated with Rutherford, who had shown in 1919 that the nitrogen nucleus could be disintegrated by alpha particles emitted during the radioactive decays of radium and thorium. In this nuclear process the nitrogen nucleus is transformed into oxygen by the collision with the alpha particle. This was

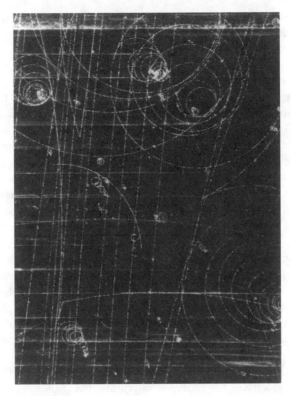

FIGURE 14.8   An electron track in a bubble chamber. The track is curved as the electron moves through a uniform magnetic field pointing into the page. The electron is slowed down by the liquid hydrogen, making the path a spiral rather that a circle. (Courtesy of Brookhaven National Laboratory.)

the first artificially produced nuclear reaction; for the first time humans had modified the structure of an atomic nucleus. However, the particles obtained as products of radioactive decays had energies of only a few MeV. The need for higher energies led scientists to invent machines that could accelerate these particles. These machines, known as particle accelerators, started in 1932 with J. D. Cockcroft and E. T. S. Walton at the Cavendish Laboratory of Cambridge University and their voltage multiplier, which was developed after a suggestion by Rutherford. Accelerating protons to energies of about 400 keV, they were able to split lithium atoms into two alpha particles.

In 1928, Rolf Wideröe in Germany had used an alternating high voltage to accelerate ions of sodium and potassium to twice the applied voltage. Ernest Lawrence was a 27-year-old professor of physics at Berkeley when Wideröe published his results. The son of an educator, Lawrence had wanted to be a physician. A physics professor at the University of Minnesota sparked his interest in physics and Lawrence decided instead to attend Yale University, obtaining his PhD in physics in 1925. He joined the physics faculty at Yale and in 1928 moved to Berkeley where he founded the world-famous Radiation Laboratory. He remained in-charge of the laboratory until his death in 1958 at the age of 57. Shortly after moving to Berkeley, while browsing through several journals in the physics library, Lawrence found Wideröe's paper in the *Archiv für Elektrotechnik*. It occurred to Lawrence that a magnetic field could steer the electrons into a circular path so that the acceleration stage could be repeated. In the summer of 1930 his graduate student M. Stanley Livingston started the design and construction of the first magnetic resonance accelerator or *cyclotron*, as part of his experimental research for a doctorate in physics. On January 2, 1931, their first cyclotron produced ions of 80 keV energy, using a borrowed magnet with a field of 13,000 gauss.

The operation of the cyclotron is based on the fact that the number of revolutions per second made by the particle (which gives us the angular velocity) does not depend on the particle's speed or on its radius of rotation. In a cyclotron, this angular velocity depends only on the charge and mass of the particle being accelerated, and on the magnetic field of the cyclotron.

Figure 14.9 shows a schematic drawing of a cyclotron. A source of particles lies between two hollow metal dees which are connected to an electrical oscillator producing an alternating potential. If a proton, for example, is injected into the cyclotron, it will be accelerated toward the dee that happens to be negative at that time. Inside the dee, the proton is shielded from the electric field but not from the magnetic field of the external magnet which forces the proton to move in a semicircle. As the proton completes a semicircle, it returns to the other dee, which turns negative as soon as the particle enters. The potential difference between the two dees causes the proton to increase its speed each time it crosses the space between the dees. Since the number of revolutions that the proton makes per second is independent of the velocity or the radius of curvature, the proton will change dees at equal intervals of time and the polarity of the dees is set to change accordingly. In a modern cyclotron, the particle may make 100 revolutions before leaving the dees with energy of several MeVs.

Owing to engineering limitations in the construction of very large magnets, cyclotrons have reached their limit, although they remain extremely useful for the range of energies under 40 MeV for protons. An accelerator that does not require a single large magnet to bend the path of the charged particles is the *synchrotron*. This type of machine uses a magnet in the form of a ring and the particles follow a single circular path instead of a spiral, as in the cyclotron. The Fermi National Accelerator

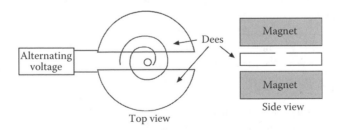

FIGURE 14.9   In a cyclotron, two metal dees lie in between the poles of a big magnet. A source of ions in the middle of the assembly injects charged particles into the dees, which are connected to an electrical oscillator. As the particles are accelerated in the space between the dees, they spiral outward and gain energy.

Laboratory (or Fermilab) in Batavia, Illinois, houses the largest synchrotron in the world. Recently, the laboratory has produced energies of the order of the TeV, or one trillion electron-volts, and for this reason, the accelerator is called the Tevatron.

MAGNETISM OF THE EARTH

As Gilbert wrote in De Magnete, the earth is a magnet. The earth has a large, dense core about 2400 kilometers in diameter, composed of heavy atoms believed to be mostly iron. Study of the seismic waves generated by earthquakes has revealed that the inner part of the earth's core is probably solid while the outer core is liquid. Very slow motions of the core, of the order of one millimeter per year, are believed to generate the magnetic field of the earth. At the earth's surface, the earth's magnetic field strength is about 5×10^{-5} T or about 0.5 gauss.

Magnetism is due to the motion of electric charges. In the earth, the slow motions in the metallic core produce electric currents in the hot, electrically conductive material. These currents flow upward and are in turn carried around by the earth's fast rotation, producing the magnetic field. This process, however, is not totally understood, and scientists were surprised to find a magnetic field in Mercury, which rotates too slowly to produce even a weak magnetic field. The source of Mercury's magnetic field is still a problem to be resolved.

The earth's magnetic field extends out into the surrounding space where it interacts with the *solar wind*, the flow of ionized atoms and electrons that constantly streams away from the atmosphere of the sun at speeds of several 100 kilometers per second. The magnetic field acts as a shield against the solar wind, slowing the supersonic particles to subsonic speeds, forming a shock wave similar to the bow wave formed by a ship moving through water (Figure 14.10). This magnetic field also protects the earth from most of the highly energetic cosmic ray particles originating in the interstellar medium.

The magnetic field of the earth traps some matter from the solar wind, creating the earth's magnetosphere, which is the region around the earth where the magnetic field plays a dominant part in controlling the physical processes taking place there. The magnetosphere was discovered in 1958 with the first American artificial satellite, the Explorer 1, which carried instruments built by James van Allen, a professor of physics at the University of Iowa. Subsequently, other planets were found to possess

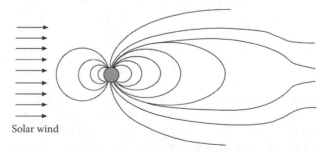

Solar wind

FIGURE 14.10 The earth's magnetic field acts as a shield against the supersonic particles that stream away from the sun.

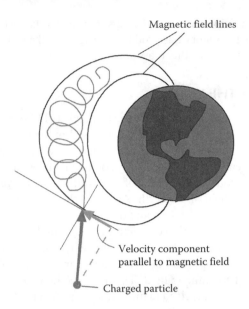

Magnetic field lines

Velocity component
parallel to magnetic field

Charged particle

FIGURE 14.11   A charged particle entering earth's magnetosphere with a velocity component parallel to the magnetic field lines moves in a spiral path.

magnetospheres. The strongest planetary magnetosphere is that of Jupiter, which completely envelops the innermost Jovian satellites.

If a charged particle—proton or electron—from the solar wind enters the earth's magnetosphere with a velocity at right angles to the magnetic field, the resulting motion of the particle will be a circle. If the particle enters the magnetosphere with a velocity component parallel to the magnetic field, it will move along the field (because of its parallel component) and also along a circle (due to its perpendicular component), so that the resulting motion is a spiral path (Figure 14.11). Since the magnetic field of the earth is stronger near the poles and weaker at the center, a particle trapped in this field will spiral around the field lines, oscillating back and forth, as in a *magnetic bottle*. The regions of space where electrons and protons from the solar wind become trapped, oscillating back and forth, are called *Van Allen belts*. The inner Van Allen belt, which extends between altitudes of 2000 and about 5000 km, contains mostly protons. The outer Van Allen belt, extending from about 13,000 to about 19,000 km, contains mostly electrons.

Solar flares are sudden gigantic explosions that take place in the atmosphere of the sun. In addition to emitting enormous amounts of energy at all wavelengths, flares also eject highly ionized matter. When this ionized matter reaches the earth, it interacts with the magnetic field producing *magnetic storms* and dumping into the atmosphere some of the particles that had been trapped in the Van Allen belts. When these high speed electrons and protons collide with gases in the upper atmosphere, atoms of nitrogen and oxygen absorb the ultraviolet radiation emitted. This results in the emission of visible light (a phenomenon called fluorescence), producing the beautiful *Northern Lights*, also known as the *aurora borealis*, or the *Southern Lights* (*aurora australis*).

PHYSICS IN OUR WORLD: AVIAN MAGNETIC NAVIGATION

Migratory birds have a built-in compass: they possess magnetite crystals near their nostrils, which allow them to detect the orientation of the magnetic field of the earth and to use it to navigate. Scientists have discovered that these birds also navigate using the positions of the stars and that in some cases they must use the magnetic and the celestial cues together to orient themselves correctly.

Peter Weindler and his research group at the J. W. Goethe University in Frankfurt studied garden warblers which breed in central Europe and migrate to Africa in the winter. In their flight to Africa, the birds take a detour west to avoid climbing over the Alps, flying instead southwest to the Iberian Peninsula and then turning southeast to Africa. Weindler's team raised a group of warbler chicks in captivity. During the summer before their first migration, they placed the birds inside cages with artificial light and a rotating dome with holes that simulated the night sky. They also isolated a group of these birds from the earth's magnetic field. When the time came for the birds' normal migration, all the warblers were placed in large cages with only celestial cues. The birds that were raised isolated from the earth's magnetic field headed south. The others oriented themselves correctly, in a southwest direction. Later in the season, when the birds should have turned southeast, both groups failed to do so. They had not been exposed to a simulation of the fall sky.

The scientists concluded that migratory birds need both magnetic and celestial information to orient themselves. The night sky provides only the general direction; south in the fall and north in the spring. The birds use the direction of the earth's magnetic field to deviate from this general orientation.

THE SOURCE OF MAGNETISM

Oersted's discovery that an electric current produces a magnetic field led Ampère to conclude in 1820 that all magnetism is due to small electric currents. Just what kind of small electric current is the cause of magnetism in a permanent magnet, for example?

In the Bohr model of the atom, an electron moves around its nucleus following a circular orbit. This motion is equivalent to a very small loop of electric current which, as we know, generates a small magnetic field (Figure 14.12a). In addition to this motion, each electron also rotates about its own axis and this rotation is again equivalent to a circular electric current that produces its own magnetic field (Figure 14.12b). The magnetic field of the electron is that of a *dipole*; that is, it looks like the magnetic field of a magnet, with north and south poles, rather than the magnetic field of a monopole. Although, the model of the atom with electrons as small spinning spheres that rotate around the nucleus is no longer considered adequate, the magnetic fields produced are what we would expect *if* the electron were spinning on its axis and revolving around the nucleus.

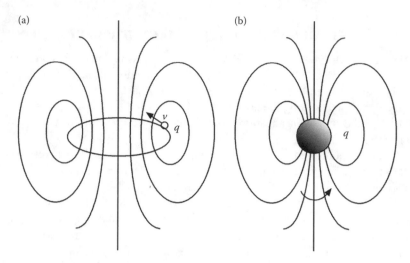

FIGURE 14.12 (a) A charged particle moving in circles produces an electric current. This loop of current generates a magnetic field. (b) A spinning charged sphere also produces a magnetic field.

Although everything contains electrons, not everything is magnetic because in most substances the magnetic fields of the electrons combine with each other and cancel out. In iron and a few other substances, these magnetic fields do not completely cancel out, and large collections of atoms align themselves, forming *magnetic domains* throughout the material. A piece of iron contains many of these magnetic domains, each formed by the alignment of perhaps millions of atoms. In a normal piece of iron, the domains are randomly oriented; unless many of these domains are aligned, the material will not be magnetic. However, the presence of a strong magnetic field in the vicinity may be enough to align many domains in the direction of this external field. You can magnetize the blade of your knife by rubbing it several times in the same direction with a large magnet. If the magnet is very strong, placing the knife on the magnet may be sufficient to align enough of its domains for it to remain magnetized for a long time.

FARADAY'S LAW OF INDUCTION

We have seen how an electric current generates a magnetic field. This suggests that perhaps a magnetic field could generate an electric current. Is this true?

The first person to answer this question was the English scientist Michael Faraday. One of ten sons of a blacksmith, Michael Faraday lacked a formal education. "My education was of the most ordinary description," he wrote, "consisting of little more than the rudiments of reading, writing and arithmetic at a common-day school." He was born in 1791 in Newington, England, and at the age of 14 became a bookbinder's apprentice in the shop of Mr. G. Riebau. This seems to have been a fortunate turn of events, because that job put him in contact with books, which the young boy decided to read. "There were plenty of books there and I read them," he later said. His master not only allowed him to read the books that were commissioned for binding, but also encouraged him to read the books in his personal library and to attend cultural events and lectures. One day, while binding a volume of the *Encyclopaedia Britannica*, Faraday encountered a 127-page article on electricity and became fascinated by it. The reading of this article stimulated his interest for science and prompted him to save enough money to buy the necessary parts with which to construct his first scientific apparatus.

In 1810, Faraday was introduced to the City Philosophical Society, which had been formed two years earlier and which met every Wednesday to discuss a paper on some topic of science presented by one of the members. At these lectures, Faraday would usually get himself a front seat to take notes which he would later rewrite and expand at home. Later, when a series of four lectures was delivered by Sir Humphrey Davy, the English chemist and Superintendent of the Royal Institution, Faraday was given a ticket to attend them. He took copious notes which he bound and later sent to Davy along with a petition for a job at the Institution. Although Davy was flattered, he could not offer Faraday a position at that time, but later, when one of Davy's employees at the Institution was involved in a fight and was dismissed, Faraday was hired as an assistant in the laboratory. Faraday was now on the road to become one of the greatest experimental scientists who ever lived.

In 1825, 12 years after his appointment as assistant, Faraday became director of the laboratory and in 1833, professor of chemistry at the Royal Institution. It was at the Royal Institution that Faraday became intrigued by Oersted's experiments showing that an electric current created a magnetic field and wanted to see it if was possible for a magnetic field to produce an electric current. With this in mind, he "had an iron ring made...[w]ound many coils of copper wire round one half, the coils being separated by twine and calico... By trial with a trough each was insulated from the other. Will call this side of the ring [which was connected to a battery] *A*." On the other side, side *B*, a second copper wire was wound (Figure 14.13). Faraday expected that the magnetic field produced by coil *A*, which would magnetize the iron ring, would in turn produce a current in the coil of copper wire wrapped around the other side of the ring, independent of the other coil. It did not happen that way. Instead, only at the instant the battery was either connected or disconnected, a current appeared in coil *B*.

It was not the presence of the magnetic field that produced a current in coil *B*, but the *change* in this magnetic field that occurred at the moment the battery was connected or disconnected. As soon as the battery was connected, the current rapidly increased from zero to a steady value and while that increase in current took place, the magnetic field produced by this current also increased. Similarly, when the battery was disconnected, the current immediately decreased to zero, producing a sudden decrease in the magnetic field. It was this increase or decrease in the magnetic field that *induced* a current in coil *B*. For this current to appear in the second coil, a voltage must exist there. Thus, we can say that a changing magnetic field induces a voltage in the second coil which in turn produces a current. This phenomenon is called *electromagnetic induction*.

Faraday performed further experiments and showed that the iron ring was not needed. Adding a switch to coil *A*, a current was induced in coil *B* whenever the switch was opened or closed. Since it was the change in the magnetic field generated

(a)   (b)

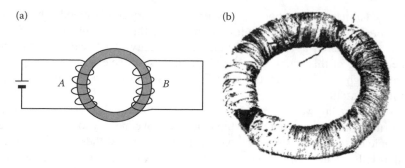

FIGURE 14.13   (a) Faraday's induction experiment. At the instant the battery was connected or disconnected so that a current would flow through coil *A*, a current was detected for a brief time in coil *B*. (b) Faraday's induction ring.

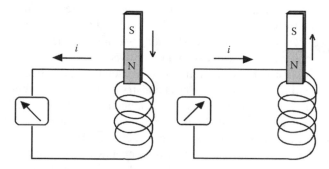

FIGURE 14.14   Moving a magnet into or out of a coil induces a current in the coil.

by the increase or decrease in the current that induced the current in the second coil, Faraday experimented with moving magnets and was able to induce electric currents in coils just by quickly moving a magnet into or out of the coil (Figure 14.14).

A coil is not necessary for electromagnetic induction to take place. A single loop of a wire or an electric circuit would work. A voltage is induced in a single loop of wire by moving a magnet in the vicinity of the loop. The faster we move the magnet, that is, the faster the magnetic field in the region of the loop changes, the greater the induced voltage. We can say that when the rate at which the magnetic field changes through the circuit is large, the induced voltage is large. In more general terms, *the induced voltage in the circuit is proportional to the rate of change of the magnetic field*. This statement is known as *Faraday's law of induction*.

In 1822, Faraday wrote in his notebook: "Convert magnetism into electricity." By 1831, he had accomplished his goal.

MOTORS AND GENERATORS

Faraday showed that a change in a magnetic field produces a current in a coil. The magnetic field threading through a spinning coil is *variable*, even if the field itself is constant (Figure 14.15). Therefore, a current would be induced in a coil rotating in a

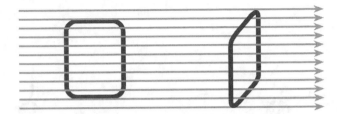

FIGURE 14.15 The magnetic field through a rotating coil changes from zero, when the plane of the coil lies parallel to the magnetic field, to a maximum value, when the plane of the coil is perpendicular to the field.

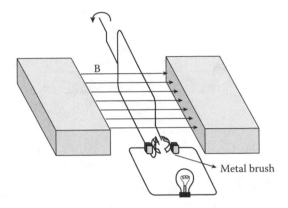

FIGURE 14.16 A loop of conducting wire with a handle generates electricity when the loop is turned in a region where there is a magnetic field. Metal brushes maintain contact with the circuit as the loop rotates. This is a simple electric generator which can provide the current for a light bulb.

region where there is a magnetic field. If you attach a handle to this coil and turn it, a current is generated in the coil for as long as you keep turning the coil. This current can be used to light a bulb, for example (Figure 14.16). This is a simple one-loop *generator*, a device to convert mechanical energy into electrical energy. A paddle wheel turned by a waterfall or by a river can provide the mechanical energy for the generator. Steam moving through a turbine can also supply the mechanical energy to turn the generator and produce electrical energy.

Suppose now that we remove the handle and place the coil inside a magnetic field. If instead of a light bulb we connect a battery so that a current flows in the coil, a torque is exerted on the loop. The reason for this is that since currents are charges in motion, and a charge in motion in a magnetic field experiences a magnetic force, an electric current placed in a magnetic field also experiences a magnetic force. This force is perpendicular to the direction of the magnetic field and to the direction of the flow of current through the conductor.

Consider the conducting loop of Figure 14.17, where the magnetic field lines in the region between the two external magnets flow from left to right. A battery is connected to the loop through a pair of metal brushes that allow two conductors in the arms of the loop to keep contact while they slide past each other. These conductors

(a)   (b)

(c)

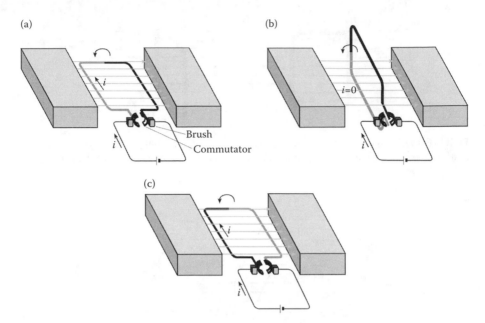

FIGURE 14.17 A simple motor. (a) The wire loop in a magnetic field is connected at one end to two semicircular conductors called the *commutator*. Current from the battery flows into the left (gray) side of the loop through the metal brush and out the other (black) side. The torque produced spins the loop counterclockwise. (b) Each brush is in contact with both commutators. Current bypasses the loop. Inertia keeps the loop rotating. (c) The black side of the loop is now on the left. Current flows into this side now and out of the gray side. The current through the loop changes direction but the loop keeps rotating clockwise.

form what is called the commutator because, as we shall see, they change the flow of current through the loop as the loop rotates.

We can now see how the loop rotates. When the brushes are aligned with the commutator segments, current flows into the left side of the loop (shown in gray in Figure 14.17a) and out of the right side. The force acting on the left side is perpendicular to the direction of the magnetic field, in this case to the right, and to the direction of the flow of current. The force on the right side of the loop is in the opposite direction, because the direction of the magnetic field is the same as before, that is, to the right, and the current is moving in the opposite direction through the loop. These two forces, which depend on the value of the current and the magnitude of the electric field, have the same magnitude, since the current flowing though the loop has the same magnitude in either direction. The two equal and opposite forces, acting at the two opposite sides of the loop, exert a torque on it which results in rotation. As the loop rotates to a position where each brush is in contact with both brushes, the current bypasses the loop, which continues to rotate due to its inertia (Figure 14.17b). As the loop rotates beyond this middle position, current again flows through the loop but in the opposite direction (it flows into what was the right side, now located on the left, and out of the other) assuring that the force exerted on whichever side of the loop happens to be on the left does not change (Figure 14.17c). The loop keeps rotating in the same direction. If a shaft and gears are attached to the loop, its rotational motion can be transmitted to a wheel which could in turn be attached to a cart. This device is a simple electric *motor*, which transforms electrical energy into mechanical energy, the reverse of a generator.

Instead of just one loop, an actual electric motor has many loops wrapped around an iron core or *armature*. In addition to increasing the torque, having many loops improves the operation, since the torque on a single loop changes continuously from zero to a maximum value, while in many consecutive loops it would stay about constant.

MAXWELL'S EQUATIONS

James Clerk Maxwell was the nineteenth-century scientist who probably made the greatest contribution to modern physics. Maxwell was born of a wealthy family in Edinburgh on June 13, 1831. Early on he showed great mathematical ability and at the age of 14 wrote a paper on geometry, "The Theory of Rolling Curves," which was accepted for reading before the Royal Society of Edinburgh. Not counting this and other short papers written while at school, Maxwell wrote over a 100 scientific papers in his short life of 48 years.

Faraday had introduced the concept of lines of force as a pictorial representation of electrical and magnetic forces. While still an undergraduate at Cambridge, Maxwell tried to give mathematical form to the physical notions of Faraday. Shortly after graduating from Cambridge, he wrote a paper entitled "On Faraday's Lines of Force," the first in a series of papers on this subject, which was read before the Cambridge Philosophical Society. Seven years later, in a second paper entitled "Physical Lines of Force," Maxwell devised a model that illustrated Faraday's law of induction. When Maxwell applied his model he realized that, in addition to Faraday's discovery that a changing magnetic field would produce an electric force, the model suggested that a changing electric field would produce a magnetic force. To take this into account Maxwell decided to modify Ampère's law.

Maxwell eventually developed a theory of electromagnetism which embraced everything that was previously known about electricity and magnetism. His four equations, known today as *Maxwell's equations*, not only summarize all the work of Coulomb, Oersted, Ampère, Faraday, and others, but also extended the relationships and symmetries between electricity and magnetism. Whereas the first three equations could be considered as restatements of previous work, his fourth equation, the extension of Ampère's law, was the key to the puzzle, the stroke of genius which unified electricity and magnetism into one single theory of electromagnetism.

Maxwell's first equation is a consequence of Coulomb's law and gives a relationship between an electric charge and the electric field it produces. The second equation describes how magnetic field lines always form closed loops; that is, they do not start or stop anywhere. In this respect, they are essentially different from electric field lines which start on positive charges and end on negative charges. For this reason, we do not observe magnetic monopoles, the magnetic equivalent of electric charges. The third equation is Faraday's law which says that a changing magnetic field creates an electric field. Maxwell's fourth equation is, as we have said, an extension of Ampère's law, and states that a changing electric field creates a magnetic field. A magnetic field then can be created by an electric current or by a changing electric field.

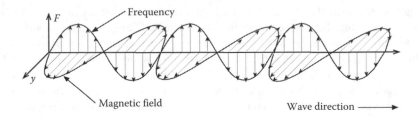

FIGURE 14.18 An electromagnetic wave consists of changing magnetic and electric fields at right angles.

According to Maxwell's third equation, a changing magnetic field creates an electric field which is also changing. The fourth equation tells us that this changing electric field would in turn create a magnetic field, which changes with time. This changing magnetic field in turn creates a changing electric field which creates a changing magnetic field, and so on. By combining his four equations into a single expression, Maxwell was able to show that even in regions where there were no electric charges or magnets, once the process started, the changing magnetic and electric fields would continue to propagate. This propagation of electric and magnetic fields through space is what we call *electromagnetic waves* (Figure 14.18).

From his equations, Maxwell was able to determine that the speed of propagation of the waves was 2.88×10^8 meters per second, almost exactly equal to the speed of light, which had been measured in his time to be 3.11×10^8 meters per second. He suggested that electromagnetic waves did exist and were observed as light. "The agreement of the results seems to shew that light … is an electromagnetic disturbance," he wrote in a paper published in 1868. Nine years after Maxwell's death, Heinrich Hertz (1857–1894) generated electromagnetic waves in his laboratory.

PHYSICS IN OUR WORLD: MICROWAVE OVENS

Microwave ovens are commonplace in our homes today. However, few of us know how they work. The physics we have learned so far should allow us to understand the basic principles of their operation.

A microwave oven is actually a small broadcasting station. A *magnetron* tube emits high frequency electromagnetic waves (microwaves). These waves are "piped" along a waveguide in the oven. A metal stirrer then directs the waves throughout all areas of the oven.

Microwave ovens designed for homes have a frequency of 2450 MHz, which corresponds to a wavelength of 122 mm. These electromagnetic waves are reflected by metals but transmitted by paper, glass, and some plastics. They are absorbed by water and sugar. As the microwaves penetrate the foods, they cause water molecules to vibrate at depths of about 5 cm, producing heat mainly as a result of the disruption of the intermolecular bonds of the water molecules.

Water molecules consist of an atom of oxygen and two atoms of hydrogen arranged so that, although the entire molecule is electrically neutral, the oxygen side of the molecule is negative, while the side with the hydrogen atoms is positive. The electric field transmitted by the microwave produces a torque on the water molecule. Since the field is oscillating, the applied torque sets the molecule into vibration. Since the water molecule in a substance such as food is bound to other molecules, retarding "frictional" forces are produced, which appear as heat in the substance.

Since the walls inside the oven are made of metal, microwaves are reflected by them. Standing waves (like the ones produced on a taut string) may be set up inside the oven. The places of maximum vibration correspond to hot spots, while the places where no vibrations of the electromagnetic field take place correspond to cold spots. To avoid the uneven cooking that these hot and cold spots would produce, foods are placed on a turntable.

Part VI

Waves

15 Wave Motion

THE NATURE OF WAVES

Wave motion exists everywhere. When a bird sings in the forest, sound waves propagate away from the bird's throat in all directions, striking the ear membranes of the other animals nearby. A leaf falling from a tree on the surface of a pond produces ripples that spread outward through the surface of the serene water. Light reaching the surface of the earth from the sun propagates through empty space as an electromagnetic wave. And as we learned in Chapter 7, even ordinary matter exhibits wave behavior.

In the examples above, sound and water waves require a medium to propagate. The bird's song could not be heard without the air through which the sound waves could travel, and the ripples spreading on the still pond are certainly impossible if the pond does not exist. An astronaut on the moon cannot attract his companion's attention by clapping his gloved hands or banging on his metal backpack since there is no atmosphere that can serve as a medium. The astronaut, however, can speak into the microphone in his helmet and the astronauts around him, the astronauts in the moon base, and the staff in Mission Control on earth, can all hear him. The electromagnetic waves generated by his radio transmitter can propagate through the vacuum of space; they do not need a medium although they can travel through certain media. Waves that require a medium to propagate are called mechanical waves. In this chapter we will study mechanical waves only, as they can be visualized.

What exactly is a mechanical wave? In Chapter 7 we learned that the elasticity of a medium allows a disturbance created at some point in space to propagate through the medium without transmitting matter. Thus, when a taut string is plucked, the hump or *pulse* produced travels down the string; but the material of the string does not travel along with the pulse, it only vibrates up and down (or sideways or at an angle, depending on the direction in which we pluck it). If matter does not travel, what is it that gets transmitted with the wave? It is energy and momentum. When the ripple produced by a pebble dropped in the water reaches a floating leaf, it lifts it; the energy required to lift the leaf is transmitted by the wave and comes from the interaction between the falling pebble and the water.

When a pulse travels along a taut string, the string vibrates in a direction that is perpendicular to the direction in which the pulse moves (Figure 15.1a). We call this wave a *transverse wave*. Pushing the end of a spring coil or "Slinky" back and forth produces a wave that travels through the coil. When the end is pushed, the first turn of the coil gets closer to the second turn, deforming the coil and forcing the second turn to move away from the approaching first turn. This results in the second turn approaching the third and the third moving toward the fourth, and so on. If the end is pulled back after it is pushed, the second turn moves back toward the first, the third toward the second, the fourth toward the third, and so on. If the end is pushed and

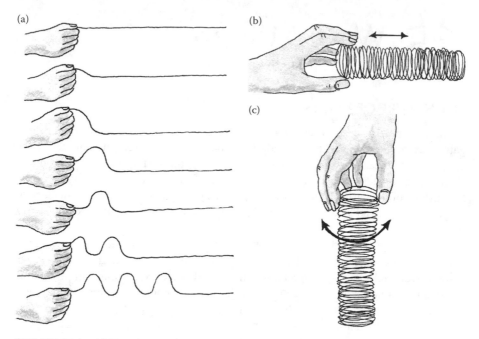

FIGURE 15.1   (a) Transverse wave. A pulse traveling to the right displaces the string in a direction at right angles to the direction in which the pulse travels. (b) A longitudinal wave produced by pushing and pulling on one end of a Slinky. (c) A torsional wave produced by repeatedly twisting one end of a coil back and forth.

pulled several times, the overall effect is seen as moving variations of the spacing between the turns; regions where the turns are closer together followed by regions where the turns are farther away from each other (Figure 15.1b). This wave is called a *longitudinal wave*, as the medium oscillates in the same direction as the wave. If we now hold the Slinky from one end so that it hangs loose in a vertical position and twist the end back and forth, a *torsional wave* will start to propagate down through the coil (Figure 15.1c).

PROPERTIES OF WAVES

When the source of a particular wave motion acts continuously rather than momentarily, a train of waves is formed instead of a single pulse. If, in addition, the source of waves acts in a cyclic way, repeating the same stages over and over like a vibration, the wave so produced is called a periodic wave. If we were to take a snapshot of a *periodic wave* as it moved through a medium, it would look like Figure 15.2. Although the plot looks like the snapshot of a transverse wave traveling along a string or cord, it can also represent a longitudinal wave. In the example with the Slinky, the regions where the turns are closer together would be represented by the crests and the places where the turns are separated, by the troughs.

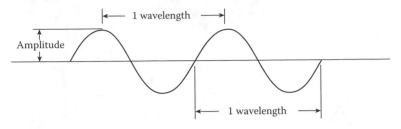

FIGURE 15.2   Graphical representation of a wave.

The distance between two adjacent peaks, between two adjacent troughs, or between any two identical points of a periodic wave is called the *wavelength*, λ (the Greek letter lambda), of the wave. If we observe one particular point in space and count the number of wavelengths that pass that point during some time interval, say one second, we would know how *frequently* the wave moves through that point. The number so obtained is called the *frequency* (*f*) of the wave motion. Frequency, then, is the number of wavelengths that pass a particular point per second and is given in cycles per second or *hertz* (Hz):

$$1Hz = 1 \text{ cycle/s}.$$

The frequency of a light wave determines perceived hue and, as we shall see in the next chapter, the frequency of a sound wave determines its perceived pitch.

Clearly, if two waves are traveling at the same speed, the wave with the shorter wavelength will have the higher frequency (Figure 15.3). Frequency and wavelength are inversely proportional for waves traveling at the same speed. A wave of twice the wavelength, for instance, would have half the frequency of another wave traveling at the same speed.

In Chapter 6 we learned that for a rotating object, the time taken to complete one revolution was called the *period, T.* In wave motion, *period* is the time (in seconds) for one complete cycle; it is therefore the inverse of the frequency, which is the number of cycles per second:

$$T = \frac{1}{f}.$$

Another important parameter in describing a wave motion is the *amplitude* (*A*) which is the maximum displacement of the medium from its equilibrium position.

The velocity (*v*) of a wave moving through a medium is determined by the medium. Velocity is equal to the displacement divided by the time interval during which the displacement occurred. Figure 15.4 shows successive snapshots of a wave traveling to the right. As one complete wavelength passes through a certain point the wave executes one cycle and this takes a time exactly equal to one period; thus,

$$v = \frac{\lambda}{T} = \lambda f \quad \left(\text{since } T = \frac{1}{f} \right).$$

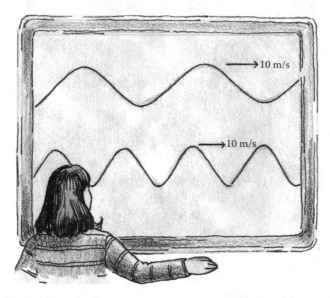

FIGURE 15.3   The observer counts more wavelengths for the lower wave than for the higher one. The frequency (the number of wavelengths that pass a point every second) is higher for the lower wave.

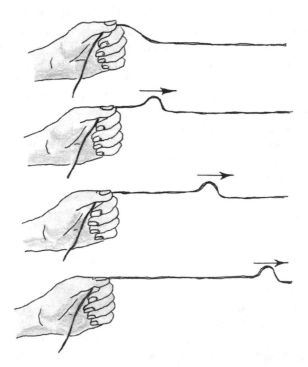

FIGURE 15.4   As a transverse wave travels along a string, a small section of the string oscillates up and down, moving in a direction that is perpendicular to the direction in which the wave propagates. It takes a time equal to one period for one wavelength to move past that section.

DIFFRACTION

One interesting property of waves is that they spread out when passing through a hole of dimensions similar to those of the wavelength. Figure 15.5 is a photograph of water waves advancing toward an obstacle with a small opening. The waves pass through the hole, propagate beyond the obstacle, and spread out around the edge of the hole. This phenomenon is called *diffraction*. A group of swimmers in the sea creates no breaks in the waves rippling onto the beach, as the waves flow around small obstacles. This is also a form of diffraction.

THE PRINCIPLE OF SUPERPOSITION

When two particles encounter each other, they collide. When two waves encounter each other, however, they pass through each other unmodified. You can hear the sound of a trumpet even while a violin is being played in the same room. The two sound waves reach your ears independently, as if they were the only ones present in the room. If you are listening to the radio while the telephone rings, you will hear the ring with the same pitch and the same loudness as if the radio was off. The same is true for water waves, waves on a string, or any other kind of wave motion.

Where the two waves overlap, their amplitudes add up algebraically; at some places the waves add up to produce larger amplitudes and at other places they combine to produce smaller amplitudes. If, for example, two waves traveling on a string meet at one point where the amplitude of one wave is 5.0 cm and the amplitude of the other is −3.0 cm, or 3.0 cm below the equilibrium position, the actual displacement of the string from the equilibrium position at that point would be 2.0 cm

FIGURE 15.5 Plane waves passing through a barrier with a small hole. The waves spread out beyond the barrier. This is known as diffraction. (From *PSSC Physics Seventh Edition*, by Haber-Schaim, Dodge, Gardner, and Shore. With permission from Kendall/Hunt Publishing Company, Dubugue, IA, 1991.)

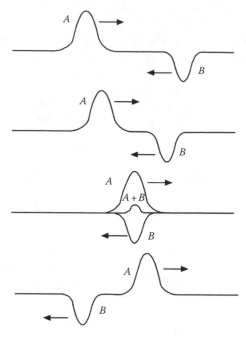

FIGURE 15.6 Superposition of two wave pulses (*A* and *B*) at a point. The waves interact destructively but then pass through each other unchanged.

(Figure 15.6). Each wave contributes to the rope's displacement regardless of the presence of the other. This property is known as the *principle of superposition* or the phenomenon of *interference*. When the displacement is enhanced, the waves are said to undergo *constructive interference*, and when the displacement is diminished, the waves undergo *destructive interference*. However, this effect is significant only when the waves have the same frequency.

CONSTRUCTIVE AND DESTRUCTIVE INTERFERENCE

Two waves of equal amplitude and frequency moving in the same direction through the same region of space interfere constructively if the crests of one wave match the crests of the other. This results in a wave with amplitude equal to twice the amplitude of either wave alone (Figure 15.7a). If the crest of one wave meets the trough of the other, there is destructive interference and the waves cancel each other out at that point (Figure 15.7b).

Figure 15.8a illustrates what happens when ripples from two nearby sources overlap. If two similar pebbles are dropped close to each other on a serene pond, circular ripples start to spread out from the place where each pebble hit the water and quickly reach the same area on the surface of the water. There will be some regions where the ripples will interfere destructively, canceling each other out, and some other regions where the interference is constructive, and the amplitude is increased. Figure 15.8a is actually a photograph of two trains of waves produced by the vibration of

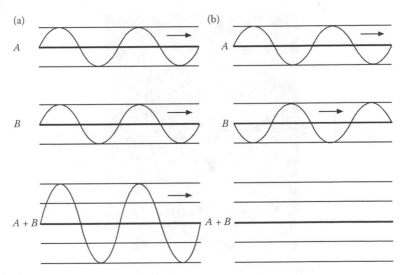

FIGURE 15.7 (a) When the crests from the two overlapping waves coincide, there is *constructive interference*; (b) when the crest and the trough coincide, there is *destructive interference* and the waves cancel each other out.

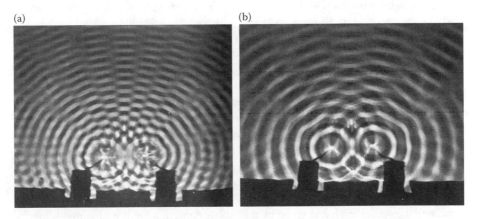

FIGURE 15.8 (a) (Left) Interference pattern produced by two sources vibrating simultaneously. (b) (Right) Interference patterns produced for a different wavelength. (From *PSSC Physics, Seventh Edition*, by Haber-Schaim, Dodge, Gardner, and Shore. With permission from Kendall/Hunt Publishing Company, Dubuque, IA, 1991.)

two rods in contact with the water and connected to the same motor. The rods vibrate synchronously and produce ripples that have the same amplitude and wavelength and that oscillate in step. They are said to be *in phase*. Waves with the same form and wavelength oscillating in phase are called *coherent* waves. The regions of constructive interference form lines that radiate from the middle of the two sources of waves and which are interspaced by lines formed by the regions of destructive interference. Here the waves are out of step and are said to be in *antiphase*.

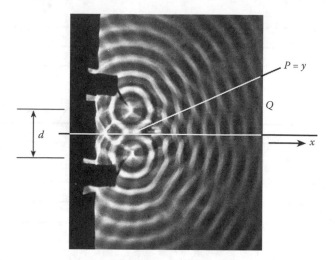

FIGURE 15.9   Two coherent sources of waves interfere constructively at P and destructively at Q. (From *PSSC Physics, Seventh Edition*, by Haber-Schaim, Dodge, Gardner, and Shore. With permission from Kendall/Hunt Publishing Company, Dubuque, IA, 1991.)

This interference pattern of regularly spaced regions of maxima and minima depends on the separation of the sources and the wavelength of the two disturbances and is characteristic of all wave motions. Although it is a simple matter to measure the wavelength of waves traveling through a rope or string or of water waves, sound waves and electromagnetic waves are not visible and the wavelength cannot be measured by observation. However, if the locations of maxima and minima can be determined, then simple geometry allows us to find the wavelength. In the photograph of Figure 15.8b, we see the different interference patterns produced by ripples of different wavelength from that of 15.8a. The separation between the sources and the positioning of the screen is the same for both cases. We can see that the spreading is more pronounced for longer wavelengths (Figure 15.8b).

Consider two coherent sources of waves separated by a distance d arranged as in Figure 15.9. If a maximum is found at some point P at a distance y above the x axis (assuming that the distance L is much greater than the separation d between the sources), the wavelength λ is given by a simple geometrical expression involving the distances y, L, and d:

$$\lambda = \frac{yd}{L}.$$

STANDING WAVES

Two identical wave motions of equal amplitude and wavelength traveling in opposite directions, toward each other, interfere in a way that depends on the location of the crests and troughs of each wave at different times (Figure 15.10). Regardless of where the individual waves are, though, there are certain points called *nodes* that do not move at all. The points in between the nodes vibrate but the wave does not move

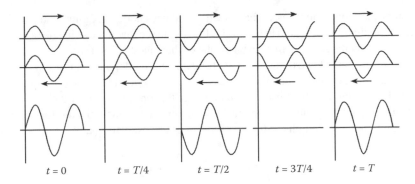

FIGURE 15.10 Two waves moving in opposite directions, shown at different times.

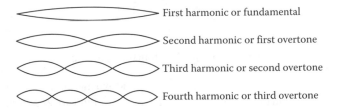

FIGURE 15.11 Standing waves on a string fixed at both ends. The fundamental mode of vibration has only two nodes at the fixed ends; the second and third harmonics have one and two additional nodes respectively.

in either direction; the nodes remain at the same locations and the overall pattern is stationary. These fixed waves are called *stationary* or *standing waves*. It can be shown that the spacing between nodes is equal to half the wavelength of the traveling waves.

Standing waves also occur on a stretched string fixed at both ends, such as in a guitar or a piano. Consider the waves generated when a guitar string is plucked. Many waves are produced which interfere with each other but only the ones with the end points fixed are sustained. These waves combine to produce an overall pattern which does not travel in any direction; the string vibrates up and down only.

One way to observe stationary waves is to tie one end of a string to a post, for example, and, holding the other end so as to keep the string taut, produce a train of waves by repeatedly shaking the free end of the string with our hand. When the waves reach the fixed end, they are reflected, and interfere with the outgoing waves that we are producing from the free end to form a standing wave pattern. When a taut string is plucked or struck, standing waves are produced and the string vibrates in different ways, called *modes of vibration*. In the *fundamental* mode or *first harmonic*, the center of the string vibrates and the only nodes are the two fixed ends. Another mode of vibration has, in addition to the fixed ends, a third node in the middle of the string; this is called the *second harmonic* or *first overtone*. Third, fourth, and other harmonics (second, third, and higher overtones) can be obtained with two, three, and more nodes in between the fixed ends of the string (Figure 15.11).

There are certain wavelengths for which standing waves appear in a stretched string such as a piano string or the string of any stringed instrument. Since the string is fixed at both ends, these two points, as we have seen, are always nodes. Examining the different modes of oscillation illustrated in Figure 15.11, we can see that if the string has a length L, the fundamental mode of oscillation takes place when one-half of a wavelength forms in between the two fixed ends; that is, in the fundamental mode, one-half of a wavelength equals the length of the string. The wavelength of the second harmonic is equal to the length of the string. For the third harmonic, we see that one and a half wavelengths equal the length of the string, and for the third, two complete wavelengths are formed in the string. Thus, the wavelengths of the first four modes of oscillation are:

$$\frac{\lambda}{2} = L \qquad\qquad \frac{2\lambda}{2} = L$$

fundamental   second harmonic

$$\frac{3\lambda}{2} = L \qquad\qquad \frac{4\lambda}{2} = L$$

third harmonic   fourth harmonic

Here we have written the wavelength of the second harmonic as $2\lambda/2$ which is just λ, and that of the fourth harmonic as $4\lambda/2$ which is equal to 2λ, so as to express all wavelengths as multiples of $\lambda/2$. We find other standing wave modes where L is equal to an integral number of wavelengths or $L = n\lambda/2$ where $n = 1, 2, 3 \ldots$ etc. The wavelength λ_n of the nth mode of oscillation is then

$$\lambda_n = \frac{2L}{n}.$$

The frequency of any mode of oscillation can be found from the formula $v = \lambda f$. For the nth mode, the frequency is

$$f_n = \frac{nv}{2L}.$$

Thus, when $n = 1$, the *fundamental frequency* is $v/2L$; when $n = 2$, we obtain the frequency for the second harmonic, which is v/L; and similarly for any other mode of oscillation. The frequencies obtained from this expression are called the *natural frequencies* of the string.

RESONANCE AND CHAOS

When we fix one end of a string and, keeping it stretched, repeatedly move the free end up and down slowly with one hand, standing waves are usually not seen right away. However, if we start increasing the frequency with which we move the free end of the string, pretty soon we reach the fundamental frequency of vibration of the string and we observe the fundamental mode. Further increasing the frequency destroys the standing wave pattern for a while until the second mode of oscillation of the string or first overtone is reached.

As we slowly increase the frequency with which we move the free end of the string, we reach other modes of oscillation. What we are doing with our rapid up and down motions is to pump energy to the string; the motion of our hand is the source of energy. We notice that it is easier to pump energy when the string is oscillating in any one of its natural frequencies. A similar situation occurs when we push a child on a swing. If the pushes are in step with the back and forth motion of the swing, it is easier to maintain the oscillations. But if the pushes are not in step with the swing, the child's motion becomes disorganized, the oscillations of the swing lose amplitude, and it is more difficult to push it. The string responds better to this influx of energy when the frequency of the driving force (our hand) matches one of the natural frequencies of oscillation of the string. This phenomenon is known as *resonance*.

On November 7, 1940, some four months after it had been inaugurated, the Tacoma Narrows Bridge collapsed after standing waves were produced by gusting winds. If the standing waves had remained stable, the bridge might not have collapsed. However, as scientists have recently discovered, a stable system may become unstable if a very small variation is introduced into its initial conditions. Even in the absence of any external random forcing, some physical systems can show regular periodic motions or apparently random motion, and the difference between the two resides in the initial conditions. A long, flexible metal beam clamped to a vibrating support would usually show a standing wave pattern; however, if started from a slightly different position, the standing wave pattern rapidly develops a chaotic motion which increases the amplitude of the oscillation (Figure 15.12). Depending on the elasticity of the beam, this chaotic condition may lead to a breaking of the beam. Instead of producing stable standing waves on the Tacoma Narrows Bridge that day, the gusting winds could have produced this chaotic motion on the bridge, causing its collapse.

The chaotic motion we are referring to here is not completely uncontrolled, helter-skelter, random motion. It is, rather, a somewhat controlled, seemingly random motion that has been discovered in the past few years. The fundamental cause of chaos is

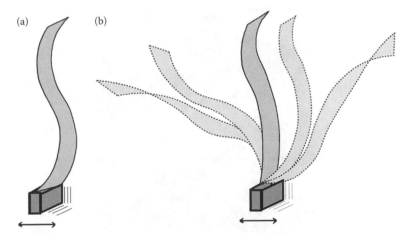

FIGURE 15.12   (a) Standing waves develop in a long metal beam attached to a vibrating clamp. (b) Depending on the initial conditions, chaotic motion can develop which may cause the beam to break.

sensitivity to initial conditions. Take, for example, a dripping faucet. A slow flow rate produces a rhythmic, periodic drip; the time interval between drops is always the same. A small increase in the flow rate may still produce a periodic drip. Further increase of the flow rate eventually produces what appears to be random, turbulent motion; the time interval between drops seems to vary randomly, without any periodicity or structure. To investigate this phenomenon, the American physicist Robert S. Shaw decided to plot the interval between drops 1 and 2 versus the interval between drops 2 and 3. When the dripping was periodic, all points fell in the same place, one on top of the other. When the flow was turbulent, the first points plotted seemed to fall all over the graph, without any pattern. As enough points accumulated, a pattern emerged; the points formed a constrained shape. Out of disorganized, turbulent behavior, an eerie order lurked; an orderly disorder, the order behind chaos (Figure 15.13).

(a)   (b)

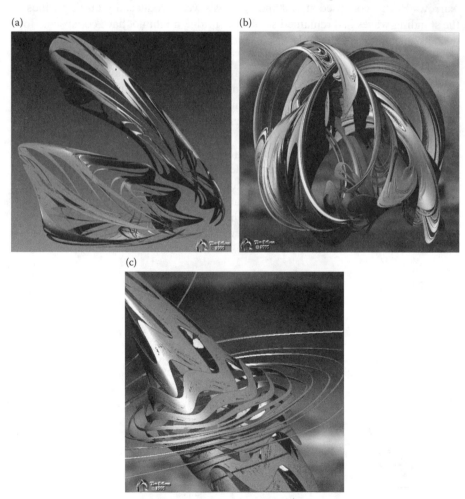

(c)

FIGURE 15.13 **(See color insert following page 268.)** Diagrams of chaotic systems. (a) Polynomial strange attractor, (b) floating point attractor, and (c) another example of a polynomial strange attractor. (Courtesy of Tim Stilson, Stanford University.)

ne of the dark lines discovered by Fraunhö.

FIGURE 6.1c This is a photograph of the spiral galaxy M-81, believed to have a similar structure to that of the Milky Way. (Courtesy of NASA, Space Telescope Science Institute.)

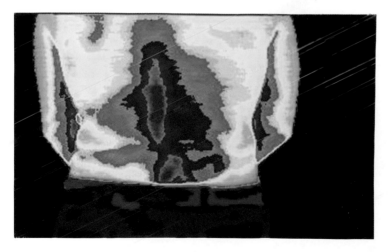

FIGURE 10.6 Thermograph showing pinched nerves, fractured vertebrae, muscle strain, and dislocations. (Courtesy of Teletherm Infrared, Dunedin, FL, Ashwin Systems International, www.thermology.com.)

(a)

(b)

(c)

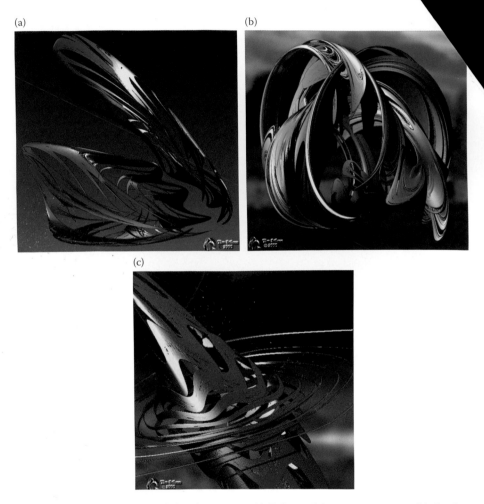

FIGURE 15.13 Diagrams of chaotic systems. (a) Polynomial strange attractor, (b) floating point attractor, and (c) another example of a polynomial strange attractor. (Courtesy of Tim Stilson, Stanford University.)

FIGURE 20.13 Artist's conception of the space-time drag near a spinning neutron star. Recent observations with NASA's Rossi X-ray Timing Explorer satellite seem to confirm the prediction of Einstein's general theory of relativity that a spinning object drags space-time along with it. (Courtesy of Joe Bergeron.)

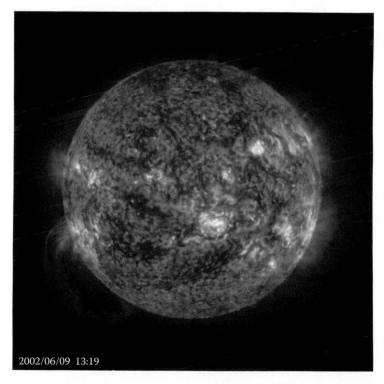

FIGURE 23.11 Nuclear fission powers the sun. (Courtesy of NASA.)

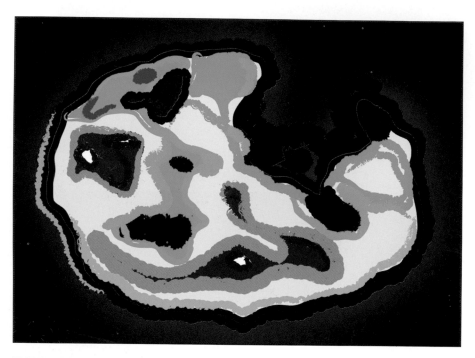

FIGURE 23.13 Radioisotope image of a cross section of a human brain, showing an area damaged by a stroke (darkened area). The bright regions indicate normal blood flow. (From Barron Storey/National Geographic Image Collection. With permission.)

This chaos appears not only in physical systems but also in living organisms. Chaotic activity is apparently responsible for certain mechanisms in the brain involved in learning. There is experimental evidence that chaotic activity switches on and off in the olfactory system of mammals with different chaotic responses for familiar and unfamiliar smells. Ventricular fibrillation, an irregular and uncontrolled contraction of the muscle fibers of the ventricles in the heart's lower chambers, is a condition that brings death within minutes. The heart is a mechanical pump and, like the dripping faucet, the condition of ventricular fibrillation may have an underlying pattern of regularity. If heart fibrillation is indeed chaotic, it might be possible to predict when it is going to appear and perhaps prevent it.

THE FRONTIERS OF PHYSICS: CHAOS IN THE BRAIN

When we see the face of a friend, hear the voice of a famous TV personality, or recognize the smell of cherry pie, we know almost instantly that what we see, hear, or smell is familiar to us. How do our brains process the complex and varied information perceived by our senses in such a short time with such accuracy? The answer seems to be with the existence of chaos in the brain.

When a person smells an odorant, for example, molecules carrying the odor are picked up by a few neuron cell receptors in the nasal cavities. The receptor cells that capture the molecules send out pulses which propagate to the olfactory center in the forebrain where the signals are analyzed and transmitted to the olfactory cortex. From there the signals are sent to other parts of the brain where they are combined with the signals from other senses and a perception is produced.

How does the brain separate one scent from all the others? How does it learn to recognize familiar scents? It seems that chaos is the property that makes perception possible.

Researchers at the University of California at Berkeley attached an array of 60 electrodes 0.5 mm apart to the olfactory bulbar surface of trained rabbits and recorded electroencephalogram (EEG) tracings as the animals sniffed. At first sight, these EEG tracings look irregular and unpredictable. The EEGs oscillate, rising and falling continuously, but when a familiar smell is detected all the waves from the array acquire some regularity, oscillating at about 40 Hz. When the differences between consecutive waves were plotted during perception of a familiar scent and during rest, the underlying order of a chaotic system was observed in the graphs.

The graphs suggested to the researchers that an act of perception consists of an explosive jump from one chaotic set of oscillations to another. They think that the olfactory bulb and cortex maintain many chaotic sets, one for each familiar scent. When a new odorant becomes familiar, another set is added to the collection.

How does the brain shift from the irregular EEGs to the more regular oscillating one of a familiar scent, sound, or sight? More recent experiments, in which the brain neurons of the visual system of anesthetized cats were tagged with a dye that fluoresces when the neuron fires, show that rather than averaging the input received by all the neurons, as had been believed, the brain actually processes all the signals. Moreover, the experiments suggest that neurons respond simultaneously to the same source rather than to many different ones.

How can the brain sort out what is important amidst all this noise? Physicist William L. Ditto of Georgia Institute of Technology thinks that this noise actually helps the brain detect weak signals. This is the idea behind the phenomenon known as stochastic resonance, in which the addition of noise enhances the detection of normally undetected signals.

Chaos in the brain, the researchers think, may be the main property that makes the brain different from a computer, even at the current stages of artificial intelligence. Future programmers will probably have to put chaos into their programs if they want their machines to achieve even a minimal level of recognition.

WATER WAVES

Water waves are an example of a fairly complicated wave phenomenon that can be understood with what we have learned about wave motion this far. As we have seen, it is energy that is transmitted in a wave, not matter. The particles of matter that make up the medium in which the wave travels vibrate around fixed positions but do not travel along with the wave. In the case of water, the particles of water move in circles, upward on the front edge and downward on the back edge (Figure 15.14).

After a disturbance is produced in water, several forces begin to act on the water. Suppose we throw a big rock or brick in the water of a lake. This makes the water next to the point of impact rise above the normal level (Figure 15.15). As when an object is thrown up into the air, the weight of the rising molecules of water slows them down until they finally stop and begin to fall. Accelerating downward, they reach the equilibrium level with a maximum speed and continue moving past this level pushing the other molecules away and creating a depression in the water. The pressure of the water itself produces a buoyant force that slows the falling water and, after stopping it, pushes it back up. Again the water overshoots and rises above

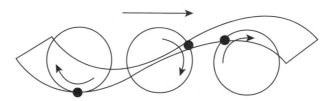

FIGURE 15.14 The particles of water move in circles when the wave passes.

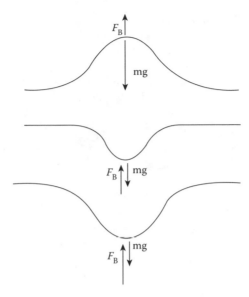

FIGURE 15.15 Two forces act on the water when a disturbance is produced; the weight of the water acting downward, and the pressure of the water itself acting upward, which produces a buoyant force F_B.

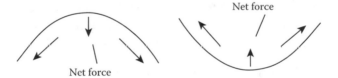

FIGURE 15.16 Ripples are propagated by forces that arise from surface tension.

the undisturbed level, and another cycle of this oscillation starts. Since water is incompressible, when the first depression is formed, the water particles next to the depression are pushed up, and a second oscillating pulse is formed. In this way the disturbance propagates away from the place where the rock was thrown.

In the case of very small waves or ripples, the role of the force of gravity is not as important as that of surface tension. When the water rises above the equilibrium level, the curved surface becomes slightly larger than the flat undisturbed surface so that the average distance between molecules increases. The forces acting between the water molecules on the surface of the disturbance pull the surface to the flat undisturbed level and the same overshooting effect takes place, producing oscillations (Figure 15.16).

SEISMIC WAVES

Earthquakes are sudden disturbances within the earth that produce waves (called *seismic waves*) that travel around the world. When an earthquake occurs, several

Detection station

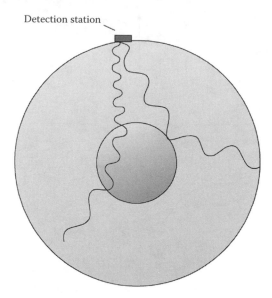

FIGURE 15.17 Waves from earthquakes on the other side of the earth, which are changed when passing through the core, are different from the waves from earthquakes that take place nearby.

kinds of waves propagate away from the epicenter at different speeds. The fastest waves, which arrive at a detection site first and consequently are called P (or primary) waves, are longitudinal and travel through the earth. Deep in the earth, the P waves travel at a speed of about 15 km/s, whereas near the surface their speed is about 5.5 km/s. The slower S (or secondary) waves, which also travel through the earth, are transverse waves; their speeds are usually 7/12 the speed of the P waves. The slowest waves are the L waves, which are surface waves with amplitudes that die out fairly rapidly with depth.

Solid regions in the earth transmit both P and S waves, while fluid regions transmit only P waves. Observation and detection of these different kinds of wave have been used to obtain information about the structure of the interior of the earth. In 1906, from analyses of P wave data, the British scientist R. D. Oldham demonstrated that the earth has a core. By observing the difference in the waves from earthquakes on the other side of the earth and those from earthquakes that had taken place at closer locations, he concluded that the waves from the other side of the earth had been altered by passing through the core (Figure 15.17). Further studies showed that the interior of the earth has two other layers besides the core: the mantle and the crust (Figure 15.18). The core appears to be composed mainly of iron, with small amounts of sulfur and oxygen; it extends 1218 km from the center of the earth. The mantle, which has a thickness of about 2900 km, consists of rocks made of iron and magnesium combined with a silicate material. It becomes fluid under high constant pressure; however, sudden changes in pressure turn it brittle.

Surrounding the mantle is the crust, a relatively thin layer of lower density rock; its thickness varies from a few kilometers under the ocean, to about 50 kilometers in

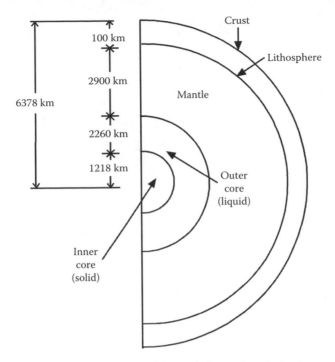

FIGURE 15.18   Structure of the interior of the earth. Detection of seismic waves has yielded information about the composition of the earth.

the continents. The lower part of the crust, together with the upper mantle, constitutes the lithosphere, a rigid layer formed by large thin plates. These plates, consisting of lower density granitic rocks, float on the slightly denser plastic mantle, carrying the continents along with them. The surface of the earth is continuously changing and the continental plates move with velocities of a few centimeters per year.

16 Sound

THE NATURE OF SOUND

Sound influences all of our activities. The human ear is capable of detecting waves carrying energies as low as 10^{-18} J. For comparison, the energy required to lift a paper clip up to a height of only one millimeter is one trillion times as great. This remarkably sensitive organ allows us to detect the immense variety of sounds in which we are constantly immersed; the snarling and barking of dogs, the mewing of cats, the chirping and singing of birds, the screaming of sea gulls, the crowing of roosters, the whispers of trees, the patter of rain, the thunder of storms, the ticking of clocks, the whir of machines, the crying of babies, the voices of singers, or the melody and harmony of music.

What is this phenomenon that we call sound? Galileo said that sound is a wave motion "produced by the vibration of a sonorous body." Some 2200 years earlier, the Greek philosopher Pythagoras had recognized that the pitch of a musical sound depends on the frequency of vibration of the object that produces the sound. Aristotle, in the fourth century BC, realized that the transmission of sound through the air is related to the motion of the air. The Roman architect and engineer Vitruvius, in the first century BC, was probably the first person to understand the concept of waves. He wrote that the propagation of sound in air was analogous to the motion of ripples on the surface of the water. In the eighteenth century, Robert Boyle showed that sound waves need a medium to travel through by pumping air out of a jar containing a bell, which could not be heard without the air.

We know today that sound is a mechanical longitudinal wave that propagates through a medium with frequencies that range from a fraction of a hertz to several megahertz. These mechanical waves are pressure waves produced by the vibrations of the molecules that make up the medium. After we strike a drumhead, it starts its return toward the position of equilibrium but, like an oscillating pendulum, it overshoots and continues moving past this position, pushing the air molecules in its path. This creates a region of higher density in the air. As the drumhead moves in the opposite direction, again toward the equilibrium position, a region of lower density air is produced. With each vibration of the drumhead, regions of higher density air followed by regions of lower density air are created. In the regions of higher density, compressed air pushes on additional air, transferring momentum, and a wave propagates through space (Figure 16.1).

Since the human ear can detect sounds between 20 and 20,000 Hz, sound of frequencies within this range are called *audio frequencies*. Sound waves with frequencies below 20 Hz are called *infrasonic*, and those with frequencies above 20,000 Hz are called *ultrasonic*.

FIGURE 16.1   After a drumhead is struck, higher and lower density regions of air propagate away from the drum.

PHYSICS IN OUR WORLD: TELEPHONE TONES

When you press a button to dial a number with a touch-tone telephone, you activate electric circuits that produce tones. In the United States, pressing the number 1 button, for example, generates a tone of frequency 697 Hz together with a tone of frequency 1209 Hz. The number 6 button generates a 770 Hz tone and a 1477 Hz tone. The sound produced by these tones is transmitted as an electromagnetic signal to the central telephone switching office where it is interpreted by electronic circuits in the switching network.

Each signal arriving at the central switching office has been combined or multiplexed together with many other signals. In the past, this multiplexing was achieved by assigning each telephone signal to a different frequency band. Today, multiplexing is performed by converting telephone signals to digital form, received by a large dedicated computer at the switching office.

If the number called is located at the same central office (the same exchange, meaning that the telephone numbers share the same initial digits), the connection is completed there. If the number called has a different exchange, the signal is sent to a second central office. For long distance calls, the call is sent to a series of interchange offices until the desired exchange is reached.

Several methods have been developed over the years to transmit telephone signals. The oldest digital system still in use today is a system in which multiplexed signals travel through copper wires that carry 24 telephone signals simultaneously. The very high frequency radio system, which carries signals across the United States in the microwave region, bounces signals across microwave towers located every 26 miles. More recently, with optical fiber (see next chapter), the telephone signals are encoded in the near-visible region of the electromagnetic spectrum. A single optical fiber can carry 30,000 simultaneous telephone signals.

THE SPEED OF SOUND

The speed of propagation of sound through a medium is a consequence of the properties of that particular medium. In a gas, the molecules are separated from each other by comparatively large distances, and the time between collisions is longer than in liquids or solids, where the molecules, being much closer together, interact through intermolecular forces rather than through collisions. Since a sound wave propagates through molecular interactions, we would expect sound to travel at a lower speed in gases. Table 16.1, lists the speed of sound in various substances.

TABLE 16.1
Speed of Sound in Some Gases, Liquids, and Solids

Substance	Speed in m/s	Substance	Speed in m/s
Gases		*Liquids at 25°C*	
Air at 0°C	331	Fresh water	1493
Air at 20°C	343	Seawater	1533
Air at 100°C	366		
Hydrogen at 0°C	1286	*Solids*	
Helium at 0°C	965	Aluminum	5100
Oxygen at 0°C	317	Iron	5130
Nitrogen at 0°C	334	Lead	2700
		Rubber	1800
		Granite	6000

Since molecular motion increases with temperature, we also expect the speed of sound to be directly proportional to the temperature of the medium through which it propagates. For sound waves moving in air, if we know the speed v_0 at 0°C, the speed is approximately

$$v = (v_0 + 0.61\ \theta)\ \text{m/s} = (331 + 0.61\ \theta)\ \text{m/s}$$

where θ is the temperature in degrees Celsius. This means that the speed of sound increases about 0.61 m/s for every one-degree rise in temperature.

Table 16.1 tells us that the speed of sound in a lighter gas, like hydrogen, is higher than the speed of sound in a heavier gas, like oxygen. Recall from Chapter 10 that the temperature of a substance depends only on the random kinetic energy per molecule, not on the number of atoms and molecules in the substance or on the kinds of atoms that form the substance. At the same temperature, then, molecules of hydrogen and oxygen have the same average kinetic energy. Since the kinetic energy of a molecule is $\frac{1}{2}\ mv^2$, at the same temperature, a hydrogen molecule, with a mass m of 2 amu, has a greater speed v than a heavier oxygen molecule of mass m equal to 32 amu.

INTENSITY OF SOUND WAVES

The sensation of loudness, although subjective, is related to an objective property, namely the *intensity* of the sound wave being perceived. Intensity is the rate at which the wave transports energy per unit of area; that is, the power delivered by the wave through a unit of area, or

$$I = \frac{P}{A}.$$

Since the units of power are watts, the units of intensity are watts per meter squared (W/m²).

The human ear can detect sounds with intensities as low as 10^{-12} W/m² and as high as 1 W/m². This is an enormous range of intensity, a factor of 10^{12} or one trillion from the lowest to the highest. Although the sound intensity of normal conversation is about 10,000 times greater than that of a whisper, we would perhaps say that normal conversation is only about 3 or 4 times as loud as a whisper. To better represent the response of the human ear to changes in sound intensity, we use the term *sound level*, expressed in units called *bels* (in honor of Alexander Graham Bell) or more commonly *decibels*, dB (1 dB is 0.1 bel). On this new scale, 0 dB corresponds to the lowest intensity that we can detect, and 120 to the maximum we can endure. Thus, when the sound intensity changes by a factor of 10, we add 10 to the sound level in dB. We can see the relation between sound intensity and sound level in Table 16.2. Sound levels for several sounds are listed in Table 16.3.

TABLE 16.2
Comparison between Sound Intensity and Sound Level

Sound Level (dB)	Sound Intensity (W/m²)	Relative Intensity
0	10^{-12}	$10^0 = 1$
10	10^{-11}	$10^1 = 10$
20	10^{-10}	$10^2 = 100$
30	10^{-9}	$10^3 = 1\ 000$
40	10^{-8}	$10^4 = 10\ 000$
50	10^{-7}	$10^5 = 100\ 000$
60	10^{-6}	$10^6 = 1\ 000\ 000$
70	10^{-5}	$10^7 = 10\ 000\ 000$
80	10^{-4}	$10^8 = 100\ 000\ 000$
90	10^{-3}	$10^9 = 1\ 000\ 000\ 000$
100	10^{-2}	$10^{10} = 10\ 000\ 000\ 000$
110	10^{-1}	$10^{11} = 100\ 000\ 000\ 000$
120	1	$10^{12} = 1\ 000\ 000\ 000\ 000$

TABLE 16.3
Some Sound Levels

Sound	Sound Level (dB)	Sound	Sound Level (dB)
Barely audible	0	Average factory	80
Rustle of leaves	10	Niagara Falls	90
Whisper at 1 m	20	Power mower	100
Quiet room	30	Rock concert (outdoors)	110
Library	40	Rock concert (indoors)	120
Average classroom	50	Threshold of pain	120
Normal conversation (1 m)	60	Jet plane at 30 m	140
Busy street	70	Jet takeoff	150

THE EAR

We learned in the previous section that the human ear can detect sounds with an intensity of 10^{-12} W/m². This corresponds to a pressure of 3×10^{-5} Pa above the atmospheric pressure, which is only six times greater than the pressure fluctuations produced by the random motion of the air molecules. If the human ear were a little more sensitive, we could hear individual molecules colliding with the eardrum!

The human ear, like the ear of other mammals, is divided into three regions: the outer ear, the middle ear, and the inner ear (Figure 16.2). The outer ear consists of the *auricle* or *pinna* and the 2.5-cm long *auditory canal*, which connects the auricle with the eardrum or *tympanum*. This has the form of a flattened cone some 9 mm

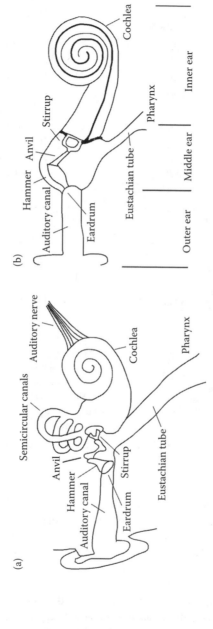

FIGURE 16.2   (a) The human ear. (b) Schematic diagram of the ear.

in diameter, stretched across the *tympanic annulus*, a small bone in the shape of a ring. Sound waves enter the ear canal through the auricle and set the eardrum into vibration.

Behind the eardrum is the middle ear which consists of a narrow, air-filled cavity where three small bones, the *hammer*, the *anvil*, and the *stirrup*, transmit the vibrations of the eardrum to the oval window of the inner ear. Leading downward from this cavity is the *Eustachian tube*, which connects the ear to the nasopharynx, the region behind the nasal passages. The purpose of this tube, about 45 mm in length, is to equalize the pressure on both sides of the eardrum.

The inner ear is contained in a bony structure known as the *labyrinth* with two openings to the middle ear: the oval window, connected to the stirrup, and the *round window*, covered by a thin membrane. This labyrinth contains three semicircular canals and a spiral tube about 30 mm in length called the *cochlea*, a word derived from the Greek world for snail (Figure 16.3). It is in the cochlea where the vibrations of the sound waves are transformed into the electrical impulses the nerves transmit to the brain. The cochlea is divided internally by three longitudinal sections: the *vestibular canal*, which ends at the oval window; the *tympanic canal*, which ends at the round window; and the *cochlear duct*, which lies in the middle of the other two and also ends at the round window.

A sound wave coming into the inner ear through the oval window travels down the vestibular canal and back along the tympanic canal. The membrane that separates these two canals contains the *organ of Corti*, a complex arrangement of some 30,000 nerve endings (Figure 16.4). The organ of Corti is covered by some five thousand sensory cells—called *hair cells* due to their hairlike appearance—which are stimulated when the membrane is set into vibration by a sound wave. The nerve fibers connected to these hair cells transmit this information to the brain in the form of electrical impulses which the brain interprets as sound.

Hair cells are about five thousandths of a millimeter thick and as fragile as a cobweb. Sound waves that carry too much energy can damage them. A few minutes exposed to a sound level above 110 dB can cause permanent damage; above this "threshold of pain," exposure for any length of time is dangerous. Each hair cell

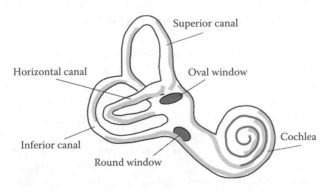

FIGURE 16.3 The inner ear.

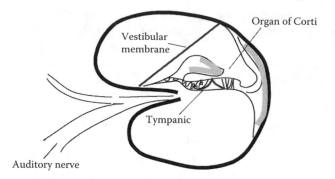

FIGURE 16.4 Cross-section of the cochlea showing the location of the organ of Corti and the vestibular and tympanic membranes.

can respond to motions as small as 100 picometers (10 billionths of a centimeter), a distance similar to the diameter of one of the larger atoms. The oscillatory pressure of the sound wave moves the hair cells and this motion is transformed into an electrical impulse at the cell which changes the electrical potential that exists between the interior and exterior of the cell, caused by different concentrations of sodium and potassium ions inside and outside the cell. When the hair cell is at rest, there is an exchange of ions through open channels in the membrane, and an equilibrium situation is reached. When the cell is moved, the number of open channels changes and the equilibrium is disturbed; this changes the electric potential of the cell from about 60 mV to about 40 mV. This change in the potential difference triggers an electrical signal that travels to the brain.

Hair cells are also the sensory receptors for the sense of balance which allows humans to walk upright. In addition to the cochlea, five other sensory organs contain hair cells; the three semicircular canals, the utricle, and the saccule. The hair cells of the utricle and the saccule are located on two thin, flat sheets, one of which is positioned vertically and the other horizontally. Close to these organs is the *otolithic* membrane which contains hundreds of thousands of small crystals, the *otoconia*, which make this membrane denser than the flat sheets that contain the utricle and the saccule. When the head is accelerated, the otoconia lags behind due to its greater inertia; this causes the hair cells of the utricle or the saccule to be displaced in the opposite direction. Displacement of the hair cells produces potential difference changes, which are communicated to the brain as electrical signals.

A similar mechanism works in the semicircular canals. In this case, the hair cells are moved when an angular acceleration takes place. The canals are filled with a fluid, the *endolymph*, which lags behind the sides of the canal when the head is rotated. This fluid exerts a pressure on the membrane where the hair cells are located, producing the displacement that causes the potential changes in the cells. As the three canals—the *superior canal*, the *horizontal canal*, and the *inferior canal*—lie along perpendicular directions, they can detect angular acceleration along three perpendicular axes in space (Figure 16.5).

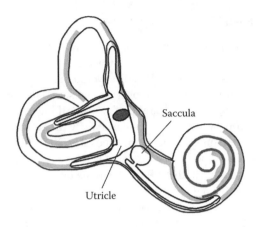

FIGURE 16.5   The labyrinth showing the three semicircular canals, the saccule, and the utricle.

THE FRONTIERS OF PHYSICS: ELECTRONIC EAR IMPLANTS

People with hearing loss caused by damage to the hair cells which cover the organ of Corti can now receive cochlear implants that replace the function of the hair cells with electrodes. These electrodes, in the form of a very thin wire less than a millimeter in diameter, are inserted through the ear canal into the cochlea to stimulate the auditory nerves electronically, allowing the deaf person to perceive sound.

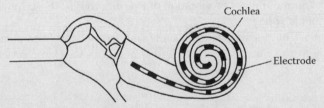

Although the first implants contained a single electrode, more recent devices use multichannel electrodes with several receptors for different frequencies. A sound processor worn behind the ear encodes the sound and transmits it as radio waves to the electrodes.

Researchers have discovered the proper position in the cochlea to place the electrodes and the range of frequencies to which each electrode should respond. High frequencies stimulate the outer spiral of the cochlea whereas low frequencies must penetrate deep inside.

The original single channel electrodes (which were approved by the Food and Drug Administration but are no longer available) gave most patients only the awareness of sound. The new multichannel devices based on the selective frequency bands give speech recognition to the patient.

THE SOUND OF MUSIC

What is the difference between music and noise? We might say that music sounds pleasant and noise generally unpleasant. But what makes a sound pleasant to our ears? Pythagoras, in the sixth century BC, discovered that when two similar stretched strings of different lengths are sounded together, the sound thus produced is pleasant if the lengths of the strings are in the ratio of two small integers.

Pythagoras was probably born in the year 560 BC in Samos, one of the Aegean islands. Little is known of his early years in Samos. In 529 BC, after traveling through Egypt and the East on advice from his mentor Thales, Pythagoras settled in the Greek colony of Crotona, in southern Italy, where he began lecturing in mathematics and philosophy. People from the upper classes attended his lectures. One of the most attentive was Theano, the beautiful daughter of his host, whom Pythagoras later married, and who, in order to attend the lectures, had to defy an ordinance which forbade women to be present at public functions. Theano later wrote a biography of her husband, now unfortunately lost.

Pythagoras discovered that if a stretched string vibrates as a whole—we would say that it vibrates in the fundamental mode—the sound produced by a second string that has half the length of the first one would be harmonious with the sound of the first string (Figure 16.6a). Strings with lengths equal to one third, one fourth, and so on, would also be harmonious with the first string. Pythagoras also realized that two stretched strings of equal length would produce pleasant sounds if one is vibrating as a whole and the other is made to vibrate in two, three, four, or more equal parts (Figure 16.6b). These modes of vibration are what we called in Chapter 15 the *overtones* of the fundamental, and together with the fundamental they form a harmonic series. Thus, Pythagoras found that the chords that sound pleasant to the ear correspond to the normal modes of vibration of a string; that is, those frequencies which form a harmonic series.

Although we have seen that the sounds with frequencies from a harmonic series are pleasant to the ear, we have not said why this is so. One characteristic of the

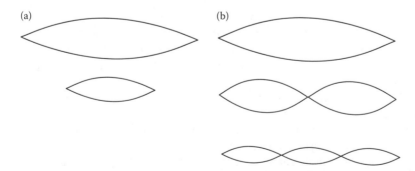

FIGURE 16.6   (a) The sound produced by a string of a certain length vibrating in the fundamental mode is harmonious with the sound of a second string that has half the length, also vibrating in the fundamental mode. (b) Two equal strings also produce harmonious sounds if one vibrates in the fundamental and the other in two, three or more equal parts (overtones).

sound of music is that the sound waves show some periodicity; we say that music has *sustained notes*. The sound waves of noise, on the contrary, are disorganized and do not show periodicity. Musicians refer to a musical tone in terms of *loudness*, *pitch*, and *quality* or *timbre*. We studied loudness earlier in the chapter; here we will consider pitch and quality.

The sensation of *pitch* is related to the frequency of sound waves; high-pitched sounds have high frequencies and low-pitched sounds have low frequencies. The human ear, as we know, can detect sounds with frequencies between 20 and 20,000 Hz; thus, sound waves outside this range have no pitch, since pitch is a subjective sensation.

The different modes of vibration of a stretched string of length L are related through their frequencies f_n by the expression

$$f_n = \frac{nv}{2L}$$

where, as we saw in Chapter 15, n refers to the mode of oscillation and v is the velocity of propagation of the waves in the string. The fundamental mode or first harmonic has $n = 1$, the second harmonic has $n = 2$, and so on. Since the frequency of vibration of the string is related to its length, we can restate Pythagoras' discovery by saying that a sound of two strings vibrating simultaneously is pleasant if the ratio of their frequencies is the ratio of two small integers. If the ratio is 2 to 1, the sounds are one octave apart. In musical terms, octaves are equal intervals of pitch.

Descriptions of the pitch relationships that exist in music are called *musical scales*. Since many relationships can be established, the number of possible musical scales is very large, and many have been used throughout history. Generally, the more advanced civilizations have developed complex scales. The most common scales are the *pentatonic* scales, which are based on five notes, and the *heptatonic* scales, based on seven notes. Western music is based on one heptatonic scale, the *diatonic* scale, with the familiar notes "do-re-mi-fa-sol-la-ti-do." The notes in this scale are named by using a LETTER (C, D, E, F, G, A, B), and each has a particular frequency. The scale itself has five whole steps (*W*) and two half steps (*H*) with two permutations or modes, *major* and *minor*. The major scale uses the interval sequence *W-W-H-W-W-W-H*, while the minor scale has the sequence *W-H-W-W-H-W-W*. A half step has a frequency ratio of 16/15 and a whole step has frequency ratios of 9/8 or 10/9. Table 16.4 shows the diatonic C-major scale.

In Table 16.4, the C with a frequency of 264 Hz is known as the *middle C* because in written music this note occurs midway between the treble and bass clefs. (*Clef* is the sign placed at the beginning of the musical staff to determine the position of the notes. The *musical staff* is the horizontal set of lines on which music is written.) The ratio between the next C, the *C above middle C*, and middle C is 528 Hz/264 Hz = 2/1. These two notes are therefore an *octave* apart. We notice that if we label middle C the first note, the C above middle C would be the eighth note, and octave means eight in Latin. The ratio between the frequency of each A and the

TABLE 16.4
Diatonic C-Major Scale

Note	Letter	Frequency (Hz)	Frequency Ratio	Interval
do	C	264		
re	D	297	9/8	Whole
mi	E	330	10/9	Whole
fa	F	352	16/15	Half
sol	G	396	9/8	Whole
la	A	440	10/9	Whole
ti	B	495	9/8	Whole
do	C	528	16/15	Half

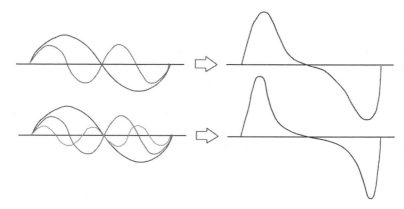

FIGURE 16.7   A stretched string vibrating in several overtones. The top diagram shows the first two harmonics and the bottom diagram the first three harmonics. On the right is the resultant waveform for each case.

preceding A or between F and the preceding F is always 2; and this is valid for any other note. We can then say that the A above the A with 440 Hz has a frequency of 880 Hz and the C below middle C has a frequency of 132 Hz.

In addition to loudness and pitch, a musical tone has *quality*. The same note when played on a violin sounds different from when it is played on a piano or sung by a soprano. The difference is in what is called the *harmonic* content of the wave; that is, the number of harmonics present in the wave. We know from the superposition principle that standing waves of different frequencies can exist in the same medium at the same time (Figure 16.7). For example, a guitar string usually oscillates in many overtones when it is plucked. A piano string when struck also oscillates in many overtones, but the relative amplitudes of these overtones are different from the ones in the guitar string. The relative oscillation energies of the different instruments determine the quality of sound they produce. Since each instrument has its own balance of overtones, a clarinet sounds different from a flute or from a soprano voice

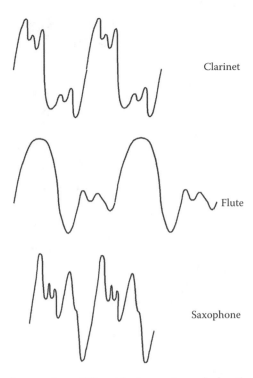

Clarinet

Flute

Saxophone

FIGURE 16.8 Sound waves from different instruments producing the same note.

when sounding the same note (Figure 16.8). The quality of the sound that we produce with our voice is also what allows us to discriminate between the different vowel sounds. When we speak, the vocal chords vibrate in the fundamental and several overtones and these *resonate* in the mouth. These resonant frequencies are called formants. When we change the shape of the mouth as we pronounce the different vowels, some harmonics are stressed more than others and a different set of formants appears, producing the different sounds.

MUSICAL INSTRUMENTS

Musical instruments consist of a source of sound and a resonator which amplifies the sound and enhances certain harmonics. Guitars and violins use strings to generate sounds and a sounding box as resonator (Figure 16.9). The vibrations of the strings are passed to the sounding box by means of the bridge. The sounding box has its own modes of vibration which resonate with the vibrations of the strings, amplifying only certain frequencies. We saw in Chapter 15 how standing waves develop on a stretched string, which make the string vibrate in several different ways or modes with frequencies given by

$$f_n = \frac{nv}{2L}.$$

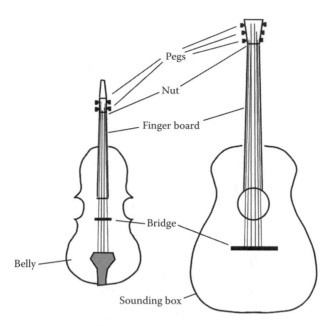

FIGURE 16.9   Guitars and violins have a sounding box or chamber that resonates with the vibrations of the strings.

The frequency of the fundamental determines the pitch of the note. Since all the strings in a guitar or a violin have the same length, the only way to change the fundamental is by changing the speed of propagation of the waves on the string. How can this speed be changed? It turns out that the speed of propagation of waves on a stretched string depends on two factors: the tension in the string and the string's density. If the tensions of the different strings in a guitar are all the same, the equation above tells us that strings of different densities vibrate with different fundamental frequencies.

Pianos have strings of different lengths and the different fundamental frequencies are achieved by varying, in addition to the length, the tension and the density of the strings. To withstand the tremendous tension of the strings, pianos have a cast-iron frame. The resonator in a piano is the *sounding board*, a flat piece of wood as big as the piano itself (Figure 16.10). When a key is depressed, a hammer, which is linked to the far end of the key, strikes the string and sets it into vibration. The vibrations of the strings are transmitted to the sounding board via the bridge over which the strings are stretched. A felt damper controlled by one of the two or three pedals present on a piano can be lifted to allow the strings to vibrate freely or lowered to stop the vibration.

In a wind instrument such as the oboe, the musician blows into the mouthpiece producing eddies which set the reeds into vibration. This causes standing waves of the air within the tube. In other instruments such as brasses, the musician's lip vibrates as he or she blows air into the mouthpiece; these vibrations also produce standing waves in the column of air in the instrument. In each

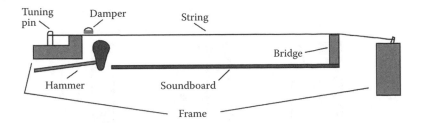

FIGURE 16.10 Schematic diagram of the hammer, damper, string, bridge, and sounding board of a piano.

case, the frequency of the note produced depends on the length of the column of air, whether the tube is open or closed, and whether it is tapered (conical) or cylindrical.

The standing waves established on the strings of stringed instruments have nodes at both ends (since they are clamped at the ends). The fundamental frequency, with nodes only at the ends, determines the pitch. The quality or characteristic sound depends on the harmonics present. The situation is similar with an air column; the closed end of the tube or pipe, such as an organ pipe, where the air is not free to move, is a *displacement* node. However, at the open end of the tube the air is free to move and a displacement *antinode* or point of maximum displacement appears there as shown in Figure 16.11. Since the distance between a node and the next antinode is one quarter of the wavelength when the air column is vibrating in the fundamental mode, then the wavelength of the fundamental must be 4 times the length L of the tube, or $\lambda_1 = 4 L$. We can obtain the frequency of the fundamental by making use of the expression relating frequency, wave speed and wavelength obtained earlier in the chapter; that is, $v = \lambda f$ or $f = v/\lambda$. Thus, the frequency of the fundamental is $f_1 = v/\lambda_1 = v/4L$ where v is the speed of sound. The next harmonic contains one additional node between the open and closed ends, as shown in Figure 16.11. Since the open end must be an antinode and the distance to the previous node is one quarter λ, the length L of the tube must be equal to three quarters the wavelength of this harmonic so that the wavelength is

$$\lambda_3 = \frac{4}{3}L; \qquad f_3 = \frac{v}{\lambda_3} = \frac{v}{(4/3)L} = \frac{3v}{4L}.$$

This frequency is three times the frequency of the fundamental; thus this harmonic is the *third* harmonic, which is the reason for the subindex 3 in the expressions above. Examining Figure 16.11, we can see that the next harmonic has two nodes in addition to the node at the end. One and one-quarter wavelengths equal the length of the tube; thus, the wavelength of this harmonic is $\lambda_5 = 4L/5$ and the frequency is $f_5 = 5v/4L$ or five times the frequency of the fundamental. Hence, this is the fifth harmonic. There are no second or fourth harmonics. Actually, there

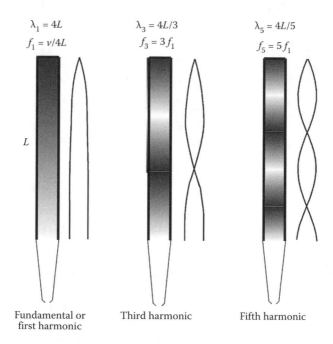

$$\lambda_1 = 4L \qquad\qquad \lambda_3 = 4L/3 \qquad\qquad \lambda_5 = 4L/5$$

$$f_1 = v/4L \qquad\qquad f_3 = 3f_1 \qquad\qquad f_5 = 5f_1$$

L

Fundamental or   Third harmonic   Fifth harmonic
first harmonic

FIGURE 16.11   Standing waves in an organ pipe closed at one end. We use the graph of a wave to indicate variations in the density of the air in the pipe caused by the vibrations. The nodes are points of higher density and minimum displacement and the antinodes or points of maxima are regions of lower density and maximum displacement.

are no even harmonics in a tube that is closed at one end; only odd harmonics are present.

In an open organ pipe both ends are antinodes and the fundamental occurs when there is a node in the middle of the pipe (Figure 16.12). In this case, the wavelength is equal to twice the length of the pipe; that is, $\lambda = 2L$. The frequency of the fundamental is then $f = v/2L$. The second harmonic has two nodes between the ends so that the wavelength is equal to the length of the pipe; the frequency is then $f_2 = v/L = 2f_1$. Examining Figure 16.12, we can see that the third harmonic has a wavelength $\lambda_3 = 2L/3$ and a frequency $f_3 = 3f_1$. As opposed to the closed pipe, in an open organ pipe all harmonics are present. This also applies to open wind instruments such as flutes, and to reed instruments with a conical bore, such as saxophones.

When an organ is played, several harmonics are present at once. The harmonic content of a pipe depends on the ratio of length to diameter, whether the pipe is closed or open, cylindrical or conical, and to some extent on the material from which the pipe is made, as well as the sound-generating edge (fipple or reed). The harmonic content determines the quality or timbre of the pipe.

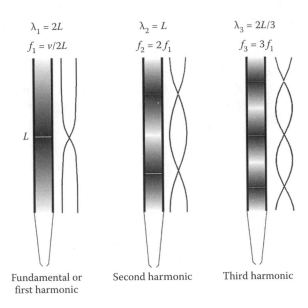

FIGURE 16.12   Fundamental, second, and third harmonics in an open organ pipe.

THE DOPPLER EFFECT

You have no doubt noticed the sudden drop in pitch from an automobile horn as the car speeds past you. We know that the pitch does not change for the driver and that it is only because the car is moving that we hear a change in pitch. The same phenomenon can be heard if you drive past a car blowing its horn; the pitch is higher when you approach the car and lower when you are leaving it behind. When a source of sound and a listener are in motion relative to one another the pitch of the sound is not the same as when the source and listener are stationary with respect to each other. This phenomenon is known as the *Doppler effect,* after the Austrian physicist Christian Doppler (1803–1853), who first explained its properties.

Suppose the horn of an automobile emits a sound of a particular frequency f and wavelength λ (= v/f) in all directions (where v is the speed of sound in air). If this automobile is standing by the side of the road and we approach it with a speed v_0, the speed of the sound waves emitted by the car appears to be increased to $v' = v + v_0$. As we move toward the car emitting the sound, we encounter more wavelengths per second. In this case, the frequency that we perceive is

$$f' = \frac{v'}{\lambda} = \frac{v + v_0}{\lambda}.$$

Although we are approaching the car, we still measure the same distance between two consecutive crests as if we were standing by the car because the car emitting the sound is at rest with respect to the air through which the sound propagates and our motion does not affect it. Since the distance between two consecutive crests is the wavelength, this means that the wavelength remains the same $\lambda = v/f$, and our expression for the frequency of the sound we hear as we approach the stationary car with a speed v_0 becomes

$$f' = \frac{v + v_0}{v/f} = f\frac{v + v_0}{v}.$$

Let us examine this expression a little more closely. Since v_0 is positive, the numerator is greater than the denominator which makes the frequency that we detect greater than the frequency perceived by a listener standing by the side of the car. A higher frequency means a higher pitch. Notice that if we decide to stop by the car with the stuck horn, then $v_0 = 0$ and $f' = f$; that is, the frequency we hear is the frequency heard by a listener at rest with the source of sound, which is what we have become by stopping. If we resume our trip and drive away from the car, then we must consider v_0 to be negative. In this case, the numerator becomes smaller than the denominator and f' is smaller than f, which implies a lower pitch. Thus, the perceived frequency is higher than the frequency of the source when the listener approaches the source and lower when the listener moves away from the source. Summarizing, the frequency perceived by a listener moving with a velocity v_0 with respect to a stationary source of sound is

$$f' = f\frac{v + v_0}{v} \qquad \text{Listener approaching the source}$$

$$f' = f\frac{v - v_0}{v} \qquad \text{Listener moving away from the source}$$

A different situation occurs when the source of sound approaches or moves away from the listener. In these two cases the wavelength changes because the source of sound is in motion with respect to the medium through which the sound propagates. The pitch we hear from the siren of an approaching fire truck is higher than the pitch heard by the driver of the fire truck (Figure 16.13a). As the fire truck moves in our direction, it gets closer to the waves that were emitted earlier and the sound waves that reach us get closer together. The siren oscillates at a constant frequency even when it is moving. At the beginning of each oscillation the fire truck has moved a little closer to us than where it was when the previous oscillation began. This makes the distance between oscillations or wavelength decrease. Since the velocity of sound remains unchanged, then the frequency has to increase. On the other hand, when the fire truck is moving away from us, the siren is a little farther away at each oscillation and the wavelength is stretched out (Figure 16.13b).

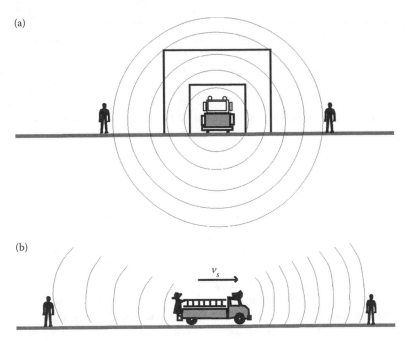

FIGURE 16.13 (a) The listeners hear the same frequency when the fire truck sounding the siren is at rest. (b) When the fire truck is moving, the listener behind it hears a lower pitch than the listener in front of the fire truck.

Suppose the siren emits a pure tone of frequency f and wavelength $\lambda = v/f$, where again v is the speed of sound in air. If the fire truck is moving with a speed v_s in the direction of the listener, the wavelength detected by the listener is shortened by $\Delta\lambda = v_s/f$ during one period of oscillation of the wave. The wavelength detected by a listener in front of the fire truck is then $\lambda' = (v - v_s)/f$ and the frequency is

$$f' = \frac{v}{\lambda'} = f\frac{v}{v - v_s}.$$

When the fire truck is moving away from the listener, v_s has opposite sign.

Summarizing, when a source of sound moves with a speed v toward or away from a stationary listener, the frequency perceived is

$$f' = \frac{v}{\lambda'} = f\frac{v}{v - v_s} \qquad \text{Source approaching listener}$$

$$f' = \frac{v}{\lambda'} = f\frac{v}{v + v_s} \qquad \text{Source moving away from listener}$$

The Doppler effect also applies to other kind of waves. It even applies to electromagnetic waves, although the mathematical expressions are somewhat different. Police radar works by measuring the Doppler shift in frequency of electromagnetic waves emitted by the radar transmitter, as they are reflected from moving cars. Certain motion-sensitive sound detectors activate some large department store or supermarket doors. Astronomers determine the motion and velocity of distant galaxies by measuring the Doppler shifts of the light they emit. The discovery of Doppler shifts in the spectra of galaxies in the 1920s enabled Hubble to discover the expansion of the universe.

SHOCKWAVES

When the source of sound moves with a speed greater than the speed of sound in the medium in which it is moving, a *shockwave* is produced. As we have seen, when the source of sound moves at a speed lower than the speed of sound, the crests of the waves get closer together in front of the moving source and farther apart behind it. If the source of sound moves exactly at the speed of sound, the waves pile up in front. The speed of sound in a particular medium depends on the elasticity of the medium; that is, on how fast and how much the molecules that make up the medium vibrate as the wave propagates. When the source of sound moves through the medium, the elastic properties of the molecules allow them to move away from the source. When an airplane moves at the speed of sound, the air molecules cannot move away fast enough and pile up in front of the plane in one large crest. If the airplane moves faster than the speed of sound, it outruns the sound waves and these pile up along the sides, producing a large crest which extends in a cone behind the plane. This large crest, formed by the constructive interference of a large number of individual crests, is the shockwave (Figure 16.14).

When an airplane flies overhead at supersonic speeds, the shockwave is heard as a very loud crack, or *sonic boom*. This carries an enormous amount of energy which

FIGURE 16.14   Shock wave formed by a source of sound moving at supersonic speeds. (Courtesy of R. J. Gomez, NASA Johnson Space Center.)

FIGURE 16.15   Bow wave formed by a duck swimming at a speed greater than the speed of propagation of water waves.

can shatter windows and produce sound levels that might reach harmful levels. For this reason, supersonic speeds have been banned over the continental United States.*

An airplane traveling at the speed of sound is said to be flying at *Mach 1*, after Ernst Mach (1838–1916), an Austrian physicist and philosopher of science who was the first to investigate the change in airflow around an object as it reached the speed of sound. Mach 2 is used when the airplane is flying at twice the speed of sound, and so on. A sonic boom is not produced only as the airplane crosses the "sound barrier." The shockwave follows the airplane and the sonic boom is heard after the plane has passed overhead. A duck moving in the water at a speed greater than the speed of propagation of surface waves on the water also produces a kind of shockwave, called a bow wave (Figure 16.15).

ULTRASOUND

When sound waves strike the surface of a solid object that is large compared with the wavelength of the sound waves, they are reflected. If the object is smaller than the wavelength of the waves impinging on it, the waves simply travel around the object almost undisturbed. A small piece of wood floating on the water does not disturb the waves produced by a speedboat; another boat does. With sound waves, when the size of the object is at least several times greater than the wavelength of the waves, the sound waves are reflected by the object; this is the familiar echo. The sound of an automobile horn with a frequency of 300 Hz has a wavelength in air equal to $\lambda = v/f = (342 \text{ m/s})/(300 \text{ Hz}) = 1.14$ m; thus, objects several meters across are large enough to reflect sound waves of this frequency. Objects only a few centimeters

---

* The sole exception is the Space Shuttle approaching the landing pad at the Kennedy Space Center. Two shock waves are heard, one from the nose and the other one, about half a second later, from the tail. Actually, all supersonic aircraft produce two sonic booms, but because they happen close to each other, they are heard as one. But the Shuttle, 37 m from nose to tail, is long enough to separate them.

across reflect only sound waves of very high frequencies, beyond the audible range of a human ear. These sound waves are known as *ultrasonic*.

Our knowledge of the speed of sound in a particular medium, say air, allows us to determine the distance to a source of sound if we know the precise time when the sound was produced. Since light travels at 3.0×10^8 m/s, we see the lightning almost instantaneously but hear the thunder a few seconds later. It takes light 10 millionths of a second to travel a distance of 3 km whereas sound waves take 9 seconds to travel the same distance. If we clap our hands and 1 second later hear the echo from a cliff in front of us, we know that the cliff is some 170 m away, since $x = vt = 342$ m/s $\times$ 1 s $= 342$ m and the sound waves traverse this distance twice. Ships and submarines use this technique in underwater rangefinding or *sonar* (*s*ound *n*avigation *a*nd *r*anging). A transmitter sends out a short-wavelength pulse of ultrasonic frequency through the water and a receiver detects the reflected pulse. From knowledge of the time it takes for the pulse to travel to the obstacle, the distance to the obstacle can be determined.

Bats also emit ultrasonic pulses of a few milliseconds duration which, after reflecting from obstacles and prey up to about four meters away, are detected by their large ears. Bats, then, can fly and hunt in complete darkness and can apparently locate their prey by distinguishing the times of arrival of the echo at the two ears. The repetition rate of the ultrasonic pulses varies from about 10 pulses per second when the obstacle or prey is farther away to some 200 pulses per second when the prey is closer. The range of frequencies of the sound emitted by bats goes from 30 kHz to nearly 120 kHz. Moths, a favorite prey of bats, seem to be able to hear those frequencies and drop quickly to the ground when they detect them. They also have another protection against their predators; their furry bodies are good sound absorbers and consequently poor sound reflectors.

Taking a cue from nature, physicists have developed a medical diagnostic and treatment tool using ultrasonic waves. An ultrasonic pulse is produced by applying a high-frequency alternating voltage to both sides of a quartz crystal which then vibrates, emitting ultrasonic waves. (Old phonographic needles use the same principle: the vibrations of the needle as it rests on the groove are transmitted to a crystal which translates them into an alternating voltage; this voltage, after it is amplified, drives the speakers.) These crystals are called piezoelectric and this technique of transforming electrical energy into mechanical energy is called the *piezoelectric effect*. After the ultrasonic pulse is reflected from the boundaries between different organs of the body, it is detected and analyzed. A two-dimensional image can be formed when multiple echoes are combined according to the time delay and intensity of the signal received. The image obtained, called a *B-scan*, is displayed on a monitor.

17 Optics

WAVES OF LIGHT

Light, as we learned in Chapter 14, propagates through space as an electromagnetic wave. Maxwell's third equation tells us that a changing magnetic field produces a changing electric field which, according to Maxwell's fourth equation, creates in turn a changing magnetic field. As Maxwell was able to show, once the process starts, the changing magnetic and electric fields continue to propagate as a wave. Although we also learned in Chapter 7 that light also behaves as a particle, the wave theory of light is adequate for many purposes. We shall occupy ourselves with the study of the wave nature of light in the next chapter. Here, we turn to two phenomena that do not depend on whether we think of light as particles or as waves and are the basis for many optical effects; the laws of *reflection* and *refraction.*

REFLECTION OF LIGHT

Since very early, people have known that light travels in a straight line. Observations of solar and lunar eclipses led the Greek thinkers to realize that light travels in straight lines at very high speed. The shadow cast by the moon on the earth during a solar eclipse or by the earth on the moon during a lunar eclipse shows that light travels in straight lines over long distances (Figure 17.1). It is often convenient to use straight lines or *rays* to represent pictorially this rectilinear motion of light. This representation will be useful in understanding some of the basic properties of light, in particular the laws of reflection and refraction and the operation of many optical devices.

When light falls upon an object, the oscillating electric field of the incoming light sets the electrons of the atoms in the object into vibration. According to Maxwell's equations, a vibrating electric charge sets up an electric field of varying strength which simultaneously creates a time-varying magnetic field. Such fields, as we know, take on a life of their own; they form the electromagnetic waves that travel through space. Thus, the electrons of the atoms or molecules of the object, in particular, the outer electrons which are less tightly bound, absorb the light's energy and start vibrating, becoming emitters of electromagnetic radiation in the process. This electromagnetic radiation, re-emitted after it is absorbed by the electrons, is the light we see when we look at the object.

The *law of reflection* describes the way in which light is reflected from the surface of an object and says that the angle at which a light ray returns from a surface, or *angle of reflection*, is equal to the angle at which the ray strikes the same surface, or *angle of incidence*. As illustrated in Figure 17.2, these angles are measured from the normal or perpendicular to the surface. The incident ray, the reflected ray, and the normal are all in the same plane. When light from a distant source falls upon an

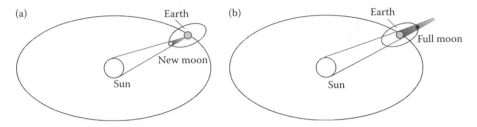

FIGURE 17.1 (a) The shadow cast by the moon during a solar eclipse or (b) by the earth on the moon during a lunar eclipse, shows that light travels in straight lines over long distances.

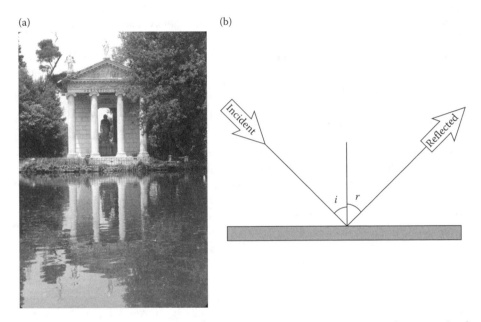

FIGURE 17.2 The angle that a reflected ray makes with the normal to a surface or angle of reflection is equal to the angle that the incident ray makes with the normal or angle of incidence. The incident and reflected rays lie in the plane that contains the normal to the surface. This is the law of reflection.

object, the incident rays of light are all nearly parallel. If the surface of the object is smooth, like a mirror or polished metal, the reflected rays are also parallel. This is called *specular* reflection, from the Latin word for mirror (Figure 17.3a). If the surface is rough, like a wall or the pages of this book, the reflected rays are not parallel to each other; they are reflected in all directions from the surface. This is called *diffuse* reflection. Diffuse reflection is what allows us to see the objects around us; light from the sun or a lamp strikes the surface of the object reflecting in all directions. The reflected rays, being reflected from the object in our direction, strike our eyes and we see the object. When we move to another location, other reflected rays strike our eyes. The same object can be seen by many people or by ourselves when

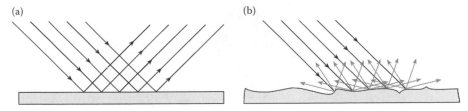

(a)   (b)

FIGURE 17.3   (a) The reflection from a smooth surface is called specular. (b) Reflection from rough surfaces is called diffuse.

we shift our position because there are reflected rays in all directions coming from the object's irregular surface (Figure 17.3b). When we look at a mirror, on the other hand, all the incident rays from a distant source are reflected in the same direction and we see the source of light not the mirror. If the rays incident on a mirror are coming from the diffuse reflection off of a rough surface, like our face, we see that surface.

REFLECTION FROM MIRRORS

When you look at yourself in a mirror, what you see is the image or likeness of your face. It looks just like you except that your right eye is still on the right of the image but the image is looking back at you and so it is the image's *left* eye. We are so used to looking at things in mirrors that we no longer pay any attention to this apparent right–left reversal. The driver we see through the rear-view mirror of our car looks fine apparently driving at the other side of the car.

Imagine that you are looking at your raised right hand in the bathroom mirror (Figure 17.4). When you think about it, it looks as if the image is saluting with the left hand. Consider one of the many rays of light that are reflected by your right thumb, one that is going to enter your eye after reflecting from the mirror. This ray appears to come from a point behind the mirror at a distance from the mirror equal to the distance from your thumb to the mirror. Now, consider another similar ray coming from your right pinky. Although this ray also enters your eye after bouncing off the mirror, it strikes the mirror at a point to the right of the ray coming from the thumb. This ray appears to come from a point behind the mirror to the right of the thumb and so it looks like a left hand instead of a right hand.

Suppose now that you are looking at the reflection of a liquid soap dispenser in the mirror (Figure 17.5). Again, the light rays appear to come from a point far behind the mirror as the dispenser is in front of the mirror. We call the point behind the mirror from where a particular ray appears to come, the *image I*, and the point on the soap dispenser where the ray actually originates, the *object O*. Labeling the distance from the object to the mirror *o*, for *object distance,* and the distance from the image to the mirror *i*, for *image distance*, we can write that for plane mirrors (to distinguish them from curved mirrors, to be considered later)

object distance (o) = image distance (i).

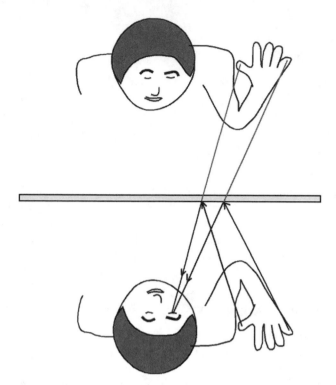

FIGURE 17.4   The image you see of your face appears reversed. Studying the ray diagrams you can see how this reversal takes place. It is in fact not reversed right to left, but front to rear.

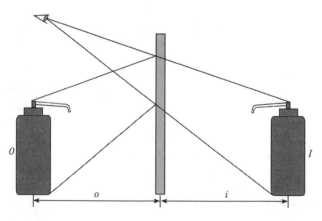

FIGURE 17.5   The distance from the object (*o*) to the mirror is the same as the distance from the image (*i*) to the mirror.

CURVED MIRRORS

Mirrors, as we know, do not have to be flat. Makeup or shaving mirrors, which are used to magnify the image of one's face, are spherical mirrors. Exterior rear-view mirrors used on the right side of some automobiles, mirrors used in some shops for security, and corner mirrors used at intersections in some small towns with narrow streets are also curved mirrors, here used to increase the viewing angle.

How do these mirrors magnify the image or increase the viewing angle? Let's look at a few light rays coming toward a curved mirror from certain specific directions. In Figure 17.6a we show a spherical mirror in which the interior surface is the reflecting surface. This type of curved mirror is called *concave*. The line joining the center of curvature C with the center of the spherical segment is called the optical axis. Consider first a ray that is parallel to the *optical axis*. This ray strikes the mirror at point P_1 and is reflected back so that the angle of reflection equals the angle of incidence. Our parallel ray is reflected back and crosses the optical axis at point F. This point, which lies midway between the center of curvature and the mirror, is called the *focal point* of the mirror. Any other ray parallel to the optical axis, like the one striking the mirror at P_2, is reflected back through the focal point.

We can then say that any ray that strikes a concave mirror parallel to the optical axis and fairly close to it is reflected through the focal point. If we now reverse the direction of the reflected and incident rays in the previous discussion, we can see that an incident ray that passes through the focal point of a concave mirror is reflected parallel to the optical axis (Figure 17.6b).

We can summarize our discussion on concave mirrors in two simple rules:

A ray that strikes a concave mirror parallel to the optical axis is reflected through the focal point.
A ray that passes through the focal point of a concave mirror is reflected parallel to the optical axis.

With these two rules we can see how a curved mirror magnifies or increases the viewing angle. We can determine the location of the image of a simple object, such as an arrow, in front of a concave mirror (Figure 17.7a). Since the tail of the arrow coincides with the optical axis, we only need to find the location of one point on the

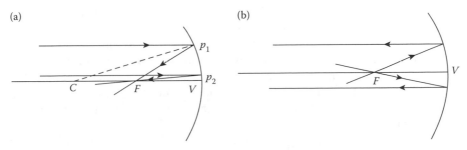

FIGURE 17.6   (a) Parallel rays are reflected by a concave mirror through the focal point. (b) A ray that passes through the focal point is reflected parallel to the optical axis.

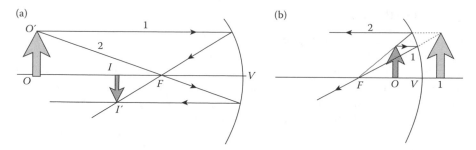

FIGURE 17.7   (a) Ray diagram to determine the image of an arrow. (b) An object placed inside the focal point of a concave mirror forms a *virtual* image behind the mirror.

object, the tip of the arrow, for instance, as we shall see shortly. We can draw several rays from this point so that they reflect from the mirror. Our task is simplified a great deal if we use simple rays that obey our two rules above. Ray 1 is parallel to the optical axis. By the first rule, ray 1 is reflected through the focal point F while ray 2 passes through the focal point and is reflected parallel to the optical axis. As we can see in Figure 17.7a, these two rays cross at one point. If we were to draw other rays from the tip of the arrow, they would also cross at this same point. When we look into the mirror, the rays of light appear to come from this last point and so this point is the image of the tip of the arrow. Notice that this point lies *below* the optical axis. We could select other points and repeat the procedure until we reconstruct most of the arrow. If we do this, we would notice that the images of these points lie *above I'*, closer to the optical axis. Finally, the image of the tail of the arrow would lie right on the axis. The image of the arrow lies below the optical axis and is inverted. Notice also that in this case the image is also smaller than the object. Since our object has a simple symmetry, there is no need to select other points; knowing the location of the image of the tip of the arrow allows us to reconstruct the entire image.

We might ask why the image of the arrow in Figure 17.7a ended up upside down and smaller, whereas the image of our face in a makeup or shaving mirror is right side up and larger. Next time you pick up one of these mirrors, place it at arm's length rather than up close. Now, your image and the images of all the other objects in the room and the walls of the room itself are upside down and in front of the mirror, not behind it. It is only up close that your face appears larger and right side up. Why? What is so special about your face? Aside from the fact that your face is really special, it is the location of the object in relation to the mirror that determines the size and appearance of the image. If the object lies between the mirror and the focal point, the image will be right side up, behind the mirror, and larger. This is why your face looks larger in a shaving mirror. They are designed so that the focal point lies slightly beyond where you would normally place the mirror to look at yourself in it. The rest of the objects in the room lie beyond the focal point and thus have smaller, inverted images.

The ray diagram in Figure 17.7b shows the image of an object placed inside the focal length of a concave mirror. Notice that this image is not only larger and right side up, as we have just discussed, but also forms behind the mirror. Images like

(a)   (b)

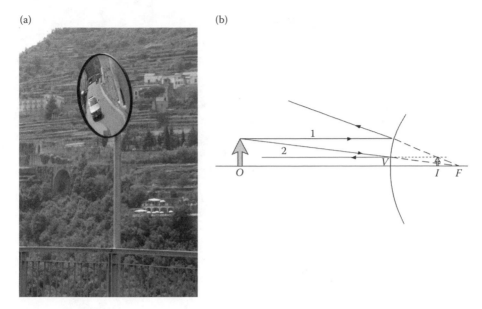

FIGURE 17.8   (a) Convex mirror positioned by a roadside at a blind corner. (b) Ray diagram for a convex mirror.

this, in which light rays do not actually pass through the image, are called *virtual images*. The name suggests that there might be other kinds of images, perhaps not as "ghostly." Indeed there are; the images formed by objects placed beyond the focal point of a concave mirror, as described above, are *real images* because light rays actually pass through them. Real images can be detected by a photographic film, for example, or formed on a projection screen placed at the location of the image. Virtual images cannot. Images formed by plane (flat) mirrors are also virtual.

A curved mirror with the reflecting surface on the outside is called a *convex* mirror (Figure 17.8). The two rules outlined above for concave mirrors also work for convex mirrors. Notice that the focal point is behind the mirror. Convex mirrors are common in tight, narrow intersections because they produce a larger field of view. For this reason, they are use widely as automobile rear-view mirrors, particularly on the right side of the vehicle.

REFRACTION OF LIGHT

As we all have noticed at one time or another, a pool of water appears to be shallower than it really is and a spoon seems to bend or even break when we use it to stir a glass of lemonade (Figure 17.9a). The reason for these seemingly strange phenomena is the fact that light travels at different speeds through different transparent media because the electromagnetic waves interact with the matter of the medium in which they travel. The speed of light in vacuum is about 300,000 km/s but decreases to about 225,000 km/s in water and 200,000 km/s in glass. The speed of light in air is about 299,900 km/s, nearly the same as in vacuum. The oscillating electric field of the

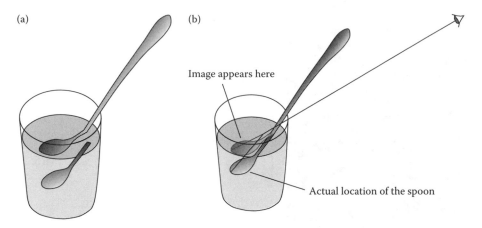

FIGURE 17.9 (a) A spoon in a glass of water seems to bend when viewed from above. When viewed from the side it appears broken. (b) The rays of light bend as light crosses the boundary between two transparent media.

electromagnetic wave sets the electrons of the atoms and molecules that make up the medium through which light is propagating into vibration. The vibrating electrons emit electromagnetic radiation of the same frequency. It is this constant absorption and re-emission of electromagnetic energy that gives the appearance that light has slowed down, although it is still traveling between interactions with the electrons at 300,000 km/s as it does in vacuum, where there are no electrons to interact with.

A ray of light from a point on a spoon under the water travels along the water in a straight line at 225,000 km/s until it encounters the boundary between water and air (Figure 17.9b). In air, it also travels in a straight line but at a greater speed, resulting in the bending of the light ray at the boundary, which our brains interpret as a bent spoon rather than a bent ray of light. We can analyze the phenomenon, as we are doing here, and understand it, but our eyes still see a bent spoon. This bending of light rays due to the different speeds of light in different media is called *refraction*. As is the case with reflection, the ray incident upon the boundary between the two media is called *incident ray* and the ray entering the new medium is called the *refracted ray*. The angle made by the incident ray and the normal to the boundary is called the angle of incidence and the angle formed by the refracted ray and the normal, the angle of refraction.

How do these changes in the speed of light result in the bending of light rays? An analogy could help us understand this phenomenon. Imagine soldiers marching in formation on a grassy field. When the soldiers come to an adjacent muddy field, they are slowed down by the mud even if they keep in step, as they have to take shorter steps. If they enter the muddy field at an angle, as shown in Figure 17.10a, the columns closer to the edge of the grassy field enter the mud first so that as each row comes to the boundary some soldiers in that row will be in the mud and the rest in the grass. The soldiers marching in the mud get slightly behind so that the next row, still on the grass, gets closer to them and the formation swivels. When

(a)   (b)

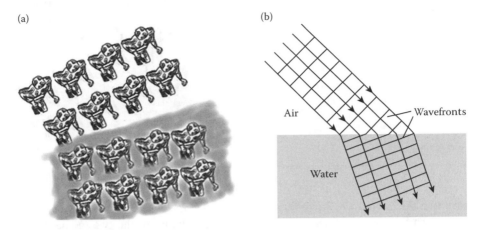

FIGURE 17.10 (a) Soldiers marching in formation slow down and change direction when they enter a muddy field. (b) Light rays slow down as they pass from air to water and this causes the rays to bend or refract.

the soldiers are all in the muddy field, they will find themselves marching in a new direction, one closer to the normal to the boundary and the rows of soldiers will be closer to each other. For light passing from air into water, we can replace the columns of soldiers by lines drawn one wavelength apart, called wave fronts (Figure 17.10b). As with the soldiers, the part of each wave front that has entered the water is slowed down in relation to the part that is still in the air so that, as light enters the water, the wave fronts are closer together and move in a direction closer to the normal. Since the distance between two consecutive wave fronts is one wavelength, we can see that the wavelength in water is shorter. The frequency, however, remains the same (in our analogy, the soldiers keep the same step). Remembering that the speed of a wave is related to its wavelength and its frequency through the expression

$$v = \lambda f$$

we can see that if the speed decreases, the wavelength also decreases in the same proportion, since the frequency remains constant. (The atoms absorb and re-emit the light at the same frequency.)

The path of a light ray through a refracting surface is reversible; that is, if the direction of a light ray is reversed, it will retrace the original path. If we reverse the direction of the rays in Figure 17.10b, the incident ray becomes the ray in the water and the refracted ray is the ray in air. In this case, the angle of refraction is greater than the angle of incidence. This is the case of the spoon in the glass of water (Figure 17.9).

The refraction of light was known to Ptolemy, who wrote a book on optics in which he discussed the relationship between the incident and refracted rays. It was, however, the Dutch scientist Willebrord van Roijen Snell (1580–1626) who first discovered the correct relationship between the angle of incidence and the angle of refraction and for this reason this relationship is known as *Snell's law*. Snell's law

TABLE 17.1

Indices of Refraction

Substance	n	Substance	n
Vacuum	1.00000	Quartz	1.54
Air	1.00029	Crown glass	1.59
Ice	1.31	Flint glass	1.75
Water	1.333	Zirconium	1.96
Acrylic plastic	1.51	Diamond	2.42

says that *the angle of refraction is in a constant relationship to the angle of incidence.* The constant in Snell's law depends on the nature of the two media. If the incident ray is in vacuum or in air (where the speed of light is almost the same as in vacuum), the constant gives the index of refraction n of the substance: that is, the ratio of the speed of light in vacuum to that in the substance, or

$$n = \frac{c}{v}$$

where c is the speed of light in vacuum, equal to 300,000 km/s, and v is the speed of light in the particular medium. The speed of light in water, for example, is 225,000 km/s. The index of refraction of water is then $n = (300{,}000 \text{ km/s}) = (225{,}000 \text{ km/s}) = 1.333$. Table 17.1 gives some values of indexes of refraction for different substances.

THE FRONTIERS OF PHYSICS: PHOTOACOUSTIC TOMOGRAPHY

Photoacoustic tomography (PAT), a combination of acoustic and optical imaging, produces high resolution images that look deeper into the body. When laser light pulses shine on the body, certain molecules that make part of the body can be used as contrast agents that absorb the light, convert it into heat, and expand. The expansion generates a wideband ultrasonic wave that can be decoded into an image.

PAT is used to monitor tumors, map blood oxygenation, and detect breast and skin cancers. It is also used to image brain function *in vivo.*

LENSES

We are all familiar with the use of lenses as magnifiers in telescopes, microscopes, binoculars, and cameras. In a lens, light is focused by refraction. A lens is a piece of transparent material in which the two refracting surfaces are curved. In most cases, the surfaces are segments of spheres. When a light ray enters a transparent medium through a curved surface, the direction of the refracted ray depends on the

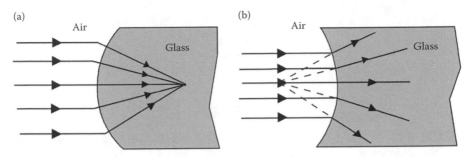

FIGURE 17.11 (a) Parallel rays incident on a convex glass surface converge at one point. (b) Parallel rays incident on a concave glass surface diverge as if they had originated at a single point.

orientation of the surface. Parallel rays incident on a spherical surface are refracted so that they converge to one point, as seen in Figure 17.11a. This type of surface is called *convex*. Parallel rays refracted by a concave surface are refracted so that they diverge and appear as if they had originated from a single point (Figure 17.11b).

An incident ray falling on a lens with two convex surfaces, so that it crosses the first surface at some point other than the center or the edge of the lens, is refracted toward the normal to the surface at that point. The ray falls on the second spherical surface, passing from glass to air. Since glass has an index of refraction greater than that of air, the ray is refracted at this second surface away from the normal (Figure 17.12). Two rays emanating from the same point on an object placed near a lens fall on the lens at different points and are refracted at slightly different angles. These rays will converge to one point on the other side of the lens. Lenses with spherical surfaces will not, in general, bring other rays emanating from the same point on the object to exactly the same point and the images formed by these lenses are not sharp, a phenomenon known as *spherical aberration* that designers of optical instruments must avoid. However, if the rays passing through the lens do not make angles greater than 10° with the *lens axis*, which is the central line perpendicular to the lens, sharp images can be obtained with spherical surfaces. Our discussion will be limited to these kinds of rays. Our discussion will also be limited to *thin lenses*, in which the thickness of the lens is very small compared with the distances to the objects and images from the lens.

All rays parallel to the lens axis are refracted so that they converge to one point on the axis (Figure 17.13). This point is called the *focal point* of the lens and the distance from the center of the lens to the focal point is known as the *focal length*. Camera lenses are identified by their focal lengths. A 50 mm lens is one with a focal length of 50 mm and is usually the normal lens for a full frame camera, one with a 36 mm by 24 mm image sensor (the same size as the older 35 mm film cameras). A 28 mm lens or a 35 mm lens is considered a wide angle lens, and a 135 mm a medium long focus. The focal length of a particular lens depends on the index of refraction of the lens material and on the curvatures of the two surfaces. As most of us have realized at one time or another, it does not matter which side of a simple magnifier we use to

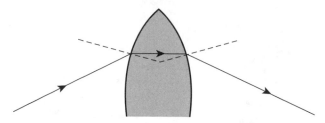

FIGURE 17.12 A ray falling on a convex surface of a lens is refracted toward the normal. At the second surface, the ray is refracted away from the normal.

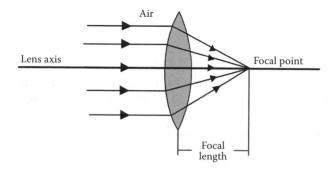

FIGURE 17.13 All rays parallel to the lens axis converge to the focal point of the lens.

look at an object; the image we see is magnified the same. We conclude that there must be a focal point on both sides of a lens and, since the image formed is the same regardless of which side faces the object, the two focal points must be equidistant from the lens, *even if the curvatures of the two sides are different.*

From our previous discussion, we can deduce simple rules that will aid us in determining the location of the image formed by a lens.

A ray that strikes a lens parallel to the lens axis is refracted to pass through the focal point.
A ray that passes through the focal point is refracted to emerge parallel to the lens axis.
A ray that strikes the lens at the center is undeviating.

A lens that converges rays to a single point is called a *converging lens*. A lens that diverges rays so that they appear to come from a single point on the side of the incident rays is called a *diverging lens* (Figure 17.14). Our three rules apply to both types of lenses. Optometrists refer to converging lenses as *positive lenses* and to diverging lenses as *negative lenses.*

Figure 17.15 shows the formation of the image of an arrow placed on the axis of a converging lens. Again, as with the mirror, we select several of the rays coming from the tip of the arrow. We are interested in particular in the three rays outlined above: ray 1, parallel to the axis; ray 2, through the front focus; and ray 3, passing through

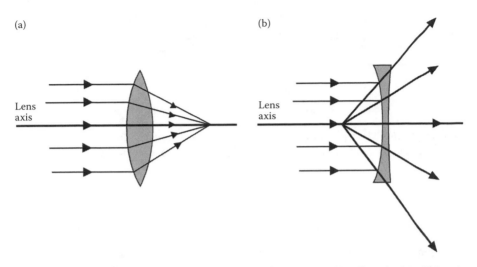

FIGURE 17.14 (a) A converging lens bends rays to a single point, while a diverging lens (b) bends rays such that they appear to diverge from a single point on the side of the incident rays.

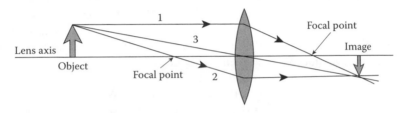

FIGURE 17.15 Ray diagram used in the location of the image formed by a converging lens.

the lens at the center (called the principal ray). These three rays converge to a point after being refracted by the lens and then diverge, so that they appear to a viewer to have originated at that point.

Lenses are used in telescopes, binoculars, and other optical instruments to magnify images. The size of the image formed by a particular lens may be larger or smaller than and even equal to the size of the object, depending on the location of the object in relation to the lens. The magnification M is the *ratio of the size of the image to the size of the object*. The magnification can be found by comparing the triangle formed by the object, the lens axis, and the undeflected ray passing through the center of the lens, with the triangle formed by this ray, the image, and the lens axis (Figure 17.16). These two triangles are similar triangles and we can write a relationship between the sides forming the right angles to give us the ratio of

$$M = \frac{\text{image size}}{\text{object size}} = \frac{\text{image distance}}{\text{object distance}}.$$

An image that is five times the distance from the lens as the object will be five times as large as the object.

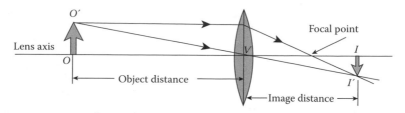

FIGURE 17.16   The triangles $O'OV$ and $I'IV$ are similar triangles. The ratio $I'I/O'O$ must equal the ratio OV/IV. Thus, image size/object size = image distance/object distance.

THE FRONTIERS OF PHYSICS: GRADIENT-INDEX LENSES

A lens with a spherical surface does not bring other rays from the same point on the object into exactly the same point, as we saw earlier in the chapter. This "spherical aberration" produces fuzzy images. Lens designers usually correct for this and other optical aberrations by grinding the lens into an aspherical shape or by adding more lenses to their design. Both solutions have their own problems. Adding more lenses increases the cost, weight, and size of the design. It also makes glare more difficult to control. Aspherical lenses are small and light but they are much more expensive to produce.

The light rays passing through a spherical lens can be brought into focus at exactly the same point regardless of where they cross the lens if the index of refraction is variable. This design would more closely resemble the way light crosses the lens in human and insect eyes. Although lens designers have attempted for many years to produce such a lens, they found themselves faced with calculations that required sophisticated computer programs.

Today's fast supercomputers have made the task possible. Several companies are now producing gradient-index lenses (GRINs) in which the index of refraction changes radially, from center to edge, or axially, from front to back. A GRIN can focus light with an optically flat surface or a simple spherical shape.

Computers can make the calculations feasible, but the process to produce the actual lenses is still complicated. A common technique is the ion-exchange process in which an ion of a heavy element in the glass is replaced by an ion of a lighter element. The areas where the lighter element is located have a lower index of refraction. With this process, it can take more than a month to have enough ions replaced in the glass. Other companies are working on similar but faster processes.

GRIN lenses with a diameter of one millimeter are currently being used in photocopiers, facsimiles, and laser printers. These lenses, shaped like tiny rods and with focal lengths of 15–30 millimeters, are arrayed in bundles of several hundred and are used as part of the scanning element in these devices.

Scientists in France are developing GRIN lenses by creating links between the components of hydrophilic polymers. These links are produced by shining light on the components. The longer the light shines, the greater the number of links and the higher the density of the polymer. Since the index of refraction of the polymer is determined by its density, the researchers can create a gradient index lens made of this polymer, which is used to make contact lenses.

TOTAL INTERNAL REFLECTION

When light passes from a medium with a high index of refraction, such as water, to one with a low index of refraction, like air, the angle of refraction is larger than the angle of incidence. Consider the rays coming from a point on a submerged object (Figure 17.17). A person outside the water would be able to see the rays from the four first locations shown. However, as the angle of incidence is increased, the rays are refracted more and more until, at a certain *critical angle*, the refracted ray merely skims along the surface. For larger angles of incidence, the ray cannot escape and the light is completely reflected, remaining in the water. This phenomenon is called *total internal reflection*.

A prism can reflect light through a right angle or even back, parallel to its initial direction (Figure 17.18). An ordinary mirror can also reflect light back to its original direction or at right angles to the initial ray. A mirror, however, even a highly polished one, does not reflect all of the light incident upon it. A newly silvered mirror might reflect up to 99% of incident light but, unless it is given a special protective coating, after a few days its reflectivity drops to about 93% due to oxidation. Total internal reflection such as with a prism is, as it says, *total*; no loss of light occurs.

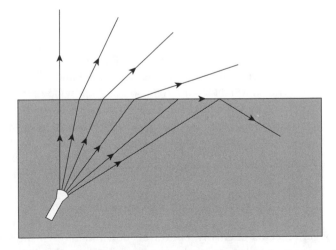

FIGURE 17.17 As the angle of incidence increases, the angle of refraction also increases until, at the critical angle, the refracted ray just skims the surface. At greater angles, the light remains inside the water. This is called total internal reflection.

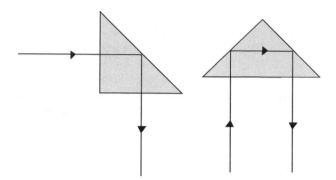

FIGURE 17.18   A prism reflects light at right angles to the original ray and can even return it parallel to the original direction.

For this reason, designers of optical equipment such as binoculars use prisms where possible, rather than mirrors.

FIBER OPTICS

The phenomenon of total internal reflection, with no loss of light, is the basis for the important area called *fiber optics*. In a thin transparent fiber, light is "piped" through a series of total internal reflections. If the fiber is thin enough, the angle of incidence is always greater than the critical angle and the light is transmitted even if the fiber is bent in any direction as long as sharp corners or kinks are avoided (Figure 17.19). To prevent damage and light losses from abrasion, the fiber is clad with transparent material having a low refractive index.

Optical fibers can be used to transmit images of objects that are otherwise inaccessible, like the interior of the human body. However, the light rays from different parts of an object would get scrambled by the multiple reflections taking place in a large-diameter fiber. For this reason, individual fine fibers, each transmitting information from a very small part of an object, are collected into a bundle with a lens at the end to form an image on the fiber bundle. The fiber bundle is surrounded by a jacket of fibers that transmit light from a light source to provide illumination of the object. At the receiving end, the image can be viewed directly by means of a lens system or displayed on a television monitor. A colon fiberscope is an example of an instrument that uses fiber optics to examine the colon without the need for surgery. Medical instruments that use fiber optics allow physicians to examine knee joints, the bladder, the vocal folds, the stomach, a fetus in the uterus, and even heart valves and major blood vessels.

Fibers that work by internal reflection are called *multimode* fibers. If the diameter of the fiber is reduced to a few micrometers, the mode of propagation changes. The light is no longer reflected internally, but travels down the fiber in what is in effect a single ray. This type of fiber is called a *single-mode* fiber. By using infrared laser light (which has a single frequency) and special glass, the light can travel 100 km or more before a repeater amplifier is needed, and because of its very high frequency,

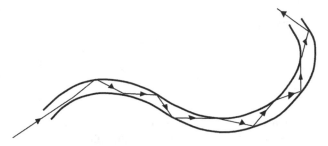

FIGURE 17.19   A light beam can be transmitted through a transparent fiber by total internal reflection, even if the fiber is bent.

a single fiber can carry thousands of telephone conversations, several television signals, and computer data, simultaneously.

OPTICAL INSTRUMENTS

Optical instruments, from a simple magnifier to the space telescope, enhance our sense of vision and allow us to study nature from a different perspective. Since the sixteenth century when the telescope and the microscope were invented (reportedly in the Netherlands), we have been able to look farther into the universe in both directions: into the very small world of the particles that are the stuff of matter, and out toward the far reaches of the universe, in the realm of the galaxies. In this last section of this chapter we will look at some examples of optical instruments.

THE CAMERA

Although modern cameras can be fairly complicated instruments, the basic optical principles behind their construction are fairly simple.

A *pinhole camera* consists of a closed box with a single pinhole through which light can enter, and uses no lenses or mirrors. Light rays from an object in front of the pinhole enter the box, producing an inverted image on the wall opposite the hole (Figure 17.20). Although it is possible to produce good photographs with this simple device, there is no provision to control the amount of light that strikes the film other than by enlarging the hole, and this produces a blurred image because every point on the object forms an image that is a disk the size of the pinhole. To let enough light through the hole so that the film is fully exposed, long exposure times are required. A pinhole camera is unsuitable for moving objects.

The simplest concept of a modern camera uses a single converging lens in front of a larger hole; this allows a great deal more light into the box (Figure 17.21). Light rays from a distant object enter the camera parallel to the lens axis and are focused at the focal point of the lens. Rays from a nearby object are focused farther back. Moving the lens toward or away from the image sensor plane allows for focusing of objects at different distances from the camera. The amount of light energy that the image sensor receives can be controlled by opening or closing the aperture of the hole with an *iris diaphragm*, a ring of metal leaves, and by changing the length of time light is allowed

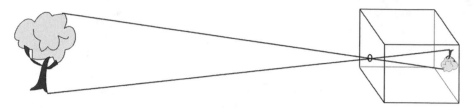

FIGURE 17.20 A pinhole camera produces an image on the wall opposite the hole by restricting the rays from each point on the object into the light-tight box.

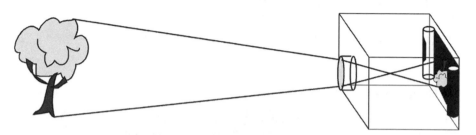

FIGURE 17.21 A simple camera consists of a converging lens that can be moved closer to or away from the image sensor plane, a variable aperture and a shutter.

to enter the camera with a *shutter*, a mechanical device which may be a curtain in front of the image sensor (a focal-plane shutter) or a set of hinged leaves in between the lens element (an intra-lens shutter). Exposure durations range from around 1/8000 s to several minutes.

Camera manufacturers calibrate the diaphragm in units called *f-stops*, defined as the ratio of the focal length of the lens (*f*),to the diameter of the aperture (*D*) allowed by the diaphragm, or

$$f\text{-}stop = \frac{f}{D}.$$

Since the amount of light reaching the lens is proportional to the area of the aperture and thus to the *square* of the diameter *D*, changing the exposure by a factor of 2 corresponds to an increase by a factor of √2 in the diameter. Camera diaphragms have scales marked with *f*-stops related by a factor of √2, i.e.

$$\frac{f}{1.4}, \frac{f}{2}, \frac{f}{2.8}, \frac{f}{4}, \frac{f}{5.6}, \frac{f}{8}, \frac{f}{11}, \frac{f}{16}, \frac{f}{22}.$$

The diameter of an *f*/1.4 aperture, for example, is twice the diameter of an *f*/2.8 aperture and therefore lets in four times as much light. If a given lens has a maximum aperture of *f*/2.8, you could achieve the same results as with an *f*/1.4 aperture by increasing the exposure time by a factor of 4, so that if the camera meter calls for 1/500 second at *f*/1.4 the exposure time for *f*/2.8 should be four times 1/500 s

FIGURE 17.22   Galileo demonstrates his telescope before the Venetian Senate in 1609. (From Archiri Alinari/Art Resource, New York. With permission.)

or 1/125 s. This relationship is called the law of reciprocity. Stated formally, it says that

$$exposure = intensity \times duration.$$

The Telescope

The telescope was invented at the turn of the seventeenth century, probably by a lens maker in Holland. In 1609, the Dutch patent office turned down a patent application for a telescope from Hans Lippershey, a Dutch lens maker, because such an instrument was already common knowledge. Word of that invention had reached Galileo that same year. Within six months, Galileo had constructed a telescope of his own with a magnifying power of ×32, which he immediately turned to the heavens (Figure 17.22). He discovered that the moon had mountains, and that the sun had spots. Studying the evolution of the sunspots, Galileo was able to show that the sun rotated on its axis, completing one rotation in 27 days. Galileo also discovered that the planet Jupiter had four large moons and worked out their periods of revolution around the planet. These moons, Io, Europa, Ganymede, and Callisto, are known today as the Galilean satellites of Jupiter. With his telescopes, Galileo also discovered that Venus had phases like the moon, evidence that Venus revolved around the sun, shining by reflected sunlight. Galileo published his discoveries in *Sidereus Nuncius* ("The Starry Messenger").

A simple telescope such as Galileo's original telescope has two lenses, the *eyepiece* and the *objective*. In this type of telescope, known as a Galilean telescope, the eyepiece is a diverging lens and the objective is a converging lens (Figure 17.23a). A modification of this design, introduced by Johannes Kepler shortly after he obtained one of Galileo's own designs, uses a converging lens as eyepiece (Figure 17.23b). These telescopes, in which lenses are used to form an image, are known as *refractors*.

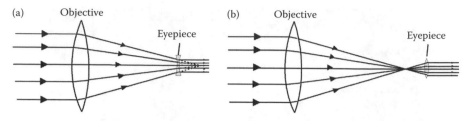

FIGURE 17.23   (a) Galilean telescope. Parallel rays from the object are focused at the focal point of the objective. These rays are intercepted by the eyepiece where they are refracted parallel. (b) In a Keplerian telescope, the eyepiece is a converging lens and the image is inverted.

Light rays from a distant object reach the objective essentially parallel to each other. These rays are focused at the focal point of the objective and then diverge, appearing to a viewer to originate there. These diverging rays are intercepted by the eyepiece. The eyepiece is placed so that its focal point coincides with the focal point of the objective; the rays diverging from this point are refracted by the eyepiece parallel to each other. Since eyepieces are small lenses with diameters of about 1 cm or less, the viewer looking into the eyepiece sees parallel rays coming from the image. In the Galilean telescope, the diverging lens used as eyepiece is placed between the objective and its focal point (Figure 17.23a), whereas in Kepler's design, the converging lens is placed at the other side (Figure 17.23b). Kepler's design produces an inverted image.

In 1663, James Gregory, a Scottish mathematician and astronomer, published the design of a telescope with a concave mirror as objective. His attempt to have it built ended in failure mainly because of the difficulty in making the mirror. In 1668, Isaac Newton designed and built a similar telescope with a concave mirror 2.5 cm in diameter as objective (Figure 17.24). Although the refracting telescope is what most people have in mind when they hear the word, all of the large research telescopes in the world are *reflectors*, as these telescopes are called. Light rays from the distant object arrive at the mirror parallel to each other, where they are reflected back and focused at the focal point of the mirror. As with refractors, the eyepiece is placed so that its focal point coincides with the focal point of the objective (Figure 17.25a). Notice that the eyepiece blocks some of the incoming light. For very large telescopes, like the 200-inch Hale telescope at Mount Palomar (Figure 17.26), this blocking does not present much of a problem. This arrangement of eyepiece and objective is called the *prime focus*. For smaller reflectors, several other arrangements have been introduced since its invention. Newton himself used the arrangement shown in Figure 17.25b, which is properly called Newtonian. Light rays reflected from the mirror are intercepted by a small plane mirror placed diagonally in front of the objective. The light rays are diverted to the eyepiece on the side. This arrangement is called the *Newtonian focus*. A popular design both for research telescopes and amateur use is the *Cassegrain* system, where the light rays are intercepted by a small convex mirror that sends the rays back to the eyepiece through a small circular hole made in the objective (Figure 17.25c). This increases the effective focal length of the system.

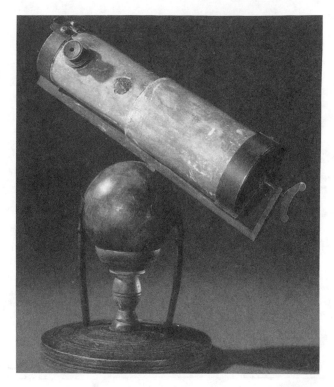

FIGURE 17.24   Newton's first telescope. (With permission from the President and Council of the Royal Society.)

THE HUMAN EYE

The human eye is a marvelous optical instrument, the result of a remarkable evolutionary process. The eye is sensitive to wavelengths from 350 nm to 750 nm, which correspond to the most intense wavelengths in sunlight. This is clearly not a coincidence but the response of evolution to the existing conditions. Had the range of intensities of sunlight been different, the eye would no doubt have evolved to be sensitive to a different range of wavelengths.

The eye is a nearly spherical gelatinous mass about 3 centimeters in diameter, slightly flattened from front to back, and surrounded by a tough membrane, the *sclera*, where the muscles that control the movements of the eye are attached. Figure 17.27 is a simplified diagram of the human eye. Light enters the eye through the *cornea*, a transparent membrane that acts as a converging lens with a focal length of about 2.4 cm. Light rays pass through the *aqueous humor*, a transparent liquid that fills the space between the cornea and the *lens*, another converging lens with a variable focal length that changes from about 4 cm to about 5 cm. The lens, a jellylike substance hard at the center and softer toward the edges, contracts and bulges under the action of the *ciliary muscles* that surround it and keep it under tension. When the ciliary muscles relax, the lens assumes a thicker shape, decreasing its radius of curvature which in turn decreases the focal length. Nearby objects are then brought into focus

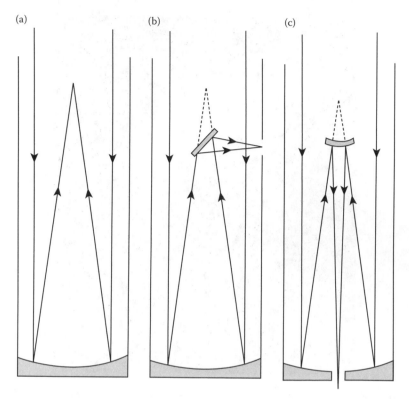

FIGURE 17.25   Reflector telescope arrangements: (a) Prime focus, (b) Newtonian focus, (c) Cassegrain focus.

on the retina. When the eye looks at a distant object, the ciliary muscles tension and the lens takes on a flatter shape, increasing the radius of curvature and consequently the focal length, and bringing the distant object to a focus on the retina. This process is called *accommodation.*

On their way to the lens, light rays pass through the *pupil.* The size of the pupil is controlled by the *iris,* a muscular ring behind the cornea that acts like the iris diaphragm in a camera lens, adjusting the amount of light that enters the eye. The iris is the colored part of the eye.

The inner chamber behind the lens is filled with a thin gelatinous substance called the *vitreous body.* This inner surface of the eyeball is covered with the millions of photoreceptor or light-sensitive cells that form the retina. There are two kinds of photoreceptor cells, *rods* and *cones.* The cones are the color-sensing cells of the retina and the rods, although not sensitive to color, are about 1000 times more sensitive to light and are therefore important for low-light vision. There are three kinds of cone cells, each responding differently to light from a colored object due to the presence of light-absorbing proteins sensitive to wavelengths in the red, green, or blue part of the spectrum.

Although the eye is capable of detecting light rays with incident angles up to 210°, vision is sharpest in a small central region about 0.25 mm in diameter called the *fovea,* which contains only cones, with an average separation of about 1 mm.

(a)   (b)

FIGURE 17.26   (a) Detail of the 200-inch Hale telescope at Mount Palomar. (b) Prime focus of the Hale. (Courtesy of Palomar Observatory.)

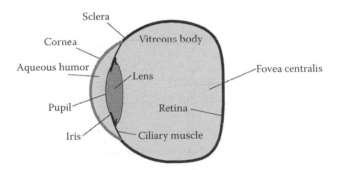

FIGURE 17.27   The human eye.

As we read a book or follow a moving object, the eye muscles move the eyeball so that light rays from the object fall on the fovea. We can see the rest of the object and our surroundings in front of us, but only with somewhat reduced resolution.

A normal eye can bring into focus objects at distances ranging from a few centimeters to infinity. The closest the eye can focus is called the *near point*, and this distance depends on the elasticity of the lens. This elasticity diminishes with age. Normally, at age 10 the near point is about 7 cm and this increases to about 10 cm at age 20, 15 cm at age 30, 25 cm at age 40, 50 cm at age 50 and over 2 m at age 60. By the late 40s or early 50s, a person's arms are usually shorter than his near point and a visit to the optician is recommended.

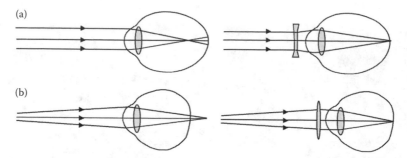

FIGURE 17.28 (a) Nearsightedness can be corrected with a diverging lens. (b) Farsightedness is corrected with a converging lens.

People who are *nearsighted* are unable to see distant objects clearly. This condition, called *myopia*, is due either to an eye that is longer than normal or to a cornea that is curved too much. Rays from a distant object are focused in front of the retina. Nearby objects are focused on the retina. Nearsightedness is corrected by a diverging lens (Figure 17.28a).

In the farsighted person, the eye is too short or the cornea insufficiently curved. People with this condition, called hyperopia, can see distant objects clearly but nearby objects appear blurred because the rays are focused behind the retina. This defect can be corrected with a converging lens (Figure 17.28b).

THE FRONTIERS OF PHYSICS: ARTIFICIAL VISION

Researchers from several universities supported by the U.S. Department of Energy Artificial Retina Project have been working on the development of an implantable artificial retina component chip that might restore sight to people suffering from retinal disorders. The chip, which is only 2 millimeters square and less than 0.02 millimeter thick, is to be implanted on the retinal surface inside the eye cavity.

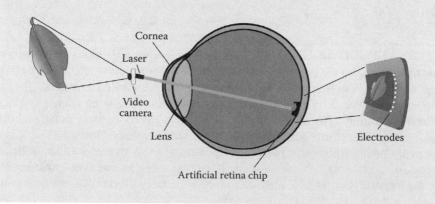

The rods and cones in the eyes of people suffering from retinosis pigmentosa or macular degeneration are defective, but the ganglion cells that line the retina are relatively intact. The new retinal chip generates electric currents that stimulate these cells which then transmit the signal to the brain, in theory enabling the patient to see. The chip is powered by an external laser, which is in turn powered by a tiny battery pack. The laser transmits the images captured by a miniature video camera to the photosensors in the chip. The laser-battery-pack video camera assembly can be mounted on regular eyeglasses.

In a breakthrough operation in 2002, a team led by Mark Humayun of the University of Southern California successfully implanted the first artificial retina into the eye of a patient who had been blind for over fifty years. In subsequent years, other patients have received artificial retinas.

18 The Nature of Light

THE WAVE NATURE OF LIGHT

As we learned in Chapter 7, light shows both particle and wave behavior. We have arrived at this conclusion through the attempts of twentieth-century physicists to understand the nature of matter. Some of the Greek philosophers believed that light consisted of particles that traveled in straight lines at high speeds and stimulated the sense of vision as they entered the eye. At the end of the fifteenth century, Leonardo da Vinci speculated that light was a wave because of the similarity between the reflection of light and the echo of a sound. In the seventeenth century, the Dutch physicist Christiaan Huygens also felt that light was a wave. However, his contemporary Isaac Newton thought that light was composed of particles. Newton's *corpuscular theory* prevailed for about two hundred years, although by the eighteenth century many optical effects had been explained in terms of the properties of waves. In the early nineteenth century, Thomas Young in England and Augustin Jean Fresnel in France demonstrated in a series of landmark experiments that light showed interference and diffraction, phenomena characteristic of waves. In Scotland, Maxwell theorized that light was the propagation of oscillating electric and magnetic fields through space: an electromagnetic wave.

Twentieth-century physicists have shown that both views are valid; that light has characteristics of particles and of waves. In this chapter, we will study the fundamentals of the wave theory of light.

THE SPEED OF LIGHT

An electric charge that is changing gives rise to a changing magnetic field, as we learned in Chapter 14. This is Maxwell's fourth equation. Faraday's Law, which became Maxwell's third equation, tells us that this changing magnetic field in turn creates a changing electric field. These changing magnetic and electric fields are inseparable: the changing magnetic field sustains the electric field, and the changing electric field sustains the magnetic field. Maxwell realized that these changing fields would sustain one another even in regions where there are no electric charges to accelerate, as in free space. These mutually sustaining fields, he predicted, propagate through space as an electromagnetic wave.

Maxwell proceeded to calculate the speed at which these electromagnetic waves travel through space. He showed that this speed was equal to the ratio of the electric and magnetic field strengths at any point in space:

$$c = \frac{E}{B}.$$

From Ampère's law, he was able to show that this speed is equal to the square root of the ratio of the electric and magnetic force constants:

$$c = \sqrt{\frac{k_E}{k_M}}.$$

When Maxwell substituted the then known values of these two constants he found that the speed of propagation of electromagnetic waves was the same as the experimentally determined values of the speed of light. He wrote,

> The velocity of transverse undulations in our hypothetical medium, calculated from the electromagnetic experiments of M. M. Kohlrausch and Weber, agrees so exactly with the velocity of light calculated from the optical experiments of M. Fizeau, that we can scarcely avoid the inference that light consists in the transverse undulations of the same medium…

Inserting up-to-date values of k_E and k_M into Maxwell's expression for the speed of electromagnetic waves, we find

$$c = \sqrt{\frac{9 \times 10^9 \, \mathrm{Nm^2 / C^2}}{1 \times 10^{-7} \, \mathrm{Tm / A}}} = 3 \times 10^8 \, \mathrm{m/s}.$$

This is the speed of light in empty space, one of the fundamental constants in nature. With the values known at the time for the two constants, Maxwell obtained 2.88×10^8 m/s.

In *Two New Sciences*, Galileo describes an experiment by which two persons on distant hills flashing lanterns can measure the speed of light. Although he concluded that the speed of light, contrary to what we might think from everyday experiences, is not infinite, he was unable to obtain a value for it. His contemporary Descartes asserted that the speed of light must be infinite. The Danish astronomer Olaus Römer experimentally showed that the speed of light is finite in 1676 by carefully measuring the times at which the Galilean satellites of Jupiter emerged from the shadow of the planet. The French astronomer Cassini had accurately measured the times of revolution of Jupiter's four known satellites so that the precise moment when a satellite was to be eclipsed by Jupiter could be calculated. Before the Academy of Sciences in Paris, Römer announced that one of Jupiter's satellites, which he had calculated should be eclipsed by Jupiter on November 9, 1676, at 5:25:45, was going to be exactly ten minutes late. Careful measurements by the skeptical astronomers of the Royal Observatory confirmed that the eclipse of the satellite had occurred at 5:35:45, in accordance with Römer's prediction. A few weeks later Römer explained to the Academy that when the earth was closer to Jupiter in its orbit, the satellites went behind Jupiter earlier than when the earth was farther away from Jupiter. He correctly deduced that the reason for this discrepancy was that the speed of light was finite. When the earth was farthest away from Jupiter, the eclipses were delayed because it took light 22 minutes to cross the earth's orbit (Figure 18.1). Römer's friend Christiaan Huygens used his data and his estimate of the diameter of the earth's orbit to obtain the first calculation of the speed of light. He obtained a value

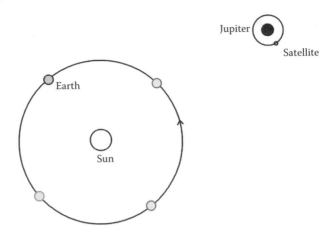

FIGURE 18.1 Römer's method to determine the speed of light. The Galilean satellites of Jupiter take longer to move behind the planet when the earth is moving away from Jupiter because the light reflected by the satellite takes longer to reach the earth.

(in modern units) of 227,000 km/s, about 24% lower than modern values. The main reason for the error was that Römer's measurement of 22 minutes for light to cross the earth's orbit is too large; it is 16 minutes approximately.

In 1729, the English astronomer James Bradley, during an attempt to detect stellar parallax, discovered the phenomenon of *stellar aberration*, which allowed him to obtain a second method of estimating the speed of light. At the time of publication of Newton's *Principia*, the motion of the earth was generally accepted but had not been experimentally measured. For about 150 years after the *Principia*, a number of scientists tried to detect stellar parallaxes with the aid of telescopes. Stellar parallax, as we might recall from Chapter 6, is the apparent shift in the position of a star due to the revolution of the earth around the sun. With this purpose in mind, Bradley mounted a vertical telescope in his chimney and measured the positions of the stars at different times of the year. Bradley did detect a tiny displacement through the year of 20.5 seconds of arc in either direction for *every* star. Since stars are at different distances from the earth, their parallaxes should all be different. Therefore, this displacement could not be due to stellar parallax. What was it, then? For about a year Bradley struggled with the problem. One day in 1728, while riding a boat on the Thames River, he noticed that the wind vane on the mast shifted direction whenever the boat put about. He realized immediately that the apparent shift in the position of every star by 20.5 seconds of arc was due to the velocity of the earth in its orbital motion about the sun. This is the same phenomenon that makes us tilt our umbrellas at an angle when we walk in the rain, even if the rain falls vertically. To observe starlight from the moving earth, we must angle the telescope very slightly in the direction in which the earth is moving which makes the star appear in a slightly different position at different times of the year (Figure 18.2a). This phenomenon is known as *stellar aberration*. Bradley realized that he could determine the speed of light from the angle of deflection and the velocity of the earth in its orbit (Figure 18.2b). He obtained a value of 304,000 km/s.

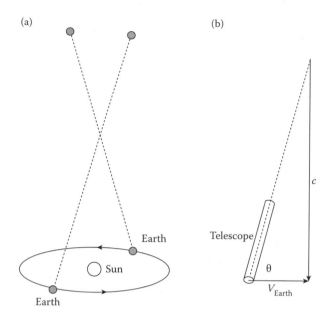

FIGURE 18.2 (a) The apparent shift in the position of a star as a result of the earth's orbital motion is known a stellar aberration. (b) By measuring the angle at which a telescope must be tilted the speed of light can be computed.

One hundred and twenty years later, in 1849, the French physicist Armand Hippolyte Louis Fizeau refined Galileo's method of flashing lights to measure the speed of light. Galileo's method of flashing lights from two adjacent hills was unsuccessful because, as we know today, the time it takes light to travel between the hills is very much shorter than the human reflex time to open and close the shutters on the lanterns. Fizeau decided to use a rapidly turning toothed wheel on one hilltop in Paris and a mirror on a hill 5.2 miles away. A source of light behind the wheel sent a beam of light between two adjacent teeth of the wheel that was reflected back along the same path by the mirror (Figure 18.3). If the wheel spun fast enough, the reflected light would pass through the next gap between the teeth. Knowing the angular frequency of rotation of the wheel in this case provided the time required for the light beam to travel the distance of 10.4 miles. Fizeau obtained a value for the speed of light of 313,300 km/s.

Shortly after this, another French scientist, Jean Bernard Léon Foucault, improved upon his friend Fizeau's method by replacing the toothed wheel by a rotating mirror. He used his new method to measure the speed of light in water and showed that it was less than in air. Foucault presented these results as his doctoral thesis. The American physicist A. A. Michelson, between 1878 and 1930, used a similar technique to measure the speed of light with great accuracy. In 1882, he reported a value of 299,853 km/s which remained the best available until his 1926 value of 299,796 ± 4 km/s, which in turn remained unchallenged until 1950.

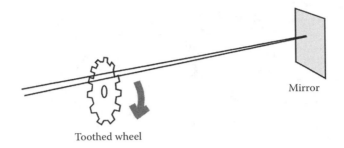

FIGURE 18.3 Fizeau's experiment to measure the speed of light.

THE ELECTROMAGNETIC SPECTRUM

Maxwell made two predictions from his theory. The first prediction was that there should be electromagnetic waves of many different frequencies all propagating through space at the speed of light. The second prediction was that electromagnetic waves exert a pressure on any surface that reflects or absorbs them. Unfortunately, Maxwell did not live to see his predictions verified, for he died of cancer at the early age of 48.

Eight years after Maxwell's death, the German physicist Heinrich Hertz used a spark-gap apparatus in which electric charge was made to oscillate back and forth, generating electromagnetic waves for the first time, thus confirming the first of Maxwell's predictions. Hertz studied physics at the University of Berlin, where he obtained his PhD in 1880 at the age of 23. In 1883 Hertz devoted himself to the study of Maxwell's theory of electromagnetism and two years later started his landmark experiments. He first made an induction coil that terminated at both ends in small metal balls separated by a small gap (Figure 18.4). The current in the primary coil A, which was connected to a battery, could be started and stopped by means of a switch. Opening and closing the switch produced a rapidly changing magnetic field in the iron core which, according to Faraday's law, induces a changing current in coil B. If the voltage was high enough, a spark would jump between the two balls. When an adjacent wire was bent so that its ends were also separated by a small gap, a spark would also jump between its two ends.

Hertz realized that each spark was actually a series of many sparks jumping back and forth between the two ends of the wire, setting changing electric and magnetic fields in the gap which, according to Maxwell's theory, propagated through space as an electromagnetic wave. When this wave reached the gap in the other wire, the electric field oscillating there produced sparks jumping back and forth as in the induction coil and this was seen as a single large spark. This second gap was a detector of electromagnetic waves.

By 1888 Hertz had measured the speed of the electromagnetic waves that he had generated with his spark-gap apparatus, obtaining a value equal to the speed of light, just as Maxwell had predicted.

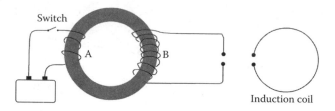

FIGURE 18.4 Hertz's experiment to produce and detect electromagnetic waves.

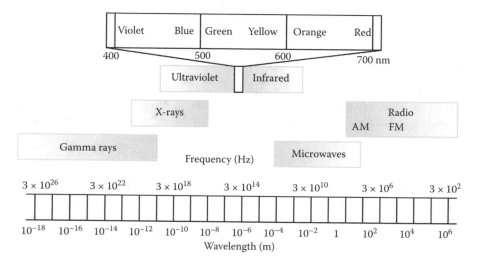

FIGURE 18.5 The electromagnetic spectrum.

Maxwell's second prediction, that electromagnetic waves exert pressure on any surface, was confirmed in 1899 by the Russian physicist P. N. Lebedev using very light mirrors in a vacuum.

The electromagnetic waves generated by Hertz had a wavelength of about one meter and were what we now call *radio waves*. The wavelengths of visible light had been measured during the first years of the nineteenth century, before the development of Maxwell's theory. Electromagnetic waves cover a wide range of wavelengths. The full range of the *electromagnetic spectrum* is shown in Figure 18.5.

COLOR

Electromagnetic waves with wavelengths from about 4×10^{-7} m to about 7×10^{-7} m constitute the visible part of the electromagnetic spectrum. We classify the different regions of the visible part of the spectrum by the names of colors. Our perception of color, however, is different from *physical* color; that is, from the infinite possible number of different electromagnetic waves with frequencies between 400 nm and 700 nm. These limits, by the way, are somewhat vague. The limits of the visible part of the spectrum are not well defined, since the sensitivity of the eye at the two ends does not stop abruptly. At 430 nm and at 690 nm, for example, the eye sensitivity has fallen to about 1% of its maximum value. If the light intensity is high enough we can see at wavelengths beyond these limits.

The present understanding of physical color stems from the beautiful experiments of Isaac Newton. It was already known in Newton's time that a beam of light passed through a prism produced a splash of the colors of the rainbow; violet, blue, green, yellow, orange, and red. This phenomenon was commonplace even at the time of Aristotle. It was Newton, however, who provided the correct explanation for the phenomenon. In his now famous letter to Henry Oldenburg, the first secretary of the Royal Society, of February 6, 1672, Newton wrote:

> . . . in the beginning of the Year 1666 (at which time I applied myself to the grinding of optic glasses of other figures than spherical) I procured me a Triangular glass-Prisme, to try therewith the celebrated *Phaenomena of Colours*. And in order thereto having darkened my chamber, and made a small hole in my window-shuts, to let in a convenient quantity of the Suns light, I placed my Prisme at its entrance, that it might be thereby refracted to the opposite wall. It was at first very pleasing divertisement, to view the vivid and intense colours produced thereby; but after a while applying my self to consider them more circumspectly, I became surprised to see them in an oblong form; which according to the received laws of Refraction, I expected should have been circular.
>
> And I saw . . . that the light, tending to [one] end of the Image, did suffer a Refraction considerably greater than the light tending to the other. And so the true cause of the length of that Image was detected to be no other, then that Light consists of *Rays differently* refrangible, which, according to their degrees of refrangibility, transmitted towards divers[e] parts of the wall.

This letter was printed in the *Philosophical Transactions* of the Royal Society for February 19, 1671–72, and became Newton's first published scientific paper. Before Newton provided the "true cause of the length of the image" and of the splash of colors produced by the prism, the accepted idea was that white light is darkened more at the thick end of the prism, so it becomes blue; is darkened less where the glass is not as thick, so it becomes green; and finally is darkened the least where it is closest to the thin end of the prism, so it becomes red. But this rationale did not explain what Newton immediately noticed, that the beam coming from a circular hole and thus having a *circular* cross section, would produce an *oblong* beam. Although this phenomenon had most certainly been noticed before Newton, it took the mind of the genius to stop and ask why this was so; the mind of a young man of twenty-three. Seeking the answer, Newton changed the size of the hole, the location of the prism, and the place where the beam hit the prism. The spectrum—a name he introduced for such a pattern of colors—did not change.

To test his idea that light was not modified by passing it through a prism, that it is physically separated into the different colors, Newton placed a prism near the hole in a shutter of his darkened room. Placing a second prism a few yards from the first, he noticed that blue light, on passing through the second prism, was refracted more than red light, as was the case in the first prism. But, more important, these colors

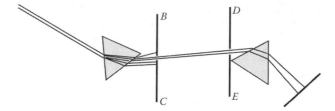

FIGURE 18.6 The *experimentum crucis*. From Newton's diagram in the *Lectiones opticae*.

were not affected by the second prism: "the purely Red rays refracted by the second Prisme made no other colours but Red & the purely blew ones no other colours but blew ones."

The crucial experiment that would leave no doubt in Newton's mind that his ideas were correct came a few years later. In his letter to Oldenburg he wrote:

> The gradual removal of these suspicions at length led me to the *Experimentum Crucis*, which was this: I took two boards [BC, DE in figure 18.6], and placed one of them close behind the Prism at the window, so that the light might pass through a small hole, made in it for that purpose, and fall on the other board, which I placed at about 12 foot distance, having first made a small hole in it also, for some of that Incident light to pass through. Then I placed another Prisme behind this second board, so that the light, trajected through both the boards, might pass through that also, and be again refracted before it arrived at the wall. This done, I took much as to make the several parts or the Image, cast on the second board, successively pass through the hole in it, that I might observe to what places on the wall the second Prisme would refract them. And I saw by the variation of those places, that the light, tending to that end of the Image, towards which the refraction of the first Prisme was made, did in the second Prisme suffer a Refraction considerably greater than the light tending to the other end. And so the true cause of the length of that Image was detected to be no other, than that *Light* consists of *Rays differently refrangible*, which, without any respect to a difference in their incidence, were, according to their degrees of refrangibility, transmitted toward divers[e] parts of the wall.

Newton demonstrated with his experimentum crucis that light consists of rays *differently refracted* which are transmitted to different parts of the wall. "When any one sort of Rays hath been well parted from those of other kinds," he further wrote, "it hath afterwards obstinately retained its colour, notwithstanding my utmost endeavors to change it." Thus, if light were modified by the prism, the second prism would also modify light, producing additional colors. Newton proved that once the colors were separated in the first prism by refraction, no further separation was possible. This was the crucial experiment.

Newton did not stop there, however. He reversed the process to prove that the colors of the spectrum could be recombined into white light. This he did first by adding a third prism in such a way as to have their spectra overlapping. They combined to form white. In 1669, the year Newton assumed the Lucasian Professorship, he repeated the experiment with a single prism and a converging lens. The spreading beam of colors struck the lens which he had placed a meter and a half from the prism, converging into a small patch of white light on the other side of the lens. The spectrum reappeared beyond this point, as the light rays diverged. In a letter to the Royal Society, shortly after he was elected a Fellow in 1672, Newton wrote:

> I have refracted it with Prismes, and reflected with it Bodies which in Day-light were of other colours; I have intercepted it with the coloured film of Air interceding two compressed plates of glass; transmitted it through coloured Mediums, and through Mediums irradiated with other sorts of Rays, and diversly terminated it; and yet could never produce any new colour out of it.
>
> But the most surprising, and wonderful composition was that of Whiteness. There is no one sort of Rays which alone can exhibit this. Tis ever compounded, and to its composition are requisite all the aforesaid primary Colours, mixed in due proportion. I have often with Admiration beheld, that all the Colours of the Prisme being made to converge, and thereby to be again mixed, reproduced light, intirily and perfectly white.
>
> Hence therefore it comes to pass, that *Whiteness* is the usual colour of *Light*; for, Light is a confused aggregate of Rays indued with all sorts of Colours, as they are promiscuously darted from the various parts of luminous bodies.

We have learned with Newton that a beam of natural white light is composed of *pure* colors, which cannot be broken down any further. The color of these pure beams is a fundamental property. Pure colors are then the simple components from which light is made. How many pure colors are there? Since a beam of white natural light can be refracted through an infinite number of angles, and each one of these refractions corresponds to a pure color, the number of possible pure colors is infinite.

As we saw in Chapter 17, the human eye contains three kinds of cone cells which are sensitive to colored light due to the presence of three types of light-absorbing molecules that change shape when they interact with light of wavelengths in the red, green, or blue region of the spectrum. Thus, all the sensory colors can be reproduced by the appropriate combination of these three pure colors. Of the infinite possible physical colors that may enter the eye, we sense only three different things. Many different combinations of physical colors may give rise to the same sensation, and we interpret this to be the same color. We can begin to see the limitations of the human eye. If the eye had more types of light-absorbing molecules, we would be able to extract more visual information from nature. We have overcome this limitation, in part, by designing and building instruments that enhance our vision.

SPECTRA: THE SIGNATURE OF ATOMS

Light passed through a prism, Newton taught us, is separated into its component colors. In analogy with the seven notes in music, Newton claimed the spectrum of white natural light contained seven colors—red, orange, yellow, green, blue, indigo, violet. He saw the spectrum as a continuous band of colors. In 1752, the Scottish physicist Thomas Melvill observed that the spectrum of a colorless alcohol flame to which a volatile substance was added was not a continuous band of colors but rather a *discrete* set of colored lines. In 1802, the English scientist William Wollaston observed the solar spectrum with the aid of a small telescope and discovered that dark lines crossed the otherwise continuous band of colors. He thought that these lines were the boundaries between the different colors and did not pursue the matter.

In 1814, the German physicist and optician Joseph von Fraunhöfer, while testing prisms made with a special glass he was studying, observed Wollaston's dark lines. He found 576 dark lines in the solar spectrum and measured the positions of the 324 most prominent ones, assigning letters of the alphabet from *A* to *K* to the more conspicuous lines. We still refer to some of the lines of the solar spectrum by the letters used by Fraunhöfer.

Later, Fraunhöfer passed the light from a star through a prism by placing it at the focal point of a telescope. He observed that the star's spectrum also showed dark lines, but these were at different locations from those in the solar spectrum. He noticed that the position of many of these lines corresponded to the colored lines of the discrete flame spectrum discovered by Melvill. He suspected that there might be a connection but his early death at the age of 39 prevented him from finding it. It was left for the German physicist Gustav Kirchhoff, some 50 years later, to find the connection. Kirchhoff was able to actually produce dark lines in his laboratory by passing a beam of white light through a glass container with gas and analyzing this light with a prism (Figure 18.7a). Kirchhoff then placed the prism to the side of the glass container to observe the glowing gas from a direction perpendicular to the beam of incident white light (Figure 18.7b). He saw a set of colored lines, like the one seen by Melvill. More interesting was his discovery that the positions of the dark lines corresponded to the positions of the bright colored lines in the spectrum seen on the side. Different gases produced different sets of dark and bright lines. Kirchhoff correctly interpreted this phenomenon by saying that the gas absorbed certain wavelengths

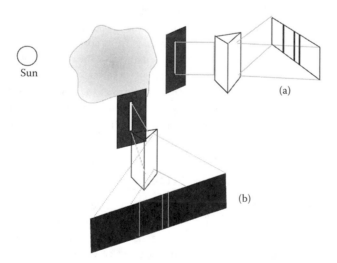

FIGURE 18.7   Kirchhoff's experiment. (a) A dark-line spectrum is observed by analyzing a beam of white light after it has passed through a gas. (b) A bright-line spectrum is seen when the prism is placed to the side of the container with the glowing gas.

from the incoming white light. These absorbed wavelengths appeared as dark lines in the front spectrum.

The spectrum crisscrossed with dark lines, like the solar spectrum, came to be known as a *dark-line* or *absorption spectrum*, whereas the spectrum of bright colored lines is called a *bright-line* or *emission spectrum*. Kirchhoff found that each particular compound when in gaseous form produces its own characteristic spectrum. The emission spectrum of sodium shows two bright yellow lines very close to each other and this spectrum is different from the beautiful set of red, orange, and yellow lines of neon, or from the red and blue lines of atomic hydrogen. Each element has its own unique spectrum, the signature of its atoms. The presence of a particular set of dark or bright lines characteristic of a certain element—its emission or absorption spectrum—is evidence of the existence of that element. Thus the spectrum of the sun tells us of the existence of certain elements in the atmosphere of the sun and the spectra of stars give us information about their composition. The most profound conclusion that has been drawn from this study is that the sun and stars are made of the same stuff as the earth.

In 1859 Kirchhoff provided a simple explanation of these phenomena. This explanation has been formulated as Kirchhoff's three laws of spectral analysis:

A heated solid or liquid emits light of all wavelengths producing a continuous spectrum.

A thin luminous gas emits an emission spectrum.

White natural light passing through a thin gas produces an absorption spectrum.

It was, however, Bohr's theory of the atom that provided a more complete explanation for these phenomena. In 1913 Niels Bohr explained that the electrons in an atom occupy certain allowed orbits about the nucleus. As we learned in Chapter 7, Bohr's theory says that when an electron drops from a higher allowed orbit to a lower one, thereby losing energy, the atom containing the electron emits light which carries off the lost energy. Since there are only certain allowed orbits, only certain electronic transitions between those orbits take place. Thus the discrete spectrum of hydrogen, for example, with its vivid red, blue-green, and blue lines, can be explained. When a beam of white light shines on a container with thin or rarefied hydrogen gas, a number of the photons that make up the beam of light will have energies that exactly match the energy difference between the lowest orbit or ground state and a particular higher orbit or excited state of the hydrogen atom. Some of these photons are absorbed by hydrogen atoms in the gas leaving them in excited states. These absorbed photons are therefore removed from the beam of white light and when we analyze this beam with a prism, the frequencies corresponding to those photons will not be there; these are the dark lines in the absorption spectrum. The billions of hydrogen atoms left in excited states are unstable and eventually their electrons make a transition to a lower energy state, emitting photons with exactly the same frequencies as the photons absorbed. These emissions are what we see as one of the characteristic hydrogen lines. A red line appears when the electron makes a transition from the third orbit to the second; a blue-green line when there is a transition from the fourth to the second orbit.

Notice that we have so far considered a *thin* or *rarefied gas*. This is because in a thin gas, the atoms making transitions to lower energy states do not interact with each other. When the gas is denser, however, some atoms begin to interact with each other while in the process of emitting light. This interaction slightly changes the energy levels of the colliding atoms. Whereas in the thin gas, where the atoms do not interact while making a transition, the energy levels of all the atoms are exactly the same and all the transitions between any two particular levels have the same energy and consequently the same frequency (recall that $E = hf$), in the denser gas some of these transitions are slightly different in energy and therefore in frequency. Since this difference is small, the lines end up very close to each other and the overall effect is the thickening of the line. In a denser gas, the collisions increase and the lines become thicker. If the denser gas is also at a high temperature, as in the interior of the sun, its atoms are moving with higher speeds. In this case, in addition to the increase in the number of collisions due to the higher kinetic energies of the atoms, there is a widening of the spectral lines due to the Doppler effect. The frequency of an emitting atom that happens to be moving away from us will be decreased slightly, whereas that of an atom moving in our direction will be increased. In the interior of the sun, both effects widen the lines so much that they actually run into each other, thus producing a continuous spectrum. It is only when these photons reach the outer cooler layers of the sun that they interact with the atoms there producing the dark lines characteristic of the sun's spectrum (Figure 18.8).

In a solid, the energy levels of the atoms that make up the solid are modified into energy bands due to the proximity of billions of other atoms (Chapter 13). For this reason a solid has a continuous spectrum. In the case of a crystal, where there are some constant energy levels, the spectrum will have some features.

The Bohr model of the atom works very well for the hydrogen-like atoms; that is, for the hydrogen atom and for other ionized atoms with one electron in their orbits. It does not work well for multi-electron atoms. It also raises many new questions for which it does not provide an answer. What causes an electron to jump down to a lower energy state, emitting light in the process? Why are there allowed orbits? In what direction is light emitted when an electron makes a transition to a lower energy state? These questions worried Einstein and other physicists. Bohr himself, in an attempt to explain the behavior of the atom with the classical theory of electromagnetism, wrote a paper in which he proposed to abandon the conservation of energy and momentum—the sacrosanct laws of physics—for atomic processes. Although at the time of the publication of this paper (1924) there was no evidence that these laws were valid at the atomic level, it soon came. In 1925, Arthur H. Compton and A. W. Simon proved that energy and momentum were conserved at the level of atoms. In July 1925 Bohr wrote: "One must be prepared for the fact that the required generalization of the classical electrodynamical theory demands a profound revolution in

FIGURE 18.8   Continuous spectrum of the sun with dark absorption lines.

the concepts on which the description of nature has until now been founded." As we shall see in Chapter 21, Bohr did not have to wait long for this "profound revolution" to take place.

YOUNG'S EXPERIMENT

The phenomena of absorption and emission of light by the electrons of atoms in a thin gas, which produce the spectra, are an example of the particle behavior of light; the photons being absorbed and emitted behave as *particles* of light. As we know, light also exhibits wave behavior. Although Newton thought that light was composed of particles, he did not reject the wave theory of light and even suggested that the different colors had different wavelengths. What Newton could not accept was the idea advanced by Huygens and Hooke that light was a spherical pressure wave propagating through a medium.

Newton had two objections to this idea. The first was that a wave did not always travel in straight lines; waves spread out in all directions and even bend around corners. Light, on the other hand, travels in straight lines as evidenced by the sharp shadows cast by a sunlit object. The second objection was related to the phenomenon of *polarization.*

The answer to Newton's first objection had to wait for Thomas Young in the early nineteenth century. The son of a Quaker banker, Young was a child prodigy who could read fluently by age 2 and by age 4 had twice read the Bible. At 14 he knew eight languages. Young studied medicine, graduating from the University of Göttingen in 1796 at the age of 23. Although he practiced medicine throughout his life, he was not a very good physician because of his poor bedside manner and probably because he was more interested in being a scientist rather than a physician. Between 1801 and 1803 he was a lecturer on science at the Royal Institution in London. He made important contributions in mechanics, acoustics, and optics. A constant that characterizes the elongation of a solid substance under tension is still called *Young's modulus.* He made considerable discoveries regarding surface tension, capillarity, and the tides. While still in medical school, he discovered the accommodation of the lens of the eye. Soon after obtaining his degree he discovered that ocular astigmatism was due to irregularities in the curvature of the cornea.

Young moved from his research on the eye to the nature of light. A century and a half before Young, the Italian physicist Francesco Maria Grimaldi had passed a beam of light through two small apertures, one behind the other. He found that the band of light on a surface behind the second aperture was slightly wider than the width of the first aperture. He realized that the beam of light had bent slightly and called the phenomenon *diffraction.* However, he could not explain why the band of light showed colored streaks at the edges. His discoveries appeared in a book published posthumously, which, for the most part, did not receive much attention. Newton was aware of Grimaldi's experiments and attributed the bending of the beam of light to interactions between the particles of light in the beam and the edge of the slit. In 1802, Young designed a similar experiment. He passed a beam of sunlight through a pinhole punched in a screen. Light spreading out from this hole passed through two other pinholes punched side by side in a second screen. Light from these two holes fell on a third screen where an interference pattern of alternating bright and dark regions or *fringes* appeared. This situation is identical to the experiment described in Chapter 15 for two coherent sources of waves. As we might recall, two waves are coherent if they have the same wavelength and a constant phase relationship. As a rule, light is incoherent because the crests and troughs of the individual waves are in random relationship to one another. In Young's experiment, coherence is assured because the light reaching the two pinholes originates from a single source.

In subsequent experiments, Young replaced the two pinholes in the second screen by two narrow parallel slits a few millimeters apart (Figure 18.9). In this case, the interference fringes are alternating dark and bright parallel bands. By measuring the distance between consecutive bright lines, Young calculated the wavelength of the light. He wrote:

> From a comparison of various experiments, it appears that the breath of the undulations [or wavelength] constituting the extreme red light must be supposed to be, in air, about one 36 thousandth of an inch, and those of the extreme violet about one 60 thousandth; the means of the whole spectrum, with respect to the intensity of light, being one 45 thousandth.

Young found that the wavelength of light was much smaller than Newton had thought. The longest wavelength in the visible spectrum, which is that of red light, is less than one-thousandth of a millimeter. When light falls on regular-sized objects, it will appear to travel in a straight line. Thus, the shadows of regular-sized objects bathed in sunlight are sharp. The shadows of pinhole-sized objects, as Grimaldi and Young saw, are not sharp.

Young's work on the wave theory of light was not well received in England mainly because it went against the particle model proposed by Newton, even after Young pointed out that Newton himself had made several statements in support of the wave theory. Young's theory had to wait until 1818, when two French scientists, Augustin Fresnel and Dominique Arago, proposed a wave theory of their own with a thorough mathematical basis. By 1850, the wave theory of light was widely accepted throughout the scientific community.

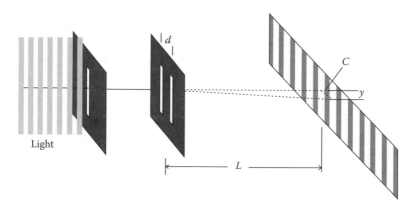

FIGURE 18.9 Young's double-slit experiment.

In Young's experiment, the two slits through which light from a single source passes are separated by a distance d (Figure 18.9). The screen where the interference phenomenon is observed is at a distance L, which, in real situations is much greater than the distance between the slits. A bright region appears on the screen at a distance y from the center C. In Chapter 15, we obtained a simple expression to calculate the wavelength of two coherent sources of waves interfering with each other, in terms of the distance between the sources (in this case, the distance between the slits), the distance L to the screen, and the distance y to a particular bright region. The expression is

$$\lambda = \frac{yd}{nL}$$

where n refers to the order in which the bright regions appear to the side of the central region; the first maximum at either side of this central region is the $n = 1$, the next one at either side is the $n = 2$ maximum, and so on.

POLARIZATION

The second of Newton's objections to the wave theory of light concerned the phenomenon of polarization. In 1669, the Danish scientist Erasmus Bartholin noticed that objects seen through a special crystal he had recently received from Iceland (now called Iceland feldspar or calcite) were doubled and that when the crystal was rotated, one image remained stationary while the other rotated along with the crystal. Assuming that the light traveling through the crystal was split into two beams, he called the beam that formed the stationary image the *ordinary beam* and that for the moving image the *extraordinary beam*. Huygens investigated this phenomenon and discovered that the two beams were further split into four beams by a second crystal unless this crystal was oriented exactly the same way or at 180° with the first, in which case no splitting occurred.

In 1717 Newton also considered this phenomenon and concluded that the beam of light is made up of particles that have two different "sides" and therefore look different when viewed from different directions. These particles are sorted out in the calcite according to the orientation they had when they entered the glass, thus producing the double images. "By those Experiments it appears," he wrote in his *Opticks*, "that the Rays of Light have different Properties in their different sides." To Newton, the particle theory explained the phenomenon of double refraction in Iceland spar crystal and that led him to reject the wave theory. "To me, at least, this seems inexplicable, if Light be nothing else than Pression or Motion propagated [as a wave]" he wrote.

The difficulty that Newton had with the wave theory of light failing to explain double refraction arose from his assumption that a wave of light had to be longitudinal, like a sound wave, rather than transverse. At the beginning of the nineteenth century, Young in England and Fresnel in France showed that light waves were transverse.

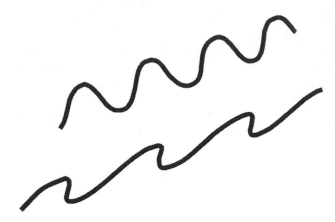

FIGURE 18.10   Vertically and horizontally polarized waves on a rope.

We know today that light is an electromagnetic wave; that is, oscillations in space and time of electric and magnetic fields that are perpendicular to each other and to the direction of propagation of the wave. It is the transverse nature of electromagnetic waves what explains polarization. If we shake a taut rope up and down, producing a train of waves, the rope vibrates in a vertical plane. In this case, the waves on the rope are polarized in a vertical plane. Shaking the rope horizontally produces horizontally polarized waves (Figure 18.10). The convention for direction of polarization of a wave of light is the direction in which the electric field oscillates. Normally, light is unpolarized because it is emitted in extremely short, randomly polarized bursts with duration in the order of nanoseconds. Since a beam of light consists of an immense number of waves, each one randomly polarized, the net result is no polarization (Figure 18.11).

In 1928 Edwin P. Land, while a physics student at Harvard, invented a transparent plastic material he called *Polaroid*. In addition to calcite, scientists had produced a synthetic crystalline material, sulfate of iodoquinine or *herapathite*, which absorbed light vibrating in the direction of orientation of the needle-shaped crystals. Since the herapathite crystals were so fragile, Land decided to embed them in a sheet of plastic. When the plastic was stretched, the crystals lined up like venetian blinds (Figure 18.12). Later, polymeric molecules composed mainly of long chains of iodine atoms replaced the herapathite crystals. Light polarized in a direction parallel to the orientation of the molecules in the Polaroid is completely absorbed because this incident light sets up vibrations in the molecules, losing most of its energy. Light polarized in a direction at right angles to the Polaroid emerges unchanged. If light is polarized at any other angle, it is partially absorbed by the molecules and therefore partially transmitted (Figure 18.13).

In a beam of unpolarized light, the electric fields of the many incoming rays are vibrating in all directions. Each electric field vector can be decomposed into a component that is parallel to, say, the orientation of the molecules in a Polaroid placed in front of the beam, and a component perpendicular to these molecules (Figure 18.14). The parallel components are all absorbed, as we know, and the

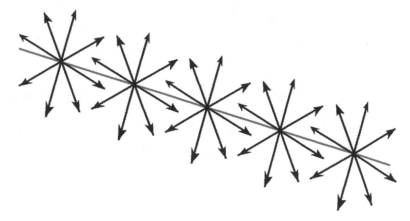

FIGURE 18.11   Several randomly polarized waves in an unpolarized beam.

(a)   (b)

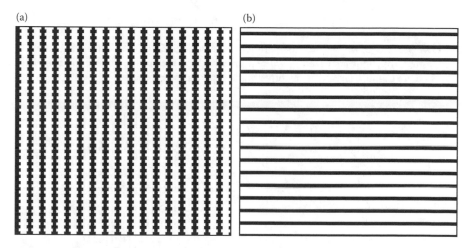

FIGURE 18.12   (a) A sheet of Polaroid in which herapathite crystals are embedded. Light polarized at right angles to these chains of crystals is transmitted. (b) Direction in which light is transmitted.

perpendicular components all pass through. Thus, when a beam of unpolarized light passes through a polarizer, half the light is absorbed and the other half is transmitted with the electric fields vibrating in one direction; that is, the light emerges polarized.

When a polarized beam of light strikes a Polaroid with its transmission direction parallel to the polarization of the beam, it will, as we have just discussed, completely pass through. If the transmission direction of the Polaroid is perpendicular to the polarization of the beam, no light is transmitted. What happens when the Polaroid is at an oblique angle to the polarization of the beam? Since each electric field vector, vibrating in the polarization direction, can be decomposed into a component parallel

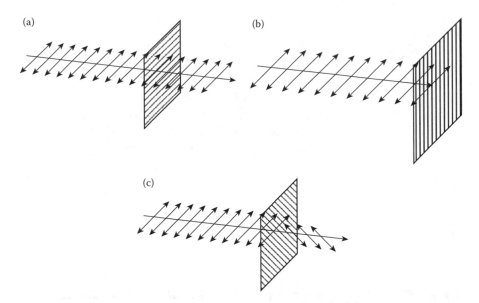

FIGURE 18.13   (a) Light polarized parallel to the orientation of the Polaroid is completely transmitted. (b) Light polarized in a direction at right angles to the Polaroid is completely absorbed. (c) Light polarized at any other angle is partially absorbed.

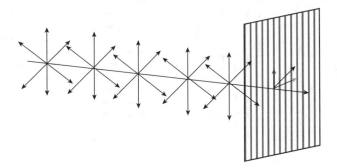

FIGURE 18.14   An unpolarized beam incident upon a Polaroid. Each electric field vector can be decomposed into a component perpendicular to the direction of the molecules in the Polaroid and another that is parallel to this direction. Only the perpendicular components are transmitted.

to the transmission direction of the Polaroid and a component that is perpendicular, part of the beam is transmitted.

When light falls upon an object, we learned in Chapter 17, the oscillating electric field of the incoming light sets the electrons of the atoms in the object into vibration. As these electrons absorb the light's energy, they start vibrating in a direction perpendicular to the incident light, becoming emitters of electromagnetic radiation in the process. This electromagnetic radiation, re-emitted after it is absorbed by the electrons, is the light we see when we look at the object. The electromagnetic radiation is re-emitted most strongly in a direction perpendicular to the direction of vibration of the electrons. Thus, light reflected by a smooth nonmetallic surface is partially polarized, with the direction of polarization parallel to the surface. The light from the morning sun reflected from the surface of a lake is partially polarized horizontally. Polarizing sunglasses, with their polarizing axis vertical, greatly reduce the glare of this reflected light. Photographers sometimes use a polarizing filter on the camera lens to reduce glare.

When light enters a calcite crystal, it becomes polarized in a very special way. Calcite and other similar crystals like tourmaline (called *noncubic* because of their particular lattice configuration), as well as certain stressed plastics such as cellophane, are *birefringent*, that is, they split a narrow beam of light into two beams which are polarized in mutually perpendicular directions. These two beams are the ordinary and extraordinary beams that Bartholin discovered. This interesting phenomenon is due to the particular atomic structure of these crystals. The ordinary and extraordinary light rays travel at different speeds through the calcite crystal because of the different ways light interacts with the atoms of the crystal. There is one particular orientation along which both rays travel at the same speed and, consequently, there is no separation of the incident ray into two rays. This direction is known as the *optic axis* of the crystal. A light ray that enters the calcite crystal along a direction other than the optical axis splits into two rays, the ordinary and extraordinary rays. The ordinary ray is polarized perpendicularly to the optic axis and the extraordinary ray is polarized in the same direction as the optic axis. Rotating the crystals makes the extraordinary ray rotate around the ordinary ray.

LASERS

The light our senses perceive consists of a great number of photons with many different frequencies, oscillating in random phases. This reflects the chaotic nature of the natural processes that give rise to them. As we know, light is emitted by atoms that have been excited to higher energy states by collisions with other atoms or by absorption of photons. Atoms in these high energy states emit radiation when they decay to their ground states. This can occur in two ways: a random, spontaneous emission; or a stimulated emission induced by the radiation emitted by other atoms of the same kind. In natural processes, spontaneous emission is dominant and the emitted light is incoherent.

Spontaneous emission occurs, for example, in fluorescent lamps. The atoms of the gas in the tube—mercury vapor mixed with an inert gas—are continuously excited to high energy states through collisions with electrons which are accelerated back and forth between two electrodes at the two ends of the tube to which an alternating voltage is applied (Figure 18.15). These excited atoms decay down to their ground states, emitting light in the ultraviolet. If some of the emitted photons have energy $h\nu$ equal to the energy difference between the ground state and an excited state in the atoms and molecules of the phosphor which coats the walls of the tube, then these atoms and molecules may absorb the photons. This absorption leaves these atoms in excited states. They decay spontaneously in several small steps, emitting light in the

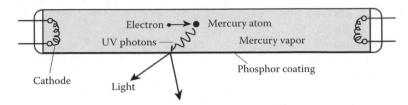

FIGURE 18.15 Electrons accelerated between two electrodes in a fluorescent tube collide with the atoms of the gas in the tube, raising them to higher energy states. As these atoms return to their ground states, they emit photons in the UV region which in turn excite atoms and molecules of the phosphor in the walls of the tube to higher states. When these atoms return to their ground states, they emit light in the visible part of the spectrum. This is the light we see.

visible part of the spectrum. Since the atoms decay at random, the emitted photons are not in phase with each other; the emitted light is not coherent.

In a *laser* (*l*ight *a*mplification by *s*timulated *e*mission of *r*adiation), the individual atoms are induced to radiate in phase. Albert Einstein predicted that if a photon of energy $h\nu = E_2 - E_1$ interacts with an atom which is in the excited state E_2, the incident photon may *stimulate* the atom to emit a second photon which not only has the same energy but also is in phase with the incident photon. Thus, stimulated emission produces light that is *coherent*.

Einstein also showed that the probability of stimulated emission from state E_2 to state E_1 is the same as the probability for absorption of a photon of energy $E_2 - E_1$ resulting in a transition from state E_1 to state E_2. In an ordinary gas, most atoms are in the ground state. It is then much more likely that a photon of the right energy will be absorbed by an atom in the ground state than it is to cause stimulated emission in an atom in the excited state. Although it is impossible to predict when a particular atom in an excited state will decay to a lower state, the *average* time of decay for a group of many atoms is of the order of 10^{-8} seconds. There are, however, certain excited states called *metastable*, in which atoms on average will stay for 10^{-3} seconds, 100,000 times longer than in an ordinary state. If there are more atoms in a metastable state than in the ground state, a situation called *population inversion*, stimulated emission will be the dominant process. A common method of achieving population inversion is by *optical pumping*, in which intense light of the right energy is used to excite many atoms to the higher state. Since the probability of absorption is the same as that for emission, the light that pumps the atoms to higher states can also pump them down to the ground state. To sidestep this problem, substances in which electrons can jump down to a third and even a fourth level are used.

A three-level laser, such as the ruby laser, has three levels that participate in laser processes; the ground state E_o, an ordinary excited state E_e, and a metastable state E_m (Figure 18.16). When a system with a large number of these atoms in the ground state is irradiated with photons of energy exactly equal to the energy difference between the ordinary excited state and the ground state; that is, $E_e - E_o$, many atoms will be excited to state E_e by absorption of one of these photons. Once excited to this level, the atoms spontaneously decay to the metastable state E_m or to the ground state in

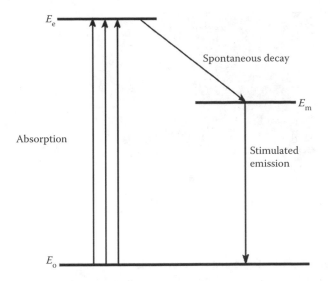

FIGURE 18.16   In a three-level laser, electrons are pumped to the excited state of energy E_e from where they spontaneously decay to a metastable state E_m or to the ground state. When enough electrons have accumulated in the metastable state, producing a *population inversion*, a source of ordinary light stimulates the electrons to decay to the ground state, starting the lasing process.

an average time of 10^{-8} s. Electrons stay in the metastable state for a much longer time. When a sufficiently large number of electrons populate the metastable state, producing a population inversion, a second source of ordinary light of energy equal to the energy difference between the metastable state and the ground state stimulates the decay of many electrons in the metastable state to the ground state, starting the lasing process.

The fundamental principles of the laser were worked out by the American physicist Charles H. Townes. In 1917, Einstein had recognized the existence of stimulated emission and, as we have seen, showed that the probability of an electron in a state E_1 of absorbing a photon of energy $E_2 - E_1$ was equal to the probability of emission of a photon of the same energy by an electron in state E_2. In 1951, Townes, then a professor of physics at Columbia University, was undertaking the problem of how to generate electromagnetic waves of great intensity in the microwave region. While sitting on a park bench one morning waiting for a restaurant to open, Townes realized that certain molecules had energy states of the right frequency for microwave emission. The problem was how to excite enough of them, keep them from decaying until a sufficiently large number of them were excited, and finally how to stimulate them to decay at once to produce a powerful enough beam. That morning, while waiting to have breakfast, Townes had all these problems solved after doing some calculations on the proverbial back of an envelope. By 1953 he and his graduate students had constructed a working prototype. Townes called his device a *maser*, for *m*icrowave *a*mplification by *s*timulated *e*mission of *r*adiation.

FIGURE 18.17   Close-up of the first laser. (From Associated Press, NY. With permission.)

Shortly after, in 1958, Townes and A. L. Schawlow showed that it was possible to construct a maser producing coherent radiation in the visible spectrum. Around the same time, the Soviet physicists A. M. Prokhorov and N. G. Basov independently worked out the principles. In 1960, the American physicist T. H. Maiman at Hughes Research Laboratories in Miami, Florida, constructed the first laser, using a synthetic ruby cylinder (Figure 18.17). Townes, Prokhorov, and Basov shared the Nobel Prize in 1964 for the discovery of the maser. Schawlow won the 1981 Nobel Prize for his work on lasers.

PHYSICS IN OUR WORLD: COMPACT DISC PLAYER

A compact disc player is a digital device; that is, the music has been stored on the disc as binary codes in the form of a sequence of pits. When the music is recorded, the changes in sound intensity are transformed by a microphone into variations in voltage. This electrical signal is sampled 44,000 times a second from each stereo channel. These values of the voltage are converted into a binary code of on-off pulses which are encoded as a sequence of pits on the surface of a polycarbonate disc 12.5 cm in diameter covered with a reflecting layer of aluminum.

In the CD player, a weak laser beam shines on the minute pits and flat reflective spaces in the spiral groove engraved on the underside of the spinning disc. When the laser beam shines on the reflective aluminum surface, it is reflected back and detected by a photocell where electrons are released. When the laser beam enters a pit in the track, the light is not reflected. Thus, the digital code of on–off signals which was encoded when the disc was produced is decoded in the CD player. As the number of samples per second is so large, the music reproduced in the player has extremely high fidelity, free of the hiss of magnetic tape and the surface noise of the old records.

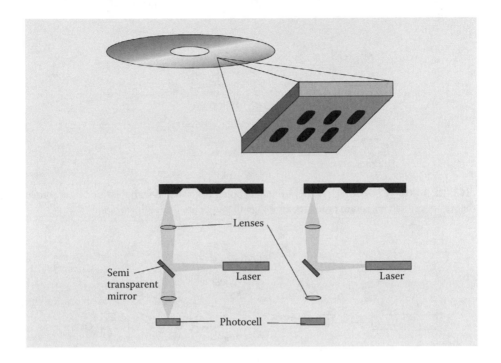

HOLOGRAPHY

Holography, although associated with lasers, was invented by the Hungarian engineer and physicist Dennis Gabor in 1947 before Maiman constructed the first laser in 1960. Gabor said that the idea for holography came to him as he waited for his turn at a tennis court in England where he was a research engineer. He wanted to improve the resolution of the electron microscope; his first paper on the subject, published in *Nature* in 1948, was entitled "A new microscopic principle." He was awarded the 1971 Nobel Prize in physics for his discovery.

Gabor coined the word *holography*, from the Greek word "holos," meaning "the whole," because it records the entire message of light, not just the intensity, as cameras or even our eyes do. A hologram is produced by the interference of two light beams, a reference beam coming directly from the light source, and the beam reflected by the object. The reflected beam varies in phase compared with the reference beam because of the interaction with the object. A stationary interference pattern is produced where the two beams overlap. A photographic film placed in this space can record this pattern: this forms the hologram. When the reference beam is passed through the developed hologram, the interference pattern diffracts the beam in such a way that the emerging beam is identical with the original object beam. If we look at the hologram from the side opposite the light source (Figure 18.18), we see an image of the object. Since the wave fronts are identical with those generated by the object, our eyes cannot distinguish between the perceived image and the real object and we see what appears to be the object itself, in three dimensions.

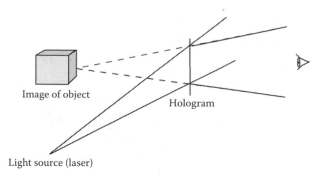

FIGURE 18.18   Looking at a hologram that is being illuminated with the same light source that produced the hologram produces an image that appears three-dimensional.

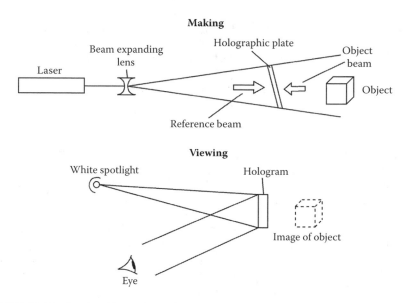

FIGURE 18.19   Denisyuk's configuration for a white-light-viewable hologram.

Gabor's original experiments permitted holograms of only very tiny thin objects (he used transparencies as objects). Anything thicker could not work because the coherence length of his mercury light source was only a fraction of a millimeter. The advent of the laser in 1960 provided a beam with a coherence length of several centimeters, and made possible holograms of three-dimensional objects. Although these holograms had to be viewed in laser light, in 1964 the Russian physicist Yu N. Denisyuk developed a technique to make holograms that were viewable in white light. Instead of positioning the reference beam and the object on the same side of the hologram, Denisyuk moved the reference beam to the opposite side (Figure 18.19). The interference pattern then formed inside the photographic film as sheets like the pages of a book. When developed, the film acted as an interference filter, which, when illuminated with white light, rejected all inappropriate wavelengths and gave single-color three-dimensional images.

Denisyuk's work, published originally in the Soviet Union, was translated into English but was unnoticed by American scientists. In 1965, Nile Hartman at Battelle Memorial Institute in Columbus, Ohio, independently discovered a similar technique to make a hologram that could be viewed in white light. In 1969, Stephen Benton of the Polaroid Corporation developed the rainbow hologram, which produces an image in colors of the spectrum. These holograms look three-dimensional only side to side, not up and down. The security holograms now embossed on some credit cards are of this type. When turned on its side, this type of hologram loses its three-dimensional appearance.

Part VII

Modern Physics

19 The Special Theory of Relativity

GALILEAN RELATIVITY

The theory of relativity is usually associated with the name Einstein. It might therefore come as a surprise that the concept of relativity did not originate with Einstein. The honor belongs to Galileo. In *Two New Sciences*, Galileo discussed the problem of the behavior of falling bodies on a moving earth, arriving at the conclusion that they fall exactly as they would appear to do if the earth were not moving. Galileo argued that you cannot tell whether the earth is moving or at rest by watching an object fall.

In *Two New Sciences*, Galileo discussed the simpler problem of uniform linear motion. According to Galileo, if we are in a ship moving along a straight line with constant speed and drop a ball from the crow's nest, the ball will fall straight down, hitting the deck at the foot of the mast, not the water. But an observer on shore will thus see the ball falling down following a curved path and not a straight line (Figure 19.1).

If we drop a ball inside a closed room, the ball will fall straight down whether the room is in our house, a cabin in a cruise ship, or the closed bathroom of an airplane. In fact, no experiment can be performed inside a closed room that will reveal to us whether or not the room is at rest or moving along a straight line at constant speed. If we are in the cabin of a ship that is moving steadily and drop a ball from the middle of the ceiling, it will drop on the middle of the floor, just as if we were doing the experiment in our room at home. Motion in a straight line at constant speed, according to Galileo, has no discoverable effects. The only way we can tell whether or not we are moving is by looking out of the window to determine whether there is relative motion between us and the earth. Sitting in an airplane we have just boarded we are sometimes fooled into thinking it is moving when the engines are running and another airplane is taxiing nearby. Unless we catch a glimpse of a building or are able to see the ground from where we sit, the vibrations from the engine and the motion of the other airplane make it impossible to decide who is moving. A similar experience occurs when, tired during a long car trip, we may find ourselves sitting low in the back seat when our driver gets stuck in a traffic jam. Seeing only the tops of the other automobiles on the other lanes, we might think we are finally moving, only to discover that the cars traveling in the opposite direction were the only ones moving.

In a sense, then, all motion is relative. In everyday situations we refer motion to the earth. Although we would say that when we are sitting in our room reading we are not moving, a hypothetical observer traveling through the solar system would

(a)   (b)

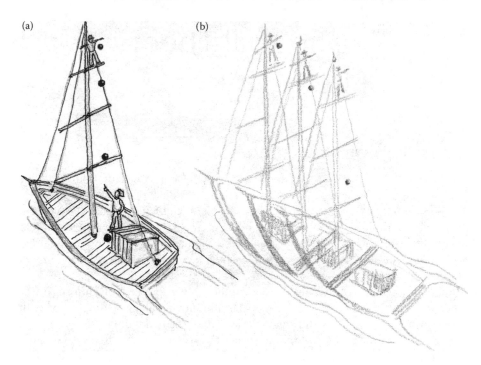

FIGURE 19.1 (a) An observer in a moving ship will see a ball falling straight down, as if the ship were not moving. (b) A person on shore will see the ball move along a curved path.

affirm that our room is actually rotating along with the earth and revolving around the sun. The description of motion depends on the particular *reference frame* to which we refer it. If we are on an airplane that is traveling at a steady 800 km/h, we would consider the book we are reading to be at rest with respect to us, the other passengers, and the plane (Figure 19.2). Of course, the book is moving along with us, the other passengers, and the plane at 800 km/h relative to the ground. Which view is the correct one? Both are. The book is at rest in the reference frame of the plane and moving at 800 km/h in the reference frame of the ground. While the plane is moving steadily, your coffee will not spill and your pen will not roll off the fold-out table. The law of inertia holds and, as we have said, nothing, other than looking out the window, will tell you that you are moving. A reference frame in which the law of inertia holds is called an *inertial reference frame*.

Suppose now that, while a plane is flying with constant velocity at 800 km/h, a flight attendant walks from the back of the plane to the front at a steady pace. Assume that she walks at a speed of 2 km/h (Figure 19.3). This, of course, is her speed in the reference frame of the plane. If the velocity of the flight attendant were to be measured from the ground, we would find it to be 802 km/h. This is hardly surprising to us. It is not uncommon to see people in a hurry walking on escalators. Their velocities with respect to the building where the escalator is located are greater than the velocity of someone who merely rides the escalator. We all have seen children walking down an "up" escalator so that they remain stationary in

FIGURE 19.2 The plane and all its contents are traveling at 800 km/h relative to the ground. The book that the passenger is reading and the passenger herself are at rest relative to the plane.

FIGURE 19.3 While the plane flies at a constant speed of 800 km/h relative to the ground, a flight attendant walks down the aisle at a speed of 2 km/h relative to the plane.

relation to the building. In this case, their velocities are the negative of the escalator velocity.

If a reference frame S' is moving with a velocity v_F relative to a second reference frame S, then the velocity v of an object relative to S is equal to its velocity v' in frame S' plus v_F; that is,

$$v = v' + v_F.$$

Returning to our example of the flight attendant, the velocity of the reference frame S' of the airplane relative to the frame of the ground is $v_F = 800$ km/h, the flight attendant's velocity relative to the plane is $v' = 2$ km/h, and her velocity relative to the reference frame S of the ground is $v = 2$km/h $+ 800$ km/h $= 802$ km/h.

Since inertial frames of reference move at constant velocities, the acceleration of an inertial frame is zero. Therefore, the acceleration of an object in one reference frame is the same as in any other inertial frame. The object we drop from the ceiling of a cabin in a steadily moving ship not only falls straight down to the ground relative to the people on the ship, it also accelerates at 9.8 m/s^2. This is the same acceleration that an observer on shore would measure for the falling object if this observer could see it. Thus, not only the law of inertia—Newton's first law—holds for inertial frames of reference but the second law and the universal law of gravitation as well. In fact all the laws of mechanics are the same for all observers moving at a constant velocity relative to each other, as Newton himself recognized. This statement is implicit in Galileo's own statement that uniform motion has no discoverable effects and is known today as the *Galilean principle of relativity*. We can state this as follows:

The laws of mechanics are the same in all inertial frames of reference.

This principle means that there is no special or absolute reference frame; all reference frames are equivalent. Thus, there is no absolute standard of rest; uniform motion has to be referred to an inertial frame.

A word of caution before we leave this section is in order. We have implicitly said that the earth is an inertial reference frame. This is not exactly true, because the earth is rotating, so any point on its surface is always accelerating. However, this acceleration is very small, and the rotational effects of the earth can be neglected. The earth can thus be considered an inertial frame of reference for our purposes.

THE MICHELSON–MORLEY EXPERIMENT

The Galilean principle of relativity was fine for mechanics. However, when Maxwell deduced the existence of electromagnetic waves that travel at the speed of light, scientists began thinking about the medium through which these waves propagate. Although Maxwell concluded that electromagnetic waves would propagate in empty space, the physicists of this period, being familiar with mechanical waves which require a medium to propagate, naturally assumed that electromagnetic waves also required a medium, and could not accept the idea of a wave propagating though empty space. On the earth, light propagated through air, water, and other transparent media. But light also came from the sun and the stars, and the space between the stars did not appear to be filled with any known substance. Clearly, they reasoned, a transmitting substance existed that filled all space; the *luminiferous æther*—or simply the "ether." The ether was a transparent medium thin enough to allow the motion of the planets but rigid enough to allow for the propagation of light with tremendous speed. Even Maxwell was convinced: "We have, therefore, some reason to believe," he wrote in 1865, "from the phenomena of light and heat, that there is an æthereal medium filling space and permeating bodies, capable of being set in motion from one part to another, and of communicating that motion to gross matter so as to heat it and affect it in various ways."

Albert A. Michelson was one of several physicists who tried to detect the motion of the earth through the ether. Like all the others, he failed. Michelson was born in Strelno, Prussia (now Strzelno, Poland), on December 19, 1852. When he was four, his parents emigrated to America and settled in San Francisco, where they went into business. There, the young boy grew up. As a teenager, Michelson entered the United States Naval Academy at Annapolis where he excelled in science, becoming a physics and chemistry instructor after graduation. In 1878, he began thinking about better ways to measure the speed of light. Realizing that he required more formal studies in optics, he traveled to France and Germany for advanced study, as was the custom in those days. Upon his return to the United States, Michelson became a professor of physics at the Case School of Applied Sciences (now known as Case Western Reserve University), and later, the first head of the physics department at the University of Chicago. In 1907, Michelson became the first American to win the Nobel Prize in the sciences.

While in Germany working in the laboratory of Hermann von Helmholtz, Michelson invented an ingenious instrument with the idea of measuring the earth's velocity with respect to the ether. *Michelson's interferometer*, as the instrument came to be known, was based on an idea first proposed by Maxwell in 1875. As the earth moves through space, Maxwell reasoned, the ether that permeates space would create a wind. If we measure the velocity of light in the direction of motion of the earth around the sun (that is, in the opposite direction to that of the ether wind) we would obtain a value equal to the speed of light with respect to the ether *minus* the speed of the ether. If we measure the speed of light in the *opposite* direction we would obtain a value equal to the speed of light with respect to the ether *plus* the speed of the ether. If we measure the speed of light in a direction at right angles to the ether wind we would obtain the actual velocity of light relative to the ether. This situation is similar to the situation of a swimmer swimming 50 meters, as measured on shore, first upstream and then the same 50 meters downstream and later swimming 50 meters back and forth at right angles to the current. The times for each round trip are going to be different. If we know at what speed the swimmer swims in still waters, we can deduce the velocity of the current.

Figure 19.4 shows a schematic representation of Michelson's interferometer. A half-silvered mirror splits a beam of light into two beams—one reflected and one transmitted—that travel in perpendicular directions. The reflected beam strikes mirror M_2 and is reflected back to the half-silvered mirror. The transmitted beam strikes mirror M_1 and is also reflected back to the half-silvered mirror. There the two beams are again partly reflected and partly transmitted and interfere. This interference pattern can be observed with a telescope.

Michelson's first attempt to detect a difference in the speed of light in any direction was unsuccessful; no difference was detected. He decided that he needed more sensitive equipment. In 1887, Michelson, now at Case Western Reserve, and his friend and collaborator, Edward W. Morley, a professor of chemistry at the University, decided

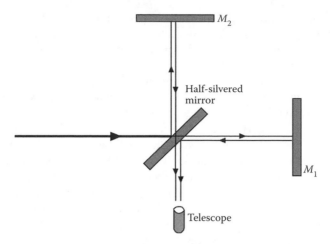

FIGURE 19.4   Schematic diagram of Michelson's interferometer. A narrow beam of light is split into two beams by a half-silvered mirror. These two beams travel in perpendicular directions to two mirrors which reflect the beams back to the partially reflected mirror where they are again partly transmitted and partly reflected. The two beams interfere with each other. This interference pattern can be observed with a telescope.

to try again. They placed the interferometer on a square slab of stone one and a half meters wide, floating on mercury to minimize vibrations as the apparatus was rotated to measure the light from the stars in different directions. When the apparatus was oriented in the direction of the earth's motion, a certain interference pattern was observed due to the difference in the optical paths of the two perpendicular beams; one parallel to the earth's velocity and the other perpendicular to it. As the interferometer was rotated 90°, the beam that was parallel to the earth's velocity was now perpendicular and the one that was perpendicular became parallel. Thus, a shift in the interference pattern was expected. Michelson and Morley calculated that this shift should have been 0.4 fringe; that is, slightly less than half the width of a fringe. The results were again negative. "The actual displacement," they reported, "was certainly less than a twentieth part of this, and probably less than a fortieth part."

Michelson was known as a meticulous experimenter and the negative results of his experiment puzzled other physicists. The Irish physicist George FitzGerald advanced an explanation in 1882. He proposed that the linear dimensions of all objects are shortened along the direction of motion of the earth. Thus, the arm of the interferometer holding the mirrors placed in the direction of motion of the earth is shortened. This contraction is exactly what is needed to compensate for the difference in the velocities of light in the direction of the ether and perpendicular to it. We do measure a smaller velocity of light in the direction opposite the ether wind as compared to the direction perpendicular to it, he proposed, but the times are the same because the length along the former direction is shorter by the right amount. Since everything on the earth, including meter sticks, is contracted in the direction of the ether wind, we cannot directly measure this contraction. In 1895, the Dutch physicist Hendrik A. Lorentz independently proposed this contraction and provided an explanation for it in terms of changes in the electromagnetic forces between the atoms. This contraction, proposed without other empirical support and only to provide an explanation for the null results of the Michelson–Morley experiment, is known today as the *Lorentz–FitzGerald contraction.*

EINSTEIN'S POSTULATES

Albert Einstein was eight years old when Michelson and Morley were attempting to measure the velocity of the ether wind. He had been late to begin to talk but was not a poor student in his early youth, as popular mythology has it. A year before this famous experiment took place, in 1886, when Einstein was seven, his mother Pauline wrote to her mother, "Yesterday Albert got his marks. Again he is at the top of his class and got a brilliant record." A year later his grandfather wrote, "Dear Albert has been back in school a week. I just love that boy, because you cannot imagine how good and intelligent he has become." At ten, Einstein entered the Luitpold Gymnasium or secondary school in Munich, a city not far from his native Ulm in Germany, where he excelled in mathematics and physics, subjects in which he was, "through self study, far beyond the school curriculum," as he wrote years later. At the age of twelve, he was given a book on Euclidian geometry which he studied with enthusiasm. "The clarity and certainty of its contents made an indescribable impression on me," he later wrote in a small autobiographical essay. Einstein, however, disliked the Gymnasium and its astringent learning atmosphere, and once remarked that its teachers were like lieutenants.

When he was fifteen, his father's business failed and the family moved to Pavia, Italy. Einstein was left behind to finish his secondary education at the Gymnasium. After six months, depressed and nervous, he persuaded his family physician to provide him with a certificate stating that owing to nervous disorders he needed the company of his family. Einstein left the Gymnasium without informing his parents and joined them in Pavia. He promised his disappointed parents that he would study on his own to prepare for the entrance examination at the prestigious Swiss Federal Polytechnic Institute, the ETH, in Zurich, where his father wanted him to study electrical engineering. Einstein was examined in political and literary history, German and French, drawing, mathematics, descriptive geometry, biology, chemistry, and physics, and was required to write an essay. He failed. However, because he had done well in mathematics and the sciences, the director of the Polytechnic suggested that he obtain a diploma at a Swiss secondary school and reapply. A year later, Einstein graduated with near-perfect grades from the cantonal school in Aarau, in the German-speaking region of Switzerland. Now armed with his high school diploma, Einstein was admitted at the ETH without further examination.

Einstein entered ETH on October 26, 1896, at the age of 17, to study physics. Under six professors—one of them Hermann Minkowski, who was to participate in giving the theory of relativity its mathematical formalism—Einstein studied mathematics. Three professors taught physics and astronomy. He also added electives such as gnomic projection, anthropology, geology, banking and stock exchange, politics, and Goethe's philosophy. He did not attend lectures regularly, however, preferring to spend his time in the physics laboratory and in the library reading the original works of Maxwell, Kirchhoff, and Hertz. He relied on good class notes taken by his friend Marcel Grossmann to cram for examinations. Einstein particularly disliked the physics courses taught by Heinrich Weber because he did not present anything about Maxwell's theory, and Maxwell's theory was "the most fascinating subject at the time that I was a student," as Einstein wrote later. The dislike was mutual, as Weber did not like Einstein's forthrightness and distrust for authority. In Europe at the time, a professor was an exalted person, respected and revered by lesser people.

Einstein graduated from the ETH in August 1900. Three other students graduated with him and the three immediately obtained assistantships at the ETH. Einstein was not offered a position. He then looked for other university positions and was rejected. In 1901, he wrote, "From what people tell me, I am not in the good graces of any of my former teachers ... I would long ago have found a [position] had not Weber intrigued against me."

In 1901, still without a job, Einstein returned to physics and wrote his first scientific paper, on intermolecular forces, which appeared in volume 4 of the prestigious scientific journal *Annalen der Physik*. He then wrote a research paper on thermodynamics which he submitted to the University of Zurich to obtain his doctoral degree. The paper was rejected by Professor Kleiner as a PhD thesis. In June 1902, with the help of a recommendation from Marcel Grossmann's father, Einstein finally obtained a job as a clerk in the Swiss Patent Office in Bern. This was not an academic job but it might have been the best job that Einstein could have had at that time, because it was undemanding and left him enough time to think about physics. It was there, at the patent office, that Einstein developed and published his Special Theory of Relativity.

1905 was Einstein's *annus mirabilis*, ranking in the annals of physics with the other *annus mirabilis* of 1666 when Newton went home in Woolsthorpe and explained how the universe works. That year of 1905 Einstein published a paper which was "very revolutionary," as he wrote to a friend. This paper was indeed revolutionary; it was the paper that laid the foundation for quantum theory, with the introduction of the concept of quanta of energy or photons, and that was eventually to earn him a Nobel Prize in physics. Less than a month after submitting this paper for publication, he sent off a second paper which was to gain him a PhD from the University of Zurich, accepted by the same Professor Kleiner who had rejected his first submission; it was the sugar paper mentioned in Chapter 1. "A New Determination of the Sizes of Molecules," he titled it. Within a month he submitted a third paper explaining the erratic, zigzag motion of a speck and helped to establish the existence of atoms. This was called Brownian motion, although Einstein did not know this. On June 30, 1905, *Annalen der Physik*, the journal where Einstein had published all his papers, received the fourth manuscript from Einstein during that year, titled "On the Electrodynamics of Moving Bodies." This was the special relativity paper submitted only 15 weeks after his first paper of that incredible year. But Einstein was not finished; he still had time that year for a fifth paper which he titled "Does the Inertia of a Body Depend upon Its Energy Content?" This was the paper containing the famous formula $E = mc^2$, which was to become synonymous with his name. Einstein was 26 that year.

Einstein's 1905 paper on special relativity starts by recognizing that Maxwell's theory of electromagnetism makes a distinction between rest and motion. He gave the example of a magnet and a wire moving relative to each other. If the magnet moves and the wire is at rest, the moving magnet generates an electric field that produces a current in the wire. However, if the wire is moving and the magnet is at rest, there is no electric field, but the moving conductor experiences a force which sets up an electric current equal to the current in the former case. This means that the states of rest and motion can be identified. Absolute motion does not exist; only relative motion matters. Galileo had discovered that. Either Galileo was wrong or Maxwell was wrong. Einstein decided that electromagnetism had to be reformulated so that the description depends only on the relative motion. He proceeded to do just that in the second part of his paper.

Einstein chose not merely to believe that only relative motion matters; he elevated this idea to the status of a postulate and called it the Principle of Relativity. Einstein introduced a second postulate "…that light is always propagated in empty space with a definite velocity c which is independent of the state of motion of the emitting body." The special theory of relativity is based on these two fundamental postulates. Let us state them here in Einstein's own words:

> Postulate 1: The Principle of Relativity: The same laws of electrodynamics and optics will be valid for all frames of reference for which the equations of mechanics hold good.
> Postulate 2: Light is always propagated in empty space with a definite velocity c, which is independent of the state of motion of the emitting body.

The first postulate is an extension of the Galilean Principle of Relativity to cover not only mechanics but also electromagnetism and optics. Actually, in extending the principle of relativity to include optics and electromagnetism, Einstein meant to include all of physics, since mechanics, optics, and electromagnetism were all of physics at that time. The Principle of Relativity can then be stated as follows:

> Postulate 1 [alternate version]: The laws of physics are the same for all observers moving in inertial (non-accelerated) frames of reference.

The second postulate seems to contradict common sense. It was Einstein's solution to a puzzle he had struggled with since he was a young boy of sixteen. What would happen if we could travel at the speed of light? Newtonian physics does not forbid us from doing so. Accelerating a spaceship continuously, for example, would eventually allow it to reach the speed of light and even any speed beyond it. Traveling at the speed of light, in the same direction as that of a light beam, we would notice that the wave pattern of the light beam disappears. Einstein realized that Maxwell's equations do not allow such a possibility. Either Maxwell's equations are wrong or traveling at the speed of light is an impossibility. Einstein decided that Maxwell equations were correct. The speed of light must be a limiting velocity; no material body can ever reach it. As such, it must be the same for all observers in inertial frames of reference.

TIME DILATION

The constancy of the speed of light for all inertial observers has important consequences for the way we measure time. Imagine two identical trains, one moving at a constant velocity v and the other standing by the railroad station. These two trains are in inertial frames of reference. Suppose further that the velocity of the moving train is extremely large, close to the speed of light. Of course, no train is capable of reaching these velocities (the spacecraft Galileo achieves velocities of about 0.0001c), but this is a thought experiment. In one of the cars in the moving train, an observer has placed a mirror on the ceiling of the car and a light source on the floor of the car. A second observer in one of the cars of the stationary train has an identical set up in her car. Imagine further that the two cars where this equipment has been installed have large windows that allow each observer to completely view the other set up but they are unable to see the ground or any other structure around the trains. For clarity, we are assuming that we are in the same reference frame as the parked train; that is, the stationary train is at rest *relative* to the ground, where we also are. Both observers agree that the distance from the mirror to the light source is the same and they label it d (Figure 19.5a).

When the two cars are facing each other, the observers fire the light sources. The observer in the moving train measures the time t' it takes for the light in his set up to make the round trip to the mirror and back. Since the distance for the round trip is 2d, the time is

$$t' = \frac{2d}{c}.$$

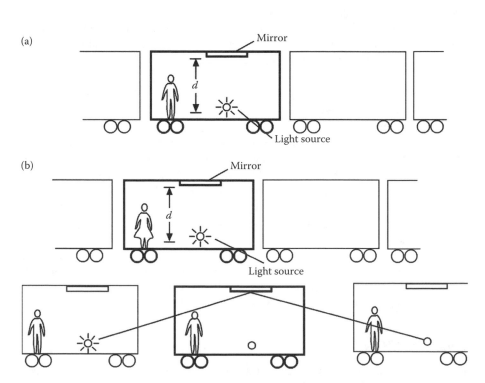

FIGURE 19.5 (a) Two identical trains with light sources and mirrors in motion relative to each other. (b) Path seen by observer on the ground.

The observer in the stationary train, however, measures a different time t for the light beam in the experiment in the moving train. Since the train is moving with a velocity v, by the time the light reaches the mirror, the mirror would have moved a certain distance (Figure 19.5b). According to the stationary observer, if the beam is to hit the mirror, it would have to move at an angle and therefore travel a longer distance. Since light travels at the same speed in both reference frames, the beam of light must travel a longer time in the reference frame of the stationary observer.

The expression for the time measured by the stationary observer for the round trip of the light beam in the moving train is

$$t = \frac{t'}{\sqrt{1 - v^2/c^2}} = \gamma t'$$

where the factor $\gamma = 1/\sqrt{1 - v^2/c^2}$ (γ is the Greek letter gamma). This expression is not difficult to derive using simple algebra (see Box). Notice that $v < c$, otherwise the quantity under the square root sign will be negative, giving us an imaginary number. This means that the factor γ is always greater than 1 (it is equal to 1 when $v = 0$), making $t > t'$; that is, the time measured by the stationary observer is always greater than the time measured by the moving observer by the factor γ.

The stationary observer sees the light beam move along the sides of an isosceles triangle of base vt, since the train is moving with speed v (Figure 19.6). If we divide this triangle into two right triangles, as shown in the figure below, we can apply the Pythagorean theorem to one of the two right triangles. According to the second postulate, the speed of the light beam traveling in the moving train as measured by the stationary observer is still c. Therefore, if the time measured by the stationary observer is t, the total distance traveled by the light beam as it moves up to the mirror and returns is ct, and the distance for the one-way trip to the mirror is then $ct/2$. This is the hypotenuse of our right triangle. Thus,

$$\left(\frac{ct}{2}\right)^2 = \left(\frac{vt}{2}\right)^2 + d^2$$

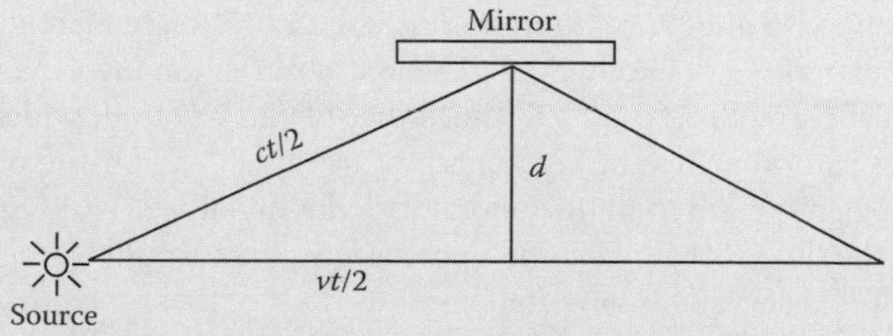

FIGURE 19.6 Right triangles used in the calculation of t.

which becomes, after multiplying through by 4,

$$c^2t^2 - v^2t^2 = 4d^2.$$

Solving for t, we obtain

$$t = \frac{2d}{\sqrt{c^2 - v^2}} = \frac{2d}{c\sqrt{1 - v^2/c^2}} = \frac{2d/c}{\sqrt{1 - v^2/c^2}}.$$

But $2d/c = t'$. Thus, the time measured by the stationary observer for the round trip of the light beam is

$$t = \frac{t'}{\sqrt{1 - v^2/c^2}} = \gamma t'$$

Since we can use the flash-mirror device as a clock by keeping track of the number of return trips, we can see that time flows more slowly in the moving frame of reference. "Not so!" the observer in the moving train would say. "I have been watching the experiment in her train," the moving observer says, "and she is the one moving. I measure a time that is longer than what she measures for her own experiment." Who is right? Both are right. That is the whole point of special relativity. The choice of reference frame does not matter, since there is no preferred frame. Our own reference frame is at rest for us but, as we have seen before, might not be for an observer able to see us from space. In fact the so-called moving train might be traveling west at the same speed as the linear speed due to the earth's rotation, and for an observer in space that train might be the stationary one. The only thing we can affirm is that the time interval between two events (the emission of the flash of light and its arrival at the mirror in our example) as measured by an observer who sees the events occur in his own reference frame is always smaller than the time interval between the same events as measured by another inertial observer. The time interval measured by an observer in his own reference frame is called the *proper time*. We can summarize our discussion by saying that:

time in the moving reference frame always flows more slowly.

This phenomenon is called *time dilation*. It is not an illusion but a real phenomenon that has been experimentally observed. The first observation involved the muon, an elementary particle with a mass 207 times that of the electron, which is found in cosmic rays, a natural source of high-energy particles. The half-life of the muon is 1.52×10^{-6} s when measured in the laboratory. Moving at nearly the speed of light, about half of the muons that reach the upper atmosphere would travel a distance of some 450 m. Yet observations have indicated that half of these muons actually travel distances of about 3200 m. This phenomenon is due to time dilation. In the rest frame of the muon, the life-time is 1.52×10^{-6} s, which is what we measure for the muons in the laboratory, when we are also in the same frame of reference. However,

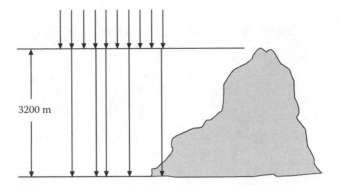

3200 m

FIGURE 19.7 In the muon's reference frame, the half-life is 1.6 μs. At nearly the speed of light, a muon would travel a distance of about 450 m. For an observer on the ground, this half-life is increased by a factor of γ to 10.8 μs, increasing the distance traveled to about 3200 m. Half of the muons detected at the top of a 3200-m mountain are detected at the foot of the mountain.

for an observer on the earth, the muon is traveling at nearly the speed of light, and the half-life as measured by this observer is increased by a factor of γ to 10.8×10^{-6} s. As seen by the observer on earth, the muons travel a greater distance due to the increased half-life (Figure 19.7).

It is not only muons that experience time dilation, but everything. Biological processes, such as the rate at which cells divide or hearts beat, are also slowed down. Time dilation implies the possibility of interstellar travel and even intergalactic travel, when the technologies of the future are able to provide us with the vehicles and engines required for acceleration to speeds close to the speed of light. Our galaxy, the Milky Way, is about one hundred thousand light-years in diameter. At close to the speed of light, it would take slightly over one hundred thousand years to travel from one end of the Galaxy to the other. This time, however, is measured by a stationary observer back on earth. A spacecraft of the future would be able to circumnavigate the entire Galaxy in less than a human lifetime, as measured on the ship.

THE FRONTIERS OF PHYSICS: INTERGALACTIC TRAVEL

Relativistic time dilation might make human intergalactic travel possible, although not with our present technology. The distance to the nearest spiral galaxy, M31, the great galaxy in Andromeda, is about one million light-years (Figure 19.8). A spaceship capable of maintaining a constant acceleration of $1g$ (9.8 m/s^2) during the entire trip, will reach M31 in only 28 years, ship time. However, the crew's families and friends back on earth—and perhaps even earth's civilization as they knew it before leaving—will be long gone way before they reach their destination, since earth time for their trip will be several million years.

FIGURE 19.8   M31, the great galaxy in Andromeda, at a distance of one million light-years from us. (Courtesy of NASA.)

A constant acceleration of 1g produces an incredible increase in speed in a relatively short time. Figure 19.9 illustrates this rapid increase in speed. We see that it takes over three years ship time to reach the nearest star, located at a distance of nearly four light-years, and some five years to reach Epsilon Eridani, a star 10 light-years away from us. It only takes 21 years, however, to reach the center of our galaxy which is 30,000 light-years away. Seven additional years, ship time, will bring this accelerating spacecraft to the vicinity of M31.

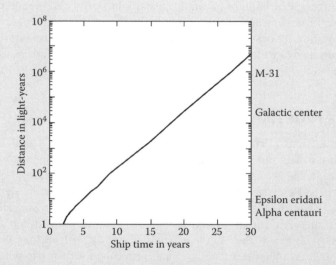

FIGURE 19.9   Ship time for an acceleration of 1g.

The ships and propulsion systems that we have developed for our nascent space exploration are but a shadow of the mighty ships needed to even consider interstellar travel. An acceleration of 1g is achievable now but cannot be maintained over a long period of time. Maintaining it for an entire trip to another star requires the engineering of a much more advanced technology.

SIMULTANEITY

In his 1905 special relativity paper, Einstein writes in one of clearest pieces of scientific prose ever written:

> If we wish to describe the motion of a material point, we give the values of its coordinates as functions of time. Now we must bear carefully in mind that a mathematical description of this kind has no physical meaning unless we are quite clear as to what we understand by "time." We have to take into account that all our judgements in which time plays a part are always judgements of simultaneous events.

But what are *simultaneous events*? Einstein gives an example which Leopold Infeld, one of Einstein's collaborators, called "the simplest sentence I have ever encountered in a scientific paper":

> If I say, for example, "the train arrives here at 7," this means: the coincidence of the small hand of my watch with the number 7 and the arrival of the train are simultaneous events.

Einstein, however, explained that although it could be possible to define "time" by substituting "the position of the small hand of my watch" for "time," such a definition is no longer satisfactory "when we have to connect in time series of events occurring at different places..." In other words, how can we say that two events, one in Paris and the other one in Los Angeles, are simultaneous? We could have a friend in Paris and another in Los Angeles call us over the telephone, we could "synchronize" our watches and be able to determine if the two events are indeed simultaneous. The problem with this approach is that the telephone conversations are transmitted using electromagnetic waves which travel at the speed of light and although, according to the second postulate, this speed is the fastest speed achievable, it is still finite. The signals then are going to take slightly different times to arrive at our location but since we know the speed of light, we can compensate for the delay in the arrival of the signal from the more distant location. Of course, if we are located midway between New York and Paris, somewhere in the Atlantic Ocean, the signals take the same time to arrive and there is no need for compensation.

The problem arises when we consider events in two different inertial systems. Consider again a train traveling with a constant velocity (Figure 19.10). On the train, a passenger is standing in the middle of a car holding two flashlights pointed toward the front and back walls which he turns on at the same time. In the inertial reference

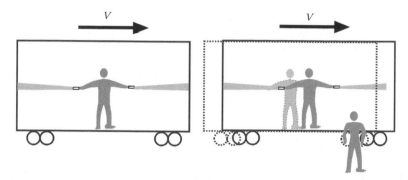

FIGURE 19.10   For the passenger in the moving train (left), the two beams that originate at the same time in the middle of the train reach the two walls simultaneously. A person standing outside the train will detect the beam striking the back wall first.

frame of the train, the two beams move at the speed of light c and reach the two walls simultaneously. For an observer standing on the station platform the two beams also move at the speed c. However, she sees the back wall of the train moving to meet the traveling light beam and the front wall moving away from the beam that is heading in the direction in which the train is moving. According to the observer standing on the platform, the beam heading toward the back wall strikes first and the two events are not simultaneous for her. Therefore, we must conclude that *events that are simultaneous in one inertial frame are not simultaneous for observers in another inertial frame.* This is a consequence of Einstein's second postulate, the constancy of the speed of light in any inertial reference frame.

LENGTH CONTRACTION

The Lorentz–FitzGerald contraction, introduced in 1882 to explain the null results of the Michelson–Morley experiment, turned out to be a consequence of the special theory of relativity. To see how this length contraction appears, let us return to our example of the two trains (Figure 19.11). One train is moving with a constant velocity v and the other is standing by the station. In the stationary train, a passenger measures the length of her car to be L_0. This length, measured by the observer in her own reference frame, is called *proper length*. The proper length is what we call the characteristic length of the object. A passenger in the moving train also measures the length of the car in the stationary train. This observer uses his precise watch to measure the time t' it takes the stationary car to pass through a reference mark he has made on the window of his own car. He calculates the length of the stationary car to be $L = vt'$. The passenger in the stationary train, which is also observing the experiment, measures the time t it takes the mark on the moving train to move at a speed v from the back to the front of her stationary car, and she is able to corroborate that the length of her car is $vt = L_0$. She knows that, because moving clocks run slow, the time t' measured by the moving observer for the passage of her car past the mark on the moving train is less that the time t she measures for that mark to move across

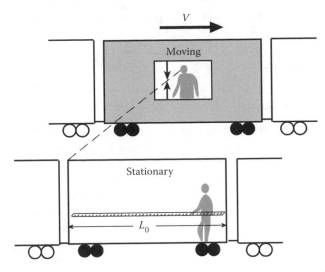

FIGURE 19.11   The passenger in the stationary train measures the length of her car to be L_0. The passenger in the moving train measures the length of the stationary train by timing its motion across a mark on his window.

her stationary car. The time t she measures, we already know, is equal to γ times the time t' measured by the moving observer:

$$t = \gamma t' = \frac{t'}{\sqrt{1 - v^2/c^2}}$$

so that $t' = t\sqrt{1 - v^2/c^2}$. The length L of the stationary car measured by the observer in the moving train is

$$L = vt' = vt\sqrt{1 - v^2/c^2}$$

and since $vt = L_0$, the proper length measured by the observer in the stationary train for her own car, the length measured by the observer in the moving car is

$$L = L_0\sqrt{1 - v^2/c^2}.$$

Since the factor $\sqrt{1 - v^2/c^2}$ is always less than 1, L is always smaller than L_0; that is, the length of any object in motion relative to an observer is always less than the length of the object measured by an observer at rest relative to the object.

Length contraction is a real effect, not an optical illusion. It is a direct consequence of the constancy of the speed of light for all inertial frames of reference and has been experimentally observed. As with time dilation, reversing the situation produces exactly the same effect. That is, if the passenger in the moving train claims that he is not moving and that the stationary train, together with the station are moving with a speed v, he would measure a proper length L_0 for his own car and, for him, the length measured by the passenger in the stationary train, moving relative to him, is shorter by a factor $\sqrt{1-v^2/c^2}$. Again, both points of view are correct since there is no preferred reference frame.

ADDITION OF VELOCITIES

In his 1905 paper on special relativity Einstein concluded that the speed of light must be a limiting velocity for any material body. If we imagine a spaceship moving at 299,000 km/s relative to us on the earth, could it not be possible for the crew on the ship to launch a reconnaissance ship at a speed of 10,000 km/s in the same direction? According to Galilean relativity, we should measure a speed of 299,000 km/s + 10,000 km/s = 309,000 km/s, greater than the speed of light of 299,792 km/s. According to special relativity, it is not possible for the reconnaissance ship to surpass the speed of light. If this is the case, how do we calculate the ship's velocity?

In his relativity paper Einstein shows that if an object moves with a velocity u' relative to an inertial frame which in turn is moving with a velocity v with respect to a second inertial frame (Figure 19.12), the velocity of the object in the second frame is

$$u = \frac{v+u'}{1+vu'/c^2}.$$

For our particular example, the ship's velocity is $v = 299,000$ km/s relative to the earth and the reconnaissance ship travels at $u' = 10,000$ km/s relative to the ship. We would measure a velocity u for the reconnaissance ship of

$$u = \frac{299,000 \text{ km/s} + 10,000 \text{ km/s}}{1+(299,000 \text{ km/s}) \times (10,000 \text{ km/s}) / (299,792 \text{ km/s})^2}$$
$$= 299,051 \text{ km/s}.$$

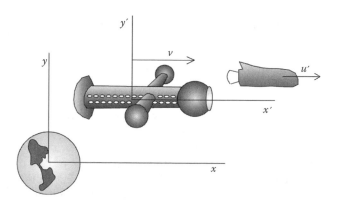

FIGURE 19.12   A reconnaissance ship traveling at a speed u' relative to a mother ship, which moves at a speed v relative to the inertial frame xy of the earth.

Notice that the numerator in Einstein's expression for the transformation of velocities is the Galilean transformation of velocities. In special relativity, this simple addition must be divided by the quantity $1 + vu'/c^2$. If the object is moving in the same direction as the moving reference frame, the quantity in the denominator is greater than 1 and the velocity u in the stationary frame is smaller than what it would be in Galilean relativity. We should also notice that when the velocities are small compared with the velocity of light, the denominator becomes almost 1, and Einstein's transformation of velocities reduces to the transformation of Galilean relativity.

$E = mc^2$

Einstein's fifth and last paper of the year 1905, "Does the Inertia of a Body Depend upon Its Energy Content?" is only three pages long. "The results of the previous investigation," writes Einstein in this short and beautiful paper, referring to his special theory, "lead to a very interesting conclusion, which is here to be deduced." What Einstein deduced was that inertial mass and energy are equivalent.

In his paper Einstein considers an atom undergoing radioactive decay and emitting light. Applying the principles of conservation of energy and momentum to the decay he showed that the atom resulting from the decay had to be less massive than the original atom. Remember that in 1905, Rutherford had not yet introduced his nuclear model of the atom and radioactivity had been discovered only some five years before. As we saw in Chapter 8, this difference in mass accounts for the binding energy of nuclei, but this problem did not exist until 1911 when Rutherford published his nuclear model. In his paper, Einstein concluded:

> If a body gives off the energy E in the form of radiation, its mass diminishes by E/c^2. The fact that the energy withdrawn from the body becomes energy of radiation evidently makes no difference, so that we are led to the more general conclusion that
>   The mass of a body is a measure of its energy-content…

"If a body gives off the energy E in the form of radiation, its mass diminishes by E/c^2." As Einstein himself later pointed out, this conclusion has general validity. An object's mass is a form of energy. Two protons colliding in a high-energy particle accelerator create a spray of new particles, as can be seen in a bubble chamber photograph (Figure 19.13), as kinetic energy is converted to mass energy. If a particle of rest mass m_0 disintegrates spontaneously, all its inertial mass is converted into energy. Since this inertial mass is $m_0 = E/c^2$, the energy released is

$$E = m_0 c^2.$$

This is Einstein's most famous equation.

This mass–energy equivalence also means that an object of mass m_0 can be created out of energy. The phenomenon of *pair production* is an example of this process. When a gamma ray photon passes near an atom, it sometimes disappears, creating an electron and its antiparticle, the positron, leaving the nearby atom unchanged (Figure 19.14). The photon's energy has been converted into mass.

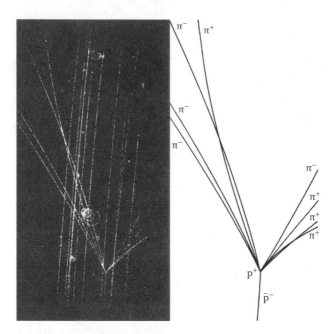

FIGURE 19.13   Bubble chamber photograph of a collision between two protons in a high-energy particle accelerator. (Courtesy of Brookhaven National Laboratory.)

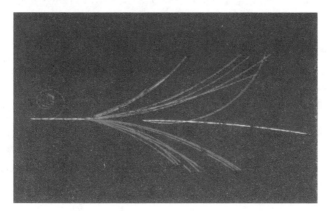

FIGURE 19.14   Pair production. A gamma ray photon of sufficient energy passes near a nucleus and disappears creating an electron and a positron whose tracks can be seen in a cloud chamber curving away from each other due to an external magnetic field. The gamma ray leaves no track. (Courtesy of Brookhaven National Laboratory.)

The total energy of a particle of mass m_0 moving with a speed v and having a kinetic energy KE is

$$E = m_0 c^2 + KE.$$

This total energy is equal to mc^2; thus,

$$mc^2 = m_0c^2 + KE.$$

The relativistic mass m of an object equals its rest mass m_0 when the object is at rest with respect to an observer; that is, when the kinetic energy is zero. When the object is in motion (relative to the observer) the relativistic mass increases. Einstein expressed this relativistic mass as

$$m = \frac{m_0}{\sqrt{1 - v^2/c^2}}.$$

This expression tells us that when the speed of the object approaches that of light its mass becomes infinite. This result is, of course, a consequence of the postulates of special relativity; it places a limit on the speed of any object. Regardless of the force applied on the object or the time this force is applied to the object, it can never reach the speed of light. Mass, as we learned early on in the book, is a measure of the inertia of a body; that is, a measure of the resistance that the body presents to a change in its state of motion by the action of an applied force. As the speed of the object increases, its resistance to the applied force or inertia increases, approaching infinity as the object's speed approaches that of light.

Mass is not the same as "quantity of matter." The relativistic increase in mass does not imply an increase in the object's size or quantity of matter. As the object increases its speed, it does not become larger or denser; it merely becomes more difficult to move. This increase in the object's inertia is what we understand by an increase in mass.

20 The General Theory of Relativity

THE PRINCIPLE OF EQUIVALENCE

Postulate 1 of the special theory, the Principle of Relativity, states that the laws of physics are the same for all *inertial* (non-accelerated) observers. This postulate, as we saw in Chapter 19, is an extension of the Galilean Principle of Relativity that includes not only mechanics but also electromagnetism and optics; that is, all of physics at the time. This postulate means that there is no absolute inertial reference frame and uniform motion has to be referred to an inertial frame. No experiment performed inside a closed chamber in an inertial reference frame can reveal to us whether the chamber is at rest or in motion.

In 1907, while preparing a comprehensive paper on the special theory of relativity, Einstein asked himself whether the postulate of relativity could be extended to non-inertial reference frames. This would mean that accelerated motion is also relative and therefore impossible to detect from inside. But we all know that accelerated motion *can* be detected from inside. If we are in a smoothly moving vehicle, we feel no motion; but if the vehicle lunges, we feel it immediately. We feel ourselves being pressed against the back of our seat when the vehicle accelerates and feel the pressure on our shoulders from the sides of the vehicle when it turns suddenly. Acceleration, Newton taught us, can be detected and therefore cannot be relative. Only uniform motion is relative. Einstein would not accept a partial relativity and kept looking for the way out. "I was sitting in a chair in the patent office in Berne, ... [t]hen there occurred to me ... the happiest thought of my life, in the following form," he wrote later in a paper intended for the journal *Nature*, which owing to its length was never published. He continued:

> The gravitational field has only a relative existence in a way similar to the electric field generated by a magnetoelectric induction. *Because for an observer falling freely from the roof of a house there exists*—at least in his immediate surrounding—*no gravitational field*. Indeed, if the observer drops some bodies then these remain relative to him in a state of rest or of uniform motion, independent of their peculiar chemical or physical nature (in this consideration the air resistance is, of course, ignored). The observer therefore has the right to interpret his state as 'at rest.'
>
> [...]
>
> The experimentally known matter independence of the acceleration of fall is therefore a powerful argument for the fact that the relativity postulate has to be extended to coordinate systems which, relative to each other, are in non-uniform motion.

Accelerated motion cannot be distinguished from inside from the effects of gravity. In another paper published that same year in which he completed the formulation of the equivalence between mass and energy, Einstein also presented a *Gedanken* or thought experiment to illustrate the relativity of accelerated frames of reference. Imagine, he proposed, a laboratory in space. We could imagine that this laboratory is inside a spaceship, far from any gravitational body (Figure 20.1a). The spaceship begins to accelerate and the scientists inside the laboratory obviously feel this acceleration. The scientists perform simple experiments inside their laboratory to determine the value of this acceleration. Let's suppose that the scientists find that the spaceship is accelerating at a rate of 9.8 m/s². The scientists in this laboratory feel the sensation of weight. One of the scientists holding a ball in her hands decides to release it. The ball is now a free body, not in contact with the accelerating spaceship and we may think it is the floor of the spaceship that rushes up to meet the ball. With respect to the spaceship the ball moves in the direction the scientists would call "down," accelerating at 9.8 m/s². The scientists perform other simple experiments, dropping objects of different masses which fall to the floor at the same rate and conclude that the spaceship continues accelerating at the constant rate of 9.8 m/s² so that everything behaves as if the laboratory were back on the ground. The following day,

(a)   (b)

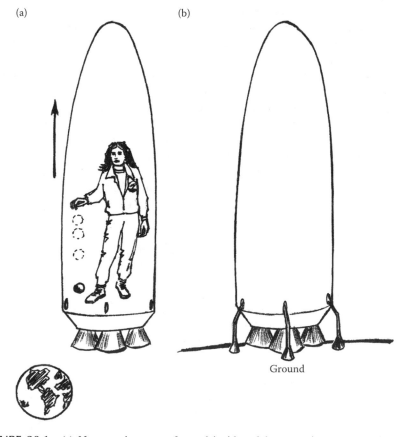

Ground

FIGURE 20.1   (a) No experiment performed inside a laboratory in space accelerating at 9.8 m/s² can allow us to distinguish it from an experiment carried out at a laboratory on the ground on earth, as shown in (b).

the scientists resume their experiments and conclude that the spaceship continues accelerating at a rate of 9.8 m/s². However, when they look outside, they are surprised to find that, while they slept, the spaceship had landed on earth and that they are actually on the ground (Figure 20.1b).

This thought experiment illustrates what Einstein called the *Principle of Equivalence*: that *it is impossible to distinguish accelerated motion from the effects of gravity*. In this sense, the effects of a uniform constant acceleration are *equivalent* to the effects of gravity. Inside the accelerating spaceship the scientists feel real gravity; it is not merely a simulation of gravity. As Einstein said, the "gravitational field has only a relative existence;" it exists while the acceleration exists. When the ball is dropped inside the laboratory in space, it ceases to be in contact with the accelerating spaceship and stops accelerating. When we drop a ball while standing on the ground on earth, it falls. But while the ball is falling, there is no gravity; the ball merely floats in space. It is not difficult to imagine the spaceship accelerating up to meet the floating ball, but to imagine the earth moving up to make contact with the ball seems ludicrous. After all, the earth does not move relative to the solar system every time an object falls to the ground. However, relative to the ball, the earth, the solar system and the entire universe are accelerating toward it. Gravity is precisely equivalent to accelerated motion.

Einstein's principle of equivalence extends the relativity principle to accelerated frames of reference. Since it generalizes the relativity of inertial frames of reference to include non-inertial frames, Einstein called the new theory the General Theory of Relativity.

Implicit in the principle of equivalence is the assumption that all objects fall to the ground with the same acceleration, regardless of their masses. If this were not the case, the effects of gravity would be distinguishable from those of acceleration. The floor of the accelerating spaceship in our thought experiment would make contact simultaneously with two objects of different masses released at the same distance from the floor. These two objects would not fall to the floor simultaneously when the spaceship is back on the ground if the acceleration due to gravity were not the same for all objects. As we saw in Chapter 3, Galileo showed experimentally that all objects fall to the ground with the same acceleration. Newton used the concept of mass in two different contexts: as the measure of inertia, the response of an object to a given force according to Newton's second law, $F = ma$, and as a measure of the gravitational effect on the object. One is the *inertial mass* and the other, the *gravitational mass*. The gravitational force acts in exactly the right proportions on objects of different masses. If the mass is doubled, the gravitational force doubles. Since the inertial mass doubles, the resistance to this double gravitational force also doubles, producing the same acceleration as before. Newton tacitly made inertial mass equal to gravitational mass by assuming Galileo's law that all objects fall with the same acceleration regardless of their composition or structure.

Einstein, with his principle of equivalence, made Galileo's law the foundation of his general theory of relativity. The principle of equivalence can be then stated in the alternative form: *gravitational mass and inertial mass are equivalent and no experiment can ever distinguish one from the other*. Between 1889 and 1922, the Hungarian physicist Roland von Eötvös performed the series of experiments that we mentioned at the end of Chapter 6 which showed that inertial mass and gravitational mass were the same to a few parts in a billion. In the 1960s and 1970s R. H. Dicke at Princeton University and Vladimir Braginsky at Moscow State University performed experiments that showed that inertial and gravitational masses were equal to 1 part in 100 billion and 1 part in a trillion, respectively. However, in 1988, a reanalysis of Eötvös' experiments seemed to indicate a slight discrepancy between inertial and gravitational mass and the existence of a fifth force, with a strength of about one-fiftieth that of the gravitational force, was postulated. More recent experiments performed by scientists from the Joint Institute for Laboratory Astrophysics in Colorado and from the Los Alamos National Laboratory in New Mexico, cast some doubts about the existence of a new force or about any discrepancy in the inertial and gravitational mass. Sophisticated experiments are currently underway that will someday clear the situation.

WARPED SPACE–TIME CONTINUUM

Sometime after introducing the principle of equivalence, Einstein realized that the usual understanding of the geometry of space should be re-examined. We may illustrate this with an example used by Einstein. Take a disc or wheel spinning around its center. From elementary geometry, we know that if we measure the circumference of the disk and divide it by its diameter, we obtain the value π. However, according to special relativity, if we measure the circumference of the disk while it is rotating, we obtain a *smaller* value than if we measure it when the disc is at rest relative to us. The reason for this discrepancy is that a line segment along the circumference of the rotating disc, in motion relative to us, will undergo a length contraction. The diameter is not changed because it is at right angles to the velocity of any point on the perimeter of the circle. Therefore, the ratio of the circumference to the diameter of a rotating circle will be less than π. This means that the geometry of space for accelerating reference frames is no longer the Euclidean geometry we learned at school, where Newton's laws hold, but a different, *curved*, geometry; Riemannian geometry.

In the formulation of the general theory of relativity, Einstein found it necessary to consider a curved four-dimensional space-time to describe the gravitational field. The curvature of a three-dimensional space is hard for us to imagine. Much harder is, of course, to visualize a curved four-dimensional space-time. However, we can start slowly and build up a good understanding of it. The idea of a four dimensional space-time in itself is not hard to understand. In fact, we already use this idea in our daily lives. Suppose that on 12th October we agree to meet a friend for lunch at a restaurant downtown in exactly one week. We first need to specify three numbers for the location in space of this restaurant. We would need to transport ourselves to 5th Street and 3rd Avenue and take the elevator to the 7th floor where the restaurant is located. This knowledge, however, is not enough for us to meet our friend. We need a fourth number. It will not do us any good to appear at the restaurant on 15th October at lunch time because our friend will not be there on that day. In addition to specifying the location in space, the *where*, we need to specify the time, the *when*. We need to appear at the restaurant at the corner of 5th Street and 3rd Avenue on the 7th floor on 19th October at noon (Figure 20.2). These four numbers constitute a four-dimensional description of the event in space-time.

FIGURE 20.2   The description in space-time of our meeting with a friend for lunch requires four numbers; three for the spatial location and one for the time of the event.

The mechanics of Newton can very well be expressed using a four-dimensional space-time formalism, but we would not gain much from doing so. The reason is that in Newtonian mechanics, time is absolute and all we would gain from this description is that the time measured by one observer is the same as the time measured by any other observer. In relativity, the time measured by two observers in motion relative to one another is different, so that treating time on a more equal footing with the space coordinates is advantageous. If our friend meeting us for lunch is actually traveling on a spacecraft at a great speed, the time interval of one week between our agreement to have lunch and the lunch date will be different for our friend. It was Hermann Minkowski, one of Einstein's professors at the ETH, who in 1907 developed the mathematical formalism of a *space-time continuum* as the underlying geometry behind the space and time relationships in the theory of relativity.

Einstein proposed that the space-time continuum is curved. To understand this curvature of space-time, let's consider first the curvature of two-dimensional space. In 1880, a noted Shakespearean scholar, Edwin A. Abbott, wrote what later became a classic science-fiction story, *Flatland*, a fantasy of strange places inhabited by two-dimensional geometrical figures. *Flatland* tells the difficulties of A Square, an inhabitant of this two-dimensional world, in imagining a third dimension; a direction perpendicular to the other two that are familiar to him (Figure 20.3). A Dutch mathematician, Dionys Bruger, wrote in 1960 a sequel to *Flatland*, *Sphereland*, a story of the difficulties of Hexagon, grandson of A Square, who also lives in a two-dimensional world. This time, however, the world is not flat but curved. Like Hexagon, we also find it very difficult to imagine a curved four-dimensional space-time continuum.

For our analogy, the two-dimensional space shall be curved, like Sphereland where Hexagon lived. The Spherelanders originally learned Euclidean geometry, which applied very well to their immediate surroundings. They never noticed that their world was not flat because Sphereland was very large and the curvature was not noticeable at small scales. Therefore the interior angles of a triangular parcel of land added to 180°,

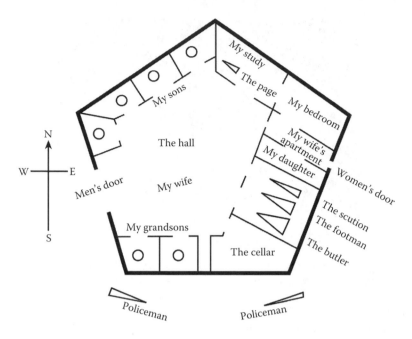

FIGURE 20.3   "Look yonder, and behold your own Pentagonal house, and all its inmates," Sphere, a visitor from the three-dimensional world tells A Square on his first trip outside Flatland.

and squares had their opposite sides parallel to each other. One day, however, a group of their scientists decided to line up a large set of very precise, straight rulers to form a very large triangle. After the scientists measured the interior angles, they were surprised to find that their sum added to 190°, not 180° as they expected. Adding more rulers to form a larger triangle yields still a greater value for the sum of the interior angles; they obtain 197° this time. A third triangle, smaller than the first gives the value of 185°. "What is going on?" they asked themselves. After much thought and discussion it is discovered that they live in a world that is curved. This curvature, however, is very difficult for them to comprehend, because it does not take place in any one of the two dimensions known to them and therefore it is completely invisible to them. It occurs in a direction perpendicular to their world, a third spatial dimension (Figure 20.4).

Our own story is very much like that of the Spherelanders, except that it occurs in three rather than in two dimensions. On a small scale, we cannot detect any deviations in our universe from Euclidean geometry. Einstein, however, reasoned that our universe is curved and that we should detect the effects of its curvature on a large scale. From his principle of equivalence, he knew that bodies of different masses and compositions must fall with the same acceleration in a gravitational field. This phenomenon has nothing to do with the falling bodies themselves and should be an intrinsic property of space, in the same way that the curved triangles observed by the Spherelanders have nothing to do with the rulers they use and are a property of their space.

In general relativity, space is distorted or curved by the presence of a gravitational field; that is, by the presence of a body. A second body that enters the space in the

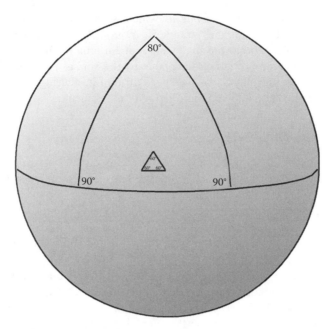

FIGURE 20.4   The people of Sphereland cannot understand why the interior angles of small triangles always add up to 180° whereas for very big triangles, the sum of the angles is larger than 180°. Eventually they invent Riemannian geometry to describe their own curved world.

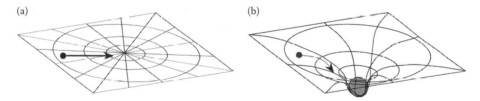

FIGURE 20.5   (a) A stretched rubber sheet is the analog of the three-dimensional space far away from any mass. (b) A billiard ball placed on it creates a dip that distorts or warps the sheet. A marble placed on this warped sheet rolls toward the billiard ball.

vicinity of the first body experiences the distortions in space-time produced by the first body. Although Einstein arrived at this conclusion in 1907 by an analysis of his principle of equivalence, the full development of a gravitational theory required complicated mathematics and several more years. However, a simple analogy will help us visualize this. Imagine a rubber sheet stretched out flat; this is the analog of our three-dimensional space without any masses in its vicinity (Figure 20.5a). A billiard ball placed on this sheet will make a dip so that a marble placed on the sheet near the billiard ball will roll toward it (Figure 20.5b). If you roll the marble on the sheet, it will follow a curved path around the dip. The billiard ball is not pulling the marble; rather it has distorted the rubber sheet in such a way as to deflect the marble. Similarly, the sun distorts or warps the space around it. In such a warped space an object merely travels along a *geodesic*, which is the path of minimum distance between two points. This path, the "straightest" possible path for a planet happens to be an ellipse. The planet orbits the sun in an elliptical orbit not because of a gravitational force of attraction exerted on the planet by the sun, as Newton affirmed, but because the mass of the sun warps the space around it, altering its geometry. For Einstein, there is no "gravitational force"; gravity is geometry.

THE BENDING OF LIGHT

Light always moves in straight lines. Since space is curved, a *straight* line is the shortest path between two points, or a geodesic. Thus, in our curved space a straight line is curved. Since the mass of an object like the sun curves the space around it, the curvature of a light ray increases as it passes near the sun (Figure 20.6).

Let's now return to our example of the accelerating spaceship. Imagine that in an optics laboratory inside the spaceship a very small round window has been installed in one of the walls. When a beam of light from a star enters the small window, it crosses the laboratory and hits the opposite wall at a particular location. The scientists inside the laboratory reasoned that since they are far away from any large mass, the space should be nearly Euclidean and the path of the light beam should be nearly straight. However, because the spaceship is accelerating, the beam does not hit the opposite wall at a point in line with the small window and the star but slightly below (Figure 20.7). The scientists undertake very precise measurements of this slight deflection of the light beam and conclude that their spaceship is accelerating at 9.8 m/s^2.

Since in the general theory of relativity, gravitational fields are locally equivalent to accelerations, the scientists expect the light beam to be bent by the same amount in the presence of the gravitational field of the earth, which produces the same acceleration. When the scientists perform the same experiment after the spaceship has landed on earth, they confirm that the deflection of the beam is exactly the same as

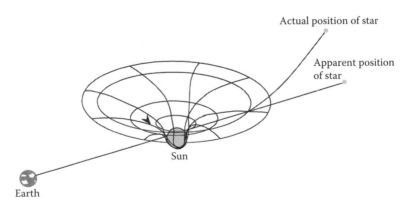

FIGURE 20.6 The path of the light from a star follows a geodesic. As it passes near the sun where the space is distorted, its curvature increases and the position of the star appears shifted when compared with its position when the light passes away from the sun.

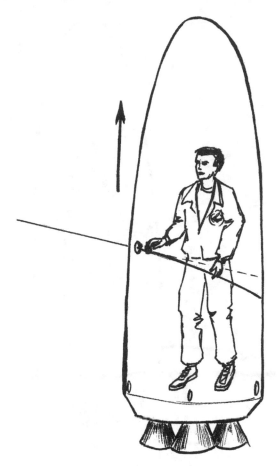

FIGURE 20.7   A beam of light follows a straight line far away from the presence of a large mass. When the beam enters the accelerating spaceship, it hits the opposite wall at a point slightly below.

when the spaceship was accelerating in space (Figure 20.8). Einstein pointed out that this is to be expected from the mass-energy equivalence given by his $E = mc^2$ of the special theory.

Einstein predicted this effect in 1907 but saw no way of designing an experiment to test it. In 1911, Einstein calculated that a light beam from a star grazing the sun would be deflected a very small angle. "A ray of light going past the Sun would accordingly undergo deflection to the amount of … 0.83 seconds of arc," he wrote toward the end of his 1911 paper "On the Influence of Gravitation on the Propagation of Light," an intermediate paper on his general theory. The difficulty was that it is not normally possible to see starlight and sunlight at the same time. But Einstein found a way. "As the fixed stars in the parts of the sky near the Sun are visible during total eclipses of the Sun," he wrote at the end of his paper, "this consequence of the theory may be compared with experience. With the planet Jupiter the displacement to be expected reaches to about 1/100 of the amount given. It would be a most desirable thing if

FIGURE 20.8   The gravitational field of the earth deflects the beam of light by the same amount as the spaceship accelerating at 9.8 m/s².

astronomers would take up the question here raised. For apart from any theory there is the question whether it is possible with the equipment at present available to detect an influence of gravitational fields on the propagation of light."

In fact, it *was* possible with the equipment of the day. In 1914 an expedition led by the German astronomer Erwin Finlay-Freundlich went to Russia to observe a total eclipse of the sun and test Einstein's prediction. The outbreak of war found the expedition already in Russia; the astronomers became prisoners of war before they could make any measurements. They were released after a few weeks, but without their equipment. Although this outcome may seem unfortunate (it certainly was unfortunate for the German astronomers), it was probably better not to have been carried through. Had they been able to make their measurements, they would probably have found a value twice as large as what Einstein had predicted in his paper. The reason is that Einstein was partially wrong in 1911.

It took Einstein four more years to develop the correct general theory of relativity. Along the way, he had to master complicated mathematics and struggle with wrong leads and dead ends. At the end, however, he obtained a physical theory of incredible power and "majestic simplicity," as Banesh Hoffmann, one of his collaborators, described it. The new theory predicts a deflection in the position of the stars whose light grazed the sun by 1.74 seconds of arc, twice his previous prediction. In 1919, Arthur Eddington, a British astronomer and member of the Royal Society, organized an expedition to the Isle of Principe in the Gulf of Guinea, off West Africa. In his *Space, Time, and Gravitation*, Eddington wrote:

> On the day of the eclipse the weather was unfavourable. When totality began the dark disc of the Moon surrounded by the corona was visible through cloud, much as the Moon often appears through cloud on a night when no stars can be seen …
>
> There is a marvellous spectacle above, and, as the photographs afterwards revealed, a wonderful prominence-flame is posed a hundred thousand miles above the surface of the Sun. We have no time to snatch a glance at it. We are conscious only of the weird half-light of the landscape and hush of nature, broken by the calls of the observers, and beat of the metronome ticking out the 302 seconds of totality …
>
> [Of the sixteen photographs obtained, only] one was found showing fairly good images of five stars, which were suitable for determination … The results from this plate gave a definite displacement, in good accordance with Einstein's theory and disagreeing with the Newtonian prediction.

This confirmation electrified the world and made Einstein into a celebrity. The telegram with the news of the confirmation was sent to Einstein by way of Lorentz in Holland who sent it to Einstein in Berlin. When the telegram arrived, Einstein read it and passed it on to a student who was in his office at the time. The student was visibly excited, but Einstein exclaimed, "I knew that the theory was correct. Did you doubt it?" The student protested and said that she thought this was a very important confirmation of his theory. What would Einstein have said, she asked him, if the results had been negative? "Then I would have been sorry for dear God. The theory is correct."

THE PERIHELION OF MERCURY

Newtonian mechanics tells us that the planets move in elliptical orbits around the sun. This is Kepler's first law. In Newtonian mechanics, space is flat. In general relativity, space is curved and when an elliptical orbit is made to fit into this curved space, it becomes deformed in such a way that the planet does not move along exactly the same path every time. Figure 20.9 illustrates what happens. Imagine that Newtonian space is represented by a flat sheet where the elliptical orbit has been drawn. To represent the warped space of general relativity, we fold or cut a wedge out of the paper and rejoin the edges. The elliptical orbit is now deformed and the planet moves along a slightly shifted ellipse. One way that astronomers use to express this is to state that the perihelion (the point in the orbit closest to the sun) precesses—that is, shifts with every revolution around the sun. For this reason, this phenomenon is called *perihelion precession.*

In 1845, the French astronomer Joseph Le Verrier analyzed the orbit of Mercury and found that it did not close. Its perihelion precessed by 574 seconds of arc every century. Although Le Verrier had taken into account the perturbations to the orbit

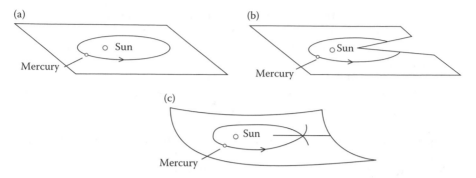

FIGURE 20.9 (a) Elliptical orbit in flat Newtonian space. (b) Cutting a wedge out of the paper removes part of the ellipse. (c) Rejoining the edges to make a curved sheet introduces a deformation in the orbit which results in precession.

of Mercury due to the other planets, especially the closest ones, Venus, Earth, and Mars, and also that of Jupiter, the most massive of the planets, the total contribution to the precession amounted to 531 seconds of arc. There were 43 seconds of arc that could not be explained with Newtonian physics. In November 1915, while finishing his masterpiece, Einstein decided that a good test for his new theory would be to calculate the orbit of Mercury. From his computation, Einstein obtained a value for the perihelion precession of 573.98 seconds of arc, almost exactly what observations had yielded. General relativity accounted for the anomalous 43 seconds of arc that Newtonian mechanics could not explain. "For a few days, I was beside myself with joyous excitement," he would write later.

More accurate experiments performed during the 1970s gave a value for the anomalous part of Mercury's perihelion precession of 43.12 ± 0.21 seconds of arc per century, in complete agreement with the prediction of general relativity of 42.98 seconds of arc. Since physics is never taken as gospel and even the masterpiece of such giants as Einstein is questioned, in recent years a controversy has appeared regarding the influence on Mercury due to the slightly oblate shape of the sun. Although the deviation from the spherical shape is only of 12 parts in a million, it may be enough to change the anomalous precession to 40 seconds of arc per century. The controversy has not yet been resolved.

THE GRAVITATIONAL TIME DILATION

According to special relativity, time in the moving reference frame always flows more slowly. The moving reference frame, however, depends on the observer. Time runs more slowly—however small the change may be—for the airplane pilot according to us on the ground. According to the pilot, it is we and the ground that are moving and it is our clocks that run slow.

Einstein showed that, according to general relativity, time runs more slowly in a gravitational field. Consider again the laboratory inside the accelerating spaceship. Imagine that there are two very precise clocks inside the laboratory, one near the floor and the other near the ceiling (Figure 20.10). These clocks are interfaced to

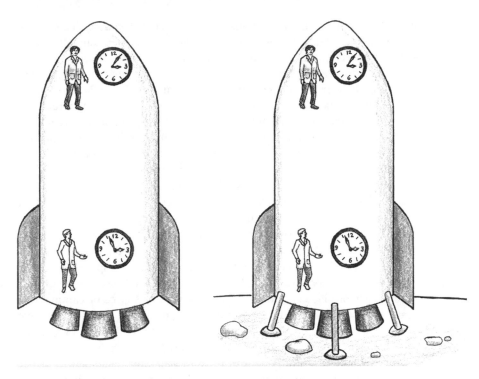

FIGURE 20.10 Comparing clocks in an accelerating laboratory in space and in a laboratory on the ground on earth.

precision instruments that control the rate at which the clocks oscillate. To compare the clocks, an electromagnetic signal oscillating in step with the clock near the floor is sent toward the clock near the ceiling. This signal, we know from special relativity, travels at the speed of light c. Since the spaceship is accelerating, the clock at the ceiling will move away from the incoming signal, which is moving at a constant speed c and the receiving equipment connected to this clock will detect a lower frequency because of the Doppler effect. The scientists in charge of this equipment conclude that the clock on the floor runs more slowly than theirs. To corroborate, they send a signal in step with their clock in the direction of the clock at the bottom. Again, due to the ship's acceleration, the receiving equipment interfaced to the clock on the floor moves toward the incoming signal, which is then Doppler-shifted toward a higher frequency. The scientists on the floor agree with the scientists running the experiment near the top; the clock at the top runs faster.

By the principle of equivalence, when the spaceship is back on the ground on earth, the clocks must behave in the same way; the clock at the top runs faster than the clock at the bottom of the laboratory. Therefore, in a gravitational field clocks run more slowly and, since Einstein taught us that time is what clocks measure, time runs more slowly. This was the first of Einstein's great predictions. Einstein also showed us how this prediction could be tested. Instead of comparing the rates of clocks in different positions in a gravitational field, we could compare the rates or frequencies of oscillation of the light emitted by atoms also in different locations. There is no distinction between a clock and the oscillations of an electromagnetic signal; modern atomic clocks are based on the emission of light with a constant frequency. In 1907, Einstein proposed to test his prediction by comparing the frequencies of the light from atoms on the sun with the frequencies of light from the same atoms on the earth. Since the gravitational field of the sun is stronger, a clock near the sun will run more slowly than on earth and the frequency of light from the sun will be lower. Since this effect is observed as a shift of the spectral lines of the sun's light toward lower frequencies—the red end of the visible spectrum—it is called the *gravitational red shift*.

The test of the gravitational red shift using Einstein's method had to wait until the 1960s and 1970s, when better instruments allowed scientists to separate this effect from other complicated effects such as convection currents of gas in the sun and the effects of pressure, all of which cause the spectral lines to shift in different ways. Modern results agreed with Einstein's prediction of about 5% shift of the spectral lines. An accurate test of the gravitational red shift had been done at Harvard in 1960 in an experiment similar to our thought experiment when the spaceship is back on the ground. The laboratory was the 24-meter tall Jefferson Tower of the physics building with a detector placed at the top of the tower and a source of gamma rays, radioactive cobalt, at the bottom of the building. When the detector was placed near the source, the gamma rays were absorbed. With the detector at the top of the building, the gamma rays were Doppler shifted toward the red due to the small change in earth's gravity and the detector did not absorb them. When the detector was moved downward to produce a blue shift that compensated for the earth's gravitational red shift, the gamma rays were absorbed. The Doppler shift measured in this experiment was extremely small; only three parts in a quadrillion (10^{15}).

The gravitational red shift, although very difficult to measure, is fairly easy to calculate. If the laboratory in the spaceship of our thought experiment accelerating at g has a height h then the electromagnetic signal takes a time $t = h/c$ to travel from the laboratory's floor to the ceiling. During this time, the elevator has increased its speed by

$$v = gt = \frac{gh}{c}.$$

Although we will not do it here, the algebraic steps necessary to extend the expression for the Doppler effect obtained in Chapter 16 to include light are not very difficult. In Doppler observations of light, it is the wavelength rather than the frequency that is more often measured and thus we replace f by c/λ. If $v < c$, the magnitude of the Doppler wavelength shift is

$$\frac{\Delta\lambda}{\lambda} = \frac{v}{c} = \frac{gh}{c^2}.$$

More direct experiments to detect the gravitational time delay have also been done using two cesium-beam atomic clocks, one on the ground and the second one on a jet plane flying overhead. In an experiment conducted in 1971 by scientists of the US Naval Observatory and Washington University in St. Louis, the results agreed with the predictions of general relativity. In 1976, a Scout D rocket carrying a hydrogen maser clock was launched by NASA for a two-hour suborbital flight to an altitude of 10,000 km. This clock is based on the transitions between two energy levels in hydrogen that emit light with frequency of 1420 MHz. After two years of analysis of the data gathered during this two-hour experiment, Robert Vessot and Martin Levine of the Smithsonian Astrophysical Observatory at Harvard and their collaborators at NASA, who had designed the experiment, reported that the frequency shifts observed agreed with the theoretical predictions of general relativity to a precision of 70 parts per million.

THE FRONTIERS OF PHYSICS: ORBITING CLOCKS

According to special relativity, the time interval between two events is different for different inertial observers. Time in the moving reference frame always flows more slowly. This phenomenon is known as time dilation. General relativity predicts that time runs more slowly in a gravitational field.

Special relativity predicts that a clock on an orbiting satellite would run more slowly than one in its ground tracking station due to the significant velocity of the satellite relative to the ground. However, because the satellite is farther away from the center of the earth, the gravitational field is weaker there and general relativity predicts that the satellite-based clock would run faster. For a satellite with an orbital radius of at about 1.5 times the earth's radius, the two effects cancel.

The NAVSTAR 2 satellites, a fleet of military satellites which are part of the US Global Positioning System, have an atomic clock on board and orbit the earth at an altitude of about 20,000 km. This orbit is about 4.2 times the earth's radius, and the time delays due to its speed relative to the ground and its position in the weaker gravitational field of the earth do not cancel out. The clocks on the NAVSTAR 2 satellites run faster than ground clocks. In a single day, the NAVSTAR 2 clocks will be 38.5 microseconds faster and the satellites clocks must be adjusted accordingly. If they were left alone, navigational errors would average about 11 km per day. The correction, calculated with both the special and general relativity predictions, allow the satellites to provide navigational positioning with an accuracy of a few meters.

BLACK HOLES

In the constellation of Orion there is a large cloud of gas, the Orion Nebula, located in the "sword" of Orion (Figure 20.11). The Orion Nebula is located within our galaxy, the Milky Way, at a distance of about 1500 light-years. In this large cloud, about 15 light-years across, stars are forming right now. The total mass of the Orion Nebula is sufficient to form as many as 100,000 stars. The density of this nebula is about 1,000 atoms per cubic centimeter. Within this cloud, regions of particularly high density

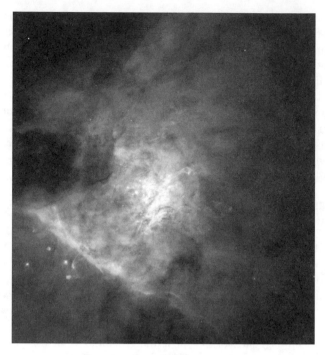

FIGURE 20.11   The Orion Nebula, a birthplace of stars. (Courtesy of C. R. O'Dell and S. K. Wong, NASA, Space Telescope Science Institute.)

fragment themselves into large balls of gas called *protostars*. Gradually, the proto-star shrinks, due to gravitational forces, becoming denser. As we know, when a gas becomes denser, its temperature rises. Eventually, the temperature in the core of the protostar is high enough for nuclear fusion to begin; the star is born.

A star the size of the sun will shine for about 10 billion years before exhausting its fuel. A larger star, one with a mass 50 times the mass of the sun, will spend its fuel much more rapidly; it will live for only a few million years. After about 10 billion years, a star like the sun, or one with a mass smaller than 4 solar masses, lacking hydrogen to fuse into helium and produce heat in its core, begins to contract because the core no longer has enough pressure to hold the overlying layers against the force of gravity. This contraction heats up the core again which starts to radiate this energy out toward the outer layers of the star. This increase in energy goes into expanding these outer layers and the star swells up. The total energy of the star is now distrib-uted over a much greater surface area with the result that the surface temperature decreases. The star becomes a *red giant*.

When the temperature of the core is hot enough, nuclear fusion of helium nuclei into carbon begins. Eventually, all the helium is used up and the core cools down once more. The red giant collapses into a *white dwarf* which will radiate energy for a few billion years. Finally, when all its energy is exhausted, the star ends up its life as a burned-out mass called a *black dwarf*. It is believed that the universe is not yet old enough for the first black dwarfs to have formed.

More massive stars have a more dramatic end. The outer layers of a star with about eight solar masses will explode into a *supernova*. In 1987, the University of Toronto astronomer Ian Shelton, working at Las Campanas Observatory in Chile, discovered a supernova in one of the satellite galaxies of the Milky Way; the Large Magellanic Cloud (Figure 11.6). Although the outer layers of the star fly apart at great speeds, the core of the star, however, collapses. Due to the enormous pressures, the electrons are pushed into the nucleus where they combine with protons to produce neutrons and neutrinos. The neutrinos escape and the core is now an incredibly dense sphere of neutrons. This is called a *neutron star*. The density of this star is so great that a neutron star with the mass of the sun will be only about 20 km across.

If a neutron star has a mass of about three solar masses, there is no force that can stop its collapse and the material of the star is compressed to a state of infinite density. At this stage, nothing is left of the star except an intense gravitational field; the star has become a *black hole*. As the star collapses, the curvature of space-time becomes increasingly pronounced so that the light rays from the star are deflected more and more. At some point during the collapse, the curvature of space-time is so severe that no light can escape and the body appears black (Figure 20.12). Since nothing can travel faster than light, nothing else can escape from the collapsed star.

At what point in the collapse of a star is the curvature of space-time severe enough to impede light from escaping? A few years after Einstein published his general theory of relativity the German astrophysicist Karl Schwarzschild calculated this critical size, now called the *Schwarzschild radius*. We can see how this radius can be calculated in terms of the concept of *escape velocity*. Recall from Chapter 4 that if an object of mass m is raised through a height h, the gravitational potential energy is increased by $PE = mgh$. This expression is only valid if the acceleration due to gravity g is assumed constant. This, in fact, is not so, as we can see from Newton's law of universal gravitation; g diminishes as $1/r^2$, where r is the distance from the center of the earth (see Chapter 6). When h is small compared to r, the expression mgh for the change in potential energy is a good approximation. If a force F is used to slowly displace an object of mass m upwards a distance h without changing its kinetic energy, this force must be equal and opposite to the weight F_{grav} of the object. Then the increase in potential energy is equal to the work done on the object, Fh; that is,

$$PE = Fh = -F_{grav}\, h$$

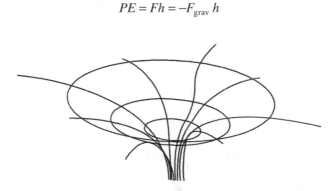

FIGURE 20.12 Graphic representation of the curvature of space-time around a black hole.

where F_{grav}, the weight of the object, is the gravitational force that the earth, mass M, exerts on the object of mass m, or

$$F_{grav} = G\frac{Mm}{r^2}.$$

Using the methods of calculus, it can be shown that the gravitational potential energy of an object of mass m at a distance r from a mass M is

$$PE = -G\frac{Mm}{r}.$$

The escape velocity v_e of an object of mass m on the surface of the earth (at a distance R from the center of the earth) is the minimum velocity that we must impart to the object so that it escapes the gravitational grasp of the earth. It the object barely moves after it leaves the gravitational field of the earth, then both its kinetic energy and potential energy must be zero. Then

$$\frac{1}{2}mv_e^2 + \left(-G\frac{Mm}{R}\right) = 0.$$

The escape velocity is then

$$v_e = \sqrt{\frac{2GM}{R}}.$$

Substituting the mass and radius of the earth, one finds that the escape velocity is 11.2 km/s. For the moon, the escape velocity is 2.3 km/s and for Jupiter, the largest planet in the solar system, 60 km/s. That is, if we throw a ball upwards with an initial velocity of 11.2 km/s (or about 40,000 km/h), the ball will leave earth and coast out toward infinity (neglecting air resistance). On the moon, we would need to impart an initial velocity of only 2.3 km/s to accomplish the same feat.

Since light always travels at speed c, we could turn our expression for the escape velocity around to find the critical radius within which the mass M of the star must be contained so that light does not escape from it; that is, the Schwarzschild radius. If we take $v = c$ in the escape velocity expression, Schwarzschild radius R_S is given by

$$R_S = \frac{2GM}{c^2}.$$

We can see that the size of the Schwarzschild radius is proportional to the mass of the star. For a star with a mass similar to that of the sun, the Schwarzschild radius is about 3 km.

Although we derived this expression from Newton's mechanics rather than from the general theory of relativity, as Schwarzschild did, the expression is still correct. The behavior of objects inside a black hole, however, cannot be explained in terms of Newton's equations. As illustrated in Figure 20.12, the space-time near a black hole becomes severely distorted. In 1965 the British physicists Stephen Hawking and Roger Penrose showed that, according to general relativity, within a black hole the curvature of space-time becomes infinite and this means that the gravitational force becomes infinite. This infinite distortion in space-time is called a *singularity*.

The distance from the point of no return or event horizon—at the Schwarzschild radius—to the singularity is not a distance in space but rather an interval of time. In a sense, then, space and time are interchanged inside a black hole. The gravitational field inside a black hole changes with time, as detected by a hypothetical observer inside the black hole. From the outside, however, it takes an infinite time for the field to change, which is another way of saying that it does not change at all; the black hole remains static for an outside observer. This strange phenomenon is due to the gravitational time delay. At the singularity, time stops. If an astronaut were to travel into a black hole and report to the space station in orbit around the black hole his location at ten-second intervals, the messages would at first arrive at ten-second intervals. Just before the astronaut crosses the event horizon, the astronauts in the space station would notice the intervals getting slightly longer. Once the astronaut has crossed the event horizon, the messages take an infinite time to arrive at the space station; that is, they never arrive. The astronaut dutifully keeps sending his messages according to his watch, but the signals cannot cross the event horizon and never reach the space station. According to the people in the space station, the astronaut's time has been gravitationally delayed. The signals have been progressively red-shifted until they appear to have an infinite wavelength; the astronaut seems to have been frozen at the event horizon, taking an infinite time to fall through. To the astronaut, time continues at its normal rate and nothing changes, except the enormous tidal forces that tear his body apart: this thought experiment would not be possible to carry through with a live astronaut.

THE FRONTIERS OF PHYSICS: SPACE-TIME DRAG

According to Einstein's general theory of relativity, a spinning object drags space-time along with it. Until recently, this prediction of general relativity had not been observed since this effect is incredibly small near the earth. Near a very massive object, like a neutron star or a black hole, the effect should be detectable. Since we can't travel to the stars, the problem in this case is how to detect the space-time drag from afar. In November of 1997, two teams of astronomers measured the periodic vibrations of the x-rays emitted by the gas falling into several rapidly spinning neutron stars and black holes and found evidence of space-time dragging.

General relativity predicts that if the gas surrounding a spinning star forms a disk that lies at an angle to the plane of rotation, the disk will wobble like a spinning top. This wobble is the cause of the oscillations observed by the astronomers. Wei Cui, of the Massachusetts Institute of Technology, studied the x-rays from several candidate black holes using NASA's Rossi X-ray Timing Explorer satellite and found greater dragging around the black holes spinning more rapidly (see Figure 20.13).

FIGURE 20.13   (See color insert following page 268.) Artist's conception of the space-time drag near a spinning neutron star. Recent observations with NASA's Rossi X-ray Timing Explorer satellite seem to confirm the prediction of Einstein's general theory of relativity that a spinning object drags space-time along with it. (Courtesy of Joe Bergeron.)

Luigi Stella of the Astronomical Observatory of Rome and his colleague, Mario Vietri of the University of Rome, studying data from the same NASA satellite, found similar oscillations in the x-rays from several rotating neutron stars. Both results were reported at the same meeting of the American Astronomical Society in Colorado.

An extremely precise measurement of the space-time drag caused by the earth as it rotates is being performed by NASA's *Gravity Probe B*, the most ambitious attempt at experimentally measure several predictions of general relativity. Launched on April 20, 2004, the *Gravity Probe B* satellite was placed into orbit at an altitude of 640 km. To measure space-time drag, the satellite carries a gyroscope that was initially aligned with a distant star. After one year and 5000 orbits, a measurement of the deviation of the orientation of the gyroscope is made. A deviation in the spin axis alignment in the direction perpendicular to the plane of the earth's rotation determines the frame dragging effect. Theoretical predictions place this value at 39 milliarcseconds or 10 billionths of a degree. In February 209 researchers announced results that matched this prediction with an uncertainty of 15%.

21 The Early Atomic Theory

THE PHYSICS OF THE ATOM

Physics underwent two great revolutions at the turn of the twentieth century, and Albert Einstein was in the middle of both. One, the theory of relativity, developed almost single-handedly by Einstein, changed our conception of space and time, as we saw in the previous two chapters. The other, quantum physics, the physics of the atom, which started with the work of Max Planck and Albert Einstein, changed our understanding of the nature of matter.

We studied the development of the early models of the atom in Chapter 7. In this and the remaining chapters, we see how this development led to our current understanding of the nature of matter, starting with the work of the two pioneers, Planck and Einstein.

BLACK BODY RADIATION

We know from experience that a hot object radiates energy. An electric stove has an orange glow when it is hot. But even cold objects radiate energy. In fact, all objects glow, but unless the temperatures are high enough, the glow is invisible. Throughout the nineteenth century, physicists knew that this type of radiation was a wave phenomenon, more properly called electromagnetic radiation. They also knew that the reason why we sometimes see the glow is because some of the frequencies of this electromagnetic radiation happen to be the same as the frequencies of light to which our eyes respond.

Just before the beginning of the twentieth century, physicists considered the problem of the radiation emitted and absorbed by objects at different temperatures. When radiation falls on an object, part is absorbed and the rest is reflected. If we place a brick on a table in the middle of a room and wait a few minutes, the brick will absorb energy from the surroundings at the same rate that it re-emits it back to the surroundings. The brick is in thermal equilibrium with the surroundings. If the brick, in thermal equilibrium with the surroundings, were to absorb more energy than it emits, it would become spontaneously warmer than its surroundings.

A light-colored object reflects more of the radiation falling upon it than a dark-colored object. Since an object in thermal equilibrium with its surroundings absorbs and emits energy at the same rate, a dark-colored object, being a good absorber of energy, is also a good emitter. An imaginary body that absorbs all radiation incident upon it is called an *ideal black body*. Ideal black bodies, then, are perfect absorbers and perfect emitters. A good approximation to a black body is a hollow block with a small hole, as shown in Figure 21.1.

When scientists measured the radiation emitted by this black body at different wavelengths, they found that the experimental results could not be explained with

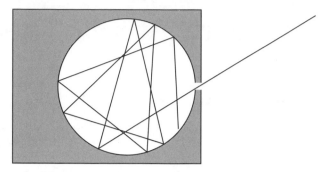

FIGURE 21.1   A cavity as an approximation to a black body. Radiation entering through the hole has a very small chance of leaving the cavity before it is completely absorbed by the walls of the cavity.

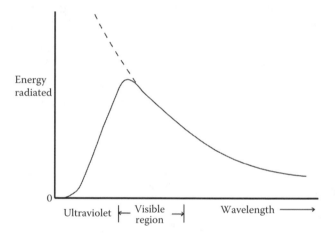

FIGURE 21.2   Black body radiation curve representing the measured values of the energy radiated at different wavelengths. The broken line shows the calculations based on what was known at the time. The calculations do not agree with the measurements at ultraviolet wavelengths.

what was known at the time about the nature of radiation. Scientists had proposed that the light emitted by an object was formed by the continuous changes in energy of charged particles oscillating within the matter. When they calculated the energy emitted in a continuous way by these oscillating particles, their results agreed with the experimental measurements only for very long wavelengths (Figure 21.2). At short wavelengths their theoretical calculations predicted that the energy radiated should become larger and larger (broken line in Figure 21.2), whereas the experiments showed that at those wavelengths the energy should approach a value of zero, an anomaly that became known as the "ultraviolet catastrophe."

In 1900, the German physicist Max Planck was able to solve the problem. He did so by making an assumption which seemed very strange, and which Planck himself could not understand initially. He assumed that the energy emitted by the charged particles in this black body could only have certain discrete values. The oscillating charged particles lose or gain energy, not in a continuous way, but in discrete jumps. In other words, the energy was radiated in little bundles or packets. Furthermore, the energy of radiation was directly proportional to the frequency of radiation. Planck called these bundles of energy quanta. His formula is

$$E = hf$$

where h is a constant, called Planck's constant, and f is the frequency. The value of Planck's constant is

$$h = 6.626 \times 10^{-34} \text{ J s} = 4.136 \times 10^{-15} \text{ eV s}.$$

According to Planck's equation, the energy of each quantum depends on its frequency of oscillation. Since the value of Planck's constant h is very small, the quantum idea becomes important only in systems where the energies are very small, as is the case with atoms. Planck's equation not only fits the observed black body radiation curves but also settles the "ultraviolet catastrophe" argument. At *high* frequencies, corresponding to short wavelengths, radiation consists of *high*-energy quanta, and only a few oscillators will have that much energy, so only a few high-energy quanta are emitted. This is the left part of the radiation curve (Figure 21.2), where the values of the energy radiated are very small, approaching zero. At *low* frequencies, corresponding to long wavelengths, the quanta have low energies and many oscillators will have these energies, resulting in the emission of many *low*-energy quanta, but they each have so little energy that even added together, the emitted radiation does not amount to much. This corresponds to the right side of the radiation curve in Figure 21.2, where the energies have small values. It is only in the middle frequencies where there are plenty of oscillators with middle wavelengths that have enough energy to emit quanta of moderate size which, when added together, produce the peak in the radiation curve. Planck's insight was to realize that different energies were associated with different wavelengths, instead of assuming that the energy was equally distributed between different wavelengths, as the previous radiation theories had proposed.

THE PHOTOELECTRIC EFFECT

About the same time that Planck was struggling with the problem of black body radiation, other physicists were trying to understand a seemingly unrelated problem, known as the photoelectric effect. It had been observed by Heinrich Hertz in 1887 that when ultraviolet light shone on certain metallic surfaces, electrons were ejected from those surfaces (Figure 21.3), *provided that the frequency of the radiation exceeded a critical threshold*. The number of electrons emitted was proportional to the intensity of the light, as was expected. However, the kinetic energies of the emitted electrons were not in any way related to the incident light intensity. This was quite puzzling.

In 1905, Albert Einstein, considered the problem of the photoelectric effect. Einstein assumed that the quanta of energy that Planck had introduced to explain the

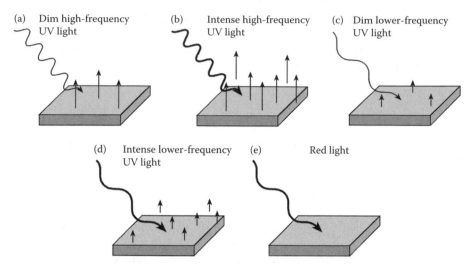

FIGURE 21.3   The photoelectric effect: (a) Dim UV light of high frequency produces few fast electrons. (b) Intense high frequency UV light produces many fast electrons. (c) Dim UV light of lower frequency produces few slow electrons. (d) Intense UV light of lower frequency produces many slow electrons. (e) Red light (much lower frequency) produces no electrons.

black body radiation were also characteristic of light rather than a special property related only to a single phenomenon. Thus, he assumed that light also consisted of discrete quanta of energy. When one of these quanta, or *photons* as he called them, struck the surface of the metal, all its energy was transmitted to an electron in the metal. Thus, an electron could only absorb energy in packets or bundles of energy, so that a light of greater intensity did not provide more energy for the electron to absorb, just more photons per second, and the chances of a single electron getting hit twice was extremely unlikely. However, a more intense beam, having more photons, would strike more electrons in the metal.

An electron would require a minimum amount of energy to break away from the surface of the metal. This minimal energy was called the *work function* φ (the Greek letter phi). Since Einstein assumed that Planck's radiation formula was universal, the energy of the photons was also given by $E = hf$, so that a minimum value of the energy implied a minimum value of the frequency. This explained the threshold frequency below which no electrons could be emitted.

Einstein's theory of the photoelectric effect accounted for all the puzzling experimental observations. According to Einstein, when a photon of energy $E = hf$ strikes an electron in a metal, the electron absorbs all the energy of the photon. The energetic electron then starts to make its way toward the surface of the metal. From the principle of conservation of energy, we know that if the energy E gained by the electron is larger than the work function φ of the metal, then the electron reaches the surface and escapes. The energy that remains is the kinetic energy of the ejected electron. That is,

energy gained from the photon = energy to leave the metal + kinetic energy,

TABLE 21.1
Work Functions of Some Elements

Element	φ (eV)
Aluminum	4.20
Calcium	2.71
Cesium	1.96
Germanium	4.61
Iron	3.91
Platinum	5.22
Potassium	2.24
Sodium	2.28
Tungsten	4.54

or

$$hf = \text{energy to free the electron} + KE.$$

Since the work function φ is the minimal energy required to free the electron, some of the electrons leave the metal with kinetic energy as great as

$$KE_{\max} = hf - \varphi$$

This expression is Einstein's equation for the photoelectric effect. In Table 21.1 we list the work functions of some elements.

The photoelectric effect has practical applications in the detection of very weak light. Some sensitive television cameras make use of photomultiplier tubes in which an incident photon ejects an electron from a photosensitive area at the front of the tube. The ejected electron is accelerated by an electric field, then strikes a metal surface, producing the emission of several more electrons, which in turn strike additional electrons from further metal surfaces. The result is a cascade of electrons down the tube (Figure 21.4). A high-gain photomultiplier tube can transform the arrival of a single photon at its face into the emission of a billion electrons, which can easily be detected.

If light consists of discrete quanta of energy or photons, then light has particle properties. But light also exhibits wave properties, as we have seen. Can light be both a particle and a wave? These two concepts are mutually contradictory. Either light is a particle or it is a wave. It cannot be both, in much the same way that a person cannot be big and small at the same time. Einstein's theory introduced a conflict in physics. It took about 20 years for this conflict to be resolved.

In this crucial year of 1905, Einstein published in three major papers his Special Theory of Relativity, his explanation of a phenomenon known as Brownian motion, and his explanation of the photoelectric effect. In light of these discoveries, the year 1905 is known in the scientific world as Einstein's *annus mirabilis*. Although by the

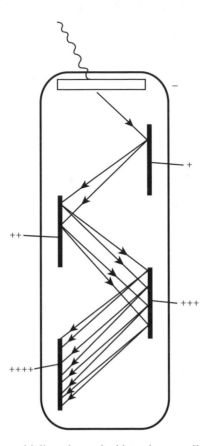

FIGURE 21.4 In a photomultiplier tube, an incident photon strikes an electron from a photosensitive area near its face. This electron in turn strikes other electrons from subsequent positively charged metal surfaces, ejecting them. A cascade of electrons is produced.

time Einstein received the Nobel Prize he had also published his General Theory of Relativity, perhaps the greatest intellectual achievement of all time, he received the Prize for his explanation of the photoelectric effect. It is believed that the Nobel committee thought at the time that the theory of relativity was too controversial.

PHYSICS IN OUR WORLD: USING PHOTONS TO DETECT TUMORS

Shining a strong light on an object produces a sharp shadow. This simple fact is the basis for transillumination, a technique used in breast examinations. A tumor inside a woman's breast casts a shadow when a strong beam of light is sent through the soft tissue of the breast. The tissue itself, however, scatters some of the light, making it difficult to detect tumors smaller than about a centimeter across.

Scientists have recently succeeded in developing a technique that might make it possible to detect very small tumors. When light enters the soft tissue of the breast, some of the photons travel in a nearly straight line, while the rest bounce off the molecules of the tissue and follow a more tortuous path. Since only the straight-line photons produce a sharp image of the tumor, scientists need to separate the straight-line photons from the rest. The straight-line photons reach the other side earlier than the rest. In other words, the direct photons will arrive at the detector slightly ahead of the other photons. The time difference between the two sets of photons is incredibly small, of the order of a few picoseconds (trillionths of a second). Physicists at the University of Michigan have developed a fast gate that opens and closes at the rate required to separate the two sets of photons when placed in front of the detector. Using this new technique, the scientists are able to detect spots 200 micrometers (thousandths of a millimeter) across inside samples of human breast tissue 3.5 millimeters thick.

The new technique has made possible the development of a method of looking at the blood flow in the breast without the familiar breast compression of regular mammograms. This technique, called Computer Tomography Laser Mammography, is currently being used for research purposes in the United States.

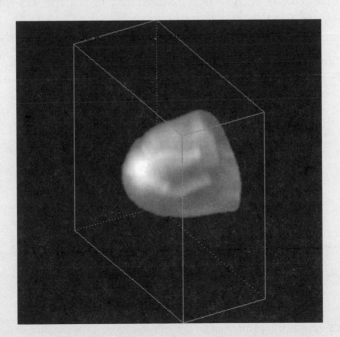

(Courtesy of Imaging Diagnostic Systems, Inc.)

THE BOHR MODEL OF THE ATOM REVISITED

As we learned back in Chapter 7, the Danish physicist Niels Bohr used Einstein's idea of the quantum of energy to construct a model of the atom which explained the observed spectrum of the hydrogen atom very well. We saw in Chapter 18 that the emission and absorption lines of the hydrogen spectrum are produced by photons emitted or absorbed in the electronic transitions between Bohr's allowed orbits. The energy of these photons is given precisely by Planck's formula $E = hf$.

If we call E_i the energy of the electron in the initial state, and E_f its energy in the final state, then $E_i - E_f = hf$ is the change in energy of an electron jumping from level i to level f. We can illustrate the absorption and emission processes with the aid of an energy-level diagram. Figure 21.5 shows the energy-level diagram for the hydrogen atom where various series of transitions are indicated by vertical arrows between the levels, named for the scientists who discovered them.

As we saw in Chapter 18, in an energy-level diagram, the letter n designates the energy level, beginning with $n = 1$, the ground state, the lowest stationary state for the electron. Since only *changes* in potential energy have any meaning, any convenient zero-level for the energy can be chosen. When $n = \infty$ the electron is so far from the nucleus that it does not feel its attraction, and we take its interaction energy to be zero. The energy of any other state, E_n, is the *difference* from this $n = \infty$ state. As the electrons jump to energy states that are closer to the nucleus, their energies *decrease* and, as we have just seen, they radiate this energy. Thus, the energies of these states are less than the energy of the $n = \infty$ state, which has been assigned the value of zero. To be less than zero, the energies of these states must be negative. Bohr showed that the energy, E_n, of a particular level n is

$$E_n = -\frac{E_{GS}}{n^2}$$

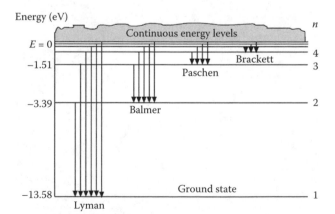

FIGURE 21.5   Energy level diagram for the hydrogen atom. Only the Balmer series is in the visible range of wavelengths.

where $E_{GS} = 2.2 \times 10^{-18}$ J $= 13.6$ eV is the lowest energy possible for an electron in a hydrogen atom, its ground state energy. This expression allows us to calculate the energy absorbed or emitted when an electron jumps from one energy level to another, in terms of the ground state energy.

Bohr's theory agrees with experiment only for the hydrogen atom, and it does not work well for more complicated atoms. Nevertheless, it was the first theory that attempted to explain the structure of the atom. It should be noted, however, that Bohr constructed his theory to agree with experiment and could not explain why the electrons did not radiate when they were in the stationary orbits allowed by his theory, or how the electrons absorbed or emitted energy when they underwent a transition between orbits. These were ad hoc assumptions. Ten years later a bright graduate student in Paris was to introduce a new idea that would become the basis for a completely different kind of physics that would explain these assumptions.

DE BROGLIE'S WAVES

Louis Victor de Broglie was a graduate student at the University of Paris looking for a dissertation topic. He was fascinated with the wave-particle duality that Einstein had introduced and reasoned that perhaps electrons could also exhibit particle and wave characteristics. This problem became his dissertation research. By 1924 de Broglie's dissertation was complete, but his advisor was apprehensive about granting him his doctorate based on a highly speculative idea. He decided to show the work to Einstein. It was only when Einstein accepted the idea as "sound" that de Broglie's degree was granted. In 1929 de Broglie received the Nobel Prize for his theory.

De Broglie's theory explained the stationary orbits in Bohr's model of the hydrogen atom by assuming that each electron orbit was a *standing wave*. Since the orbits in Bohr's model are circles, the electron wave must be a circular standing wave that closes in on itself, as shown in Figure 21.6.

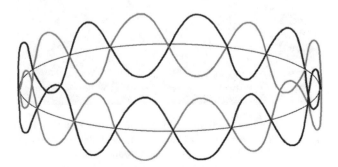

FIGURE 21.6   De Broglie's standing waves for the allowed orbits.

The frequency and wavelength of electron waves in de Broglie's theory are

$$f = \frac{E}{h} \quad \text{and} \quad \lambda = \frac{h}{mv}$$

where m is the mass of the electron and v its velocity. In 1927, the American physicists C. J. Davisson and L. H. Germer observed interference patterns with electron beams, as if the electron beam had wave properties (see Figure 21.7). De Broglie's prediction was thus confirmed.

The wave nature of electrons is the basis for the electron microscope, invented in 1932 by Ernst Ruska in Germany. The resolution of a microscope (that is, its ability to reproduce details in a magnified image) is inversely proportional to the wavelength of the light used to illuminate the object. Electron wavelengths are much shorter than those of visible light. Since the energy of the electrons is proportional to their frequency and inversely proportional to their wavelength, fast electrons have shorter wavelengths. "Illuminating" an object with electron waves in an electron microscope, therefore, produces higher magnifications than ordinary light. Some of the most powerful electron microscopes achieve magnifications of over ×1,000,000 allowing the observation of objects as small as 0.05 nm. Figure 21.8 shows Apollo 14 lunar dust samples at ×2000 magnification.

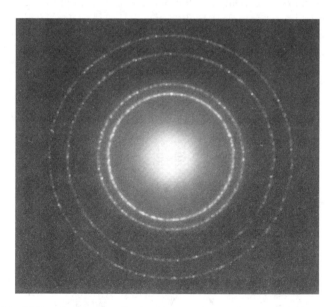

FIGURE 21.7 Diffraction pattern obtained by passing an electron beam through a crystal. (From *PSSC Physics Seventh Edition*, by Haber-Schaim, Dodge, Gardner, and Shore. With permission from Kendall/Hunt Publishing Company, Dubuque, IA, 1991.)

FIGURE 21.8 Apollo 14 lunar particles at 2000 magnification. The spherical object is a glass spherule, a bead rich in iron which was formed by small impacts on the lunar surface. (Courtesy of P. Hintze, NASA Kennedy Space Center, Corrosion Technology Laboratory.)

QUANTUM MECHANICS

After de Broglie introduced his matter-wave hypothesis, things began to move fast. The major problem that physicists were trying to solve was that of the wave nature of the electron. In 1926 the Austrian physicist Erwin Schrödinger presented an equation that described the wave behavior of the electron in the hydrogen atom. Independently of Schrödinger, a 24-year-old professor of physics at Göttingen by the name of Werner Heisenberg undertook a purely mathematical treatment of atomic particles. Heisenberg's approach was later proven equivalent to Schrödinger's wave equation theory.

This new theory became known as *quantum mechanics*. In the years that followed, many discussions took place between Einstein, Bohr, Heisenberg, Schrödinger, and most of the great physicists of the time, about the interpretation of quantum mechanics, and especially about the meaning of the wave equation. Soon they agreed that the waves associated with the atomic particles like the electron did not represent the actual motion of the electron. They quickly realized that the electron neither acts like a wave nor was smeared out throughout the wave. The physicist Max Born suggested instead that these waves were waves of *probability*; that is, they represented the *probability* of finding the electron in some region of space.

What is then the actual motion of the electrons around the nucleus? Quantum mechanics tells us that this question cannot be answered because, as Bohr pointed out, the electron is not simply a particle, in the macroscopic sense. Thus, it is meaningless to ask how it moves from one point to another. The only thing we can ask now is about the probability of finding the electron at a certain location around the nucleus. After an observation, it is possible to determine that the electron occupied some position in space. Another observation may locate the electron at some other point. The path that the electron takes in moving between those two points cannot be known. In fact, an electron has a well defined position only when it is being observed. *Position is not an intrinsic property of an electron.*

As we can see, quantum mechanics has weird consequences. In fact, *weird* is an adjective that is very often found associated with quantum mechanics. That doesn't make it wrong, however. Many real devices, like computer chips for example, are based on our knowledge of quantum mechanics. Quantum mechanics gives us a very accurate picture of the atomic world. The only problem is that this picture is highly mathematical and impossible to translate into concrete images that we can visualize. Furthermore, although most physicists agree that quantum mechanics is correct (in the sense that the scientific method allows a theory to be correct), its interpretation is still subject to animated discussions not only among physicists and other members of the scientific community, but also among philosophers. The best we can say is that quantum mechanics works, but nobody *really* understands it.

22 Quantum Mechanics

THE BEGINNINGS OF QUANTUM MECHANICS

Einstein's 1905 paper on the photoelectric effect, in which he considered Planck's idea of quanta of energy to be a universal characteristic of light, marks the beginning of quantum theory. For over ten years Einstein stood alone on the issue of energy quantization of light; and although in 1915 experimental confirmation of his photoelectric effect came from painstaking experiments performed by Millikan, acceptance of the idea that light was composed of particles was far from universal. Millikan himself wrote in 1915: "Despite … the apparent complete success of the Einstein equation, the physical theory of which it was designed to be the symbolic expression is found so untenable that Einstein himself, I believe, no longer holds to it." Einstein *did* hold to it. It was only in the mid 1920s, when Heisenberg, Schrödinger, Dirac, and others developed the first forms of quantum theory, that serious discussions concerning the particle nature of light and the meaning of the new theory began. Discussions about the meaning of quantum theory continue today.

THE NEW MECHANICS OF THE ATOM

In the summer of 1922, Niels Bohr delivered a series of invited lectures at the University of Göttingen in Germany. Attending these lectures was a 22-year-old graduate student from the University of Munich, Werner Heisenberg. Bohr's lecture left a profound impression upon the young student. "I shall never forget the first lecture," Heisenberg wrote later. "The hall was filled to capacity. The great Danish physicist … stood on the platform, his head slightly inclined, and a friendly but somewhat embarrassed smile on his lips. Summer light flooded in through the wide-open windows. Bohr spoke fairly softly, with a slight Danish accent … each of his carefully formulated sentences revealed a long chain of underlying thoughts, of philosophical reflections, hinted at but never fully expressed. I found this approach highly exciting; what he said seemed both new and not quite new at the same time."

The young student "dared to make a critical remark" during one of the lectures. Bohr immediately recognized the brilliant mind of the student and invited him after the lecture for a walk over the Hain Mountain to further discuss his objections. At the end of their long walk that afternoon Bohr suggested that Heisenberg join his Institute in Copenhagen after graduation. "You must pay us a visit in Copenhagen," Bohr told him, "perhaps you could stay with us for a term, and we might do some physics together." "My real scientific career only began that afternoon," wrote Heisenberg later, "suddenly the future looked full of hope and new possibilities, which, after seeing Bohr home, I painted to myself in the most glorious colors all the way back to my lodgings."

The Niels Bohr Institute in Copenhagen was a place where the young brilliant physicists from Europe, America, and the Soviet Union would gather to study the problems of the atom. There Heisenberg found the atmosphere he needed to unleash his profound intellect. He had obtained his PhD in 1924, at the age of 23, and had immediately joined Bohr at his Institute. After almost a year with Bohr, Heisenberg returned to Göttingen as an assistant to Max Born, then director of the Physics Institute. At Göttingen, Heisenberg worked on the problem of atomic spectral lines, trying to find a mathematical expression for the line intensities of the hydrogen spectrum. This attempt proved unsuccessful, leading to "an impenetrable morass of complicated mathematical equations, with no way out." Nevertheless, something useful came out of this. Heisenberg decided that the problem of the actual orbits of the electrons inside the atom should be ignored; they have no meaning. Only the frequencies and amplitudes associated with the spectral lines have any meaning, since they can be observed directly. Rather than the "morass of complicated mathematical equations," Heisenberg turned to a much simpler system: a simple pendulum. The oscillations of a pendulum could be used as a model for the atomic vibrations.

Toward the end of May 1925 a bout of spring hay fever made Heisenberg so ill that he asked Born for two weeks leave of absence. He went straight for Heligoland, a small rocky island in the North Sea, near Hamburg, with fresh sea breezes, far from the pollen of the meadows of the mainland. There Heisenberg had enough time to reflect upon his problem. "A few days were enough to jettison all the mathematical ballast that invariably encumbers the beginning of such attempts, and to arrive at a simpler formulation of my problem." Lightning had struck. In a few more days, he had some preliminary calculations which looked quite promising. What he needed was the proof that his new formalism presented no contradictions and, in particular, that it obeyed the principle of conservation of energy. One evening, he reached a point where the proof could be obtained. Heisenberg, too excited to stop, worked well into the night. By three o'clock in the morning, he had invented a new mechanics.

Returning to Göttingen, Heisenberg wrote up his paper, finishing it in July 1925. In his paper, he described the energy transitions of the atom as an array of numbers and found rules that these arrays must obey in order to calculate atomic processes. Unsure that his paper was ready for publication, he showed it to Born, who immediately recognized these arrays as matrices, and the rules that they obeyed as the rules for matrix algebra. Born sent the paper to *Zeitschrift für Physik*. Max Born enlisted the assistance of his student Pascual Jordan and together they further developed Heisenberg's theory. Matrices, it turned out, were the correct tools to describe atomic processes.

The fundamental property of Heisenberg's new mechanics is that measurable quantities like position and momentum are no longer represented by numbers but by matrices. And matrices do not obey the commutative law of multiplication. That is, if we multiply matrix p by matrix q, the answer we get is different from what we get when we multiply matrix q by matrix p. In Newtonian mechanics, if $q = 3$ is the position of a particle and $p = 2$ its momentum, the product of these two *numbers* does not depend on the order in which we multiply them; that is, $p \times q = 2 \times 3 = 6$ and $q \times p = 3 \times 2 = 6$. In a paper Born and Jordan published together, they show that the difference between the two matrix products **pq** and **qp** is proportional to Planck's constant h. In Newtonian physics, this difference is equal to zero since $p \times q - q \times p = 6 - 6 = 0$. Since our universe is discrete, not continuous, Planck's constant is not zero and the difference between the two matrix products is not zero. Since h is very small, we do not notice this discreteness in the macroscopic world. The discrete nature of our world forces us to forgo the use of simple numbers to represent quantities like position and momentum; they must be represented by matrices. That Heisenberg was able to realize this in the few days at Helgoland can only be explained as the insight of a genius.

In July 1925, Heisenberg gave a lecture at the Cavendish Laboratory in Cambridge. In the audience was another brilliant physicist, Paul Dirac, eight months younger than Heisenberg. Although Heisenberg did not mention his new theory during the lecture, he did discuss it with a few scientists afterward and even left a prepublication copy of his paper with Dirac's supervisor, who passed it on to the young scientist. Dirac immediately understood the importance of Heisenberg's work and began working on it himself. A few weeks later, Dirac wrote a lucid paper where he showed that Heisenberg's new mechanics was a complete theory that replaced Newtonian mechanics. Meanwhile, Heisenberg, Born, and Jordan arrived at the same conclusion using a different path. After only a few months of concentrated work by these four scientists, the first coherent theory of the atom emerged. It was called matrix mechanics and is one of the several forms of quantum mechanics, the mechanics of the atom.

WAVE MECHANICS

Einstein in his 1905 paper on the photoelectric effect had theorized that light was a particle—the photon—and in 1909 suggested that a theory of light should incorporate both the particle and wave theories of light. In 1923, as we saw in Chapter 21, Louis de Broglie suggested that all matter—an electron, for example—should also display wavelike behavior. De Broglie proceeded to calculate the wavelength of the electron and to suggest that his prediction could be confirmed experimentally if a diffraction pattern of electrons were observed. This work was de Broglie's doctoral thesis. In 1927, Davisson and Germer observed interference patterns in electron beams.

Erwin Schrödinger, a professor of physics at the University of Zurich, read in one of Einstein's papers of 1925 a positive reference to de Broglie's work. Einstein, we recall, had commented favorably on de Broglie's ideas when Paul Langevin, de Broglie's thesis director, sent Einstein a copy of the thesis. According to de Broglie's ideas, a wave of some sort must be associated with an electron. When the electron orbits the nucleus in Bohr's model of the atom, a *resonance condition* that gives rise to Bohr's stationary states must exist, in a similar fashion to the appearance of stationary configurations produced by standing waves on a string. For a circular orbit, de Broglie reasoned, an integral number of wavelengths must coincide exactly with the circumference of a stationary-state orbit (Figure 22.1). After studying de Broglie's papers, Schrödinger used the mathematics of waves to calculate the allowed energy levels, obtaining values that did not agree with the observed patterns

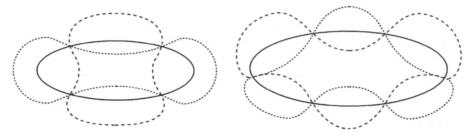

FIGURE 22.1   De Broglie's standing waves for circular stationary orbits.

of atomic spectra. Not being successful with his initial calculation, Schrödinger put the work aside for several months. Peter Debye, head of the research group at Zurich, having heard of de Broglie's work, asked Schrödinger to give a seminar on the subject. The discussion at the seminar stimulated Schrödinger to return to his work. In a few weeks during the Christmas holidays, Schrödinger developed a wave equation which governed the propagation of electron waves.

With his wave equation, Schrödinger correctly obtained the light spectrum of hydrogen. Before direct experimental verification of electron interference, de Broglie's strange idea that the electron was a wave was quantitatively vindicated by Schrödinger. His paper was published in January 1926. *Wave mechanics*, as Schrödinger's theory was initially called, was a second completely general way of formulating the mechanics of the atom. As Dirac was later on to prove, the two approaches—matrix mechanics and wave mechanics—are equivalent, and both are generalizations that include Newton's theory as a special case.

What are these waves in Schrödinger's theory? The first answer came from Schrödinger himself. Perhaps electrons and all other subatomic particles are not really particles but waves. The entire universe is nothing but a great wave phenomenon. This interpretation was quickly rejected by Max Born because it would not explain the fact that you could count individual particles with a Geiger counter and see their tracks in a cloud chamber. Max Born offered a better solution to this puzzle. In 1924, before Schrödinger developed his theory, Bohr had written a paper in which he had attempted to solve the contradiction between the particle picture and the wave picture introduced by de Broglie. In that paper Bohr had considered the concept of *probability waves*. He interpreted the electromagnetic waves not as real waves but as waves of probability the intensity of which determines the probability that the atom is going to undergo a transition from one energy level to another. Born extended Bohr's concept and developed a clear mathematical definition of the wave functions in Schrödinger's theory which he interpreted as probability waves. These waves were not three-dimensional waves like electromagnetic waves, but waves of many dimensions; they were, then, abstract mathematical entities. According to Bohr's interpretation, the Schrödinger wave function of an electron gives the probability of locating an electron in a particular region of space. More precisely, the square of the wave amplitude at some point in space is the probability of finding the electron there.

PIONEERS OF PHYSICS: SCHRÖDINGER'S INSPIRED GUESS

Schrödinger reasoned that if an electron has a wavelike behavior, then it should be possible to describe this behavior in space and time by a wave equation. He devised an equation—a wave equation—that controls the behavior of the wave and that specifies the connection between this behavior and the behavior of the particle. His starting point was the principle of conservation of energy. Applying de Broglie's relation between momentum and wavelength and Planck's $E = hf$ relation, Schrödinger was able to guess the mathematical form of the wave equation. We must stress that Schrödinger's equation cannot be derived; it was obtained by a postulate. Its justification is not that it has been deduced from previous known facts, but that its predictions can be experimentally verified. "Where did we get that [equation] from?" asked Richard Feynman, "Nowhere. It is not possible to derive it from anything you know. It came from the mind of Schrödinger."

HEISENBERG'S UNCERTAINTY PRINCIPLE

In Helgoland, we recall, Heisenberg had discovered that observable or measurable quantities were to be represented by matrices. In 1927, Heisenberg showed that if two matrices **q** and **p** represented two physical properties of a particle, like position and momentum or time and energy, which obeyed the noncommutative rule, that is, that the difference between the product **pq** and **qp** was proportional to Planck's constant, then one could not measure both properties simultaneously with infinite precision. This statement is known as *Heisenberg's Uncertainty Principle.*

To illustrate this principle, let's consider a possible experiment with electrons. In Chapter 18, we discussed the diffraction of light in Young's double-slit experiment in which a beam of light passed initially through a single narrow slit and then through a pair of slits that formed two coherent beams. These beams interfered with each other, producing an interference pattern on a screen. We can understand this interference pattern if we think of light as a wave. In fact, this was the experimental proof given by Young that light was a wave phenomenon. Light, Einstein told us, also behaves as a particle. Can we think of light as made up of particles and still understand Young's double-slit experiment? What if we use electrons instead of photons?

Consider a source of electrons, like the electron gun in a regular television tube, a thin metal plate with two very narrow slits in it, and a phosphor-coated screen that produces a flash when an electron collides with it, like a television screen (Figure 22.2a). What we observe is that the flashes on the screen indicate that each electron hits the screen in just one point. If we cover either one of the slits and let the experiment run for some time, counting the number of flashes at any given position on the screen, what we obtain for the probability of arrival of the electron at various locations on the screen is the pattern shown in Figure 22.2b. Uncovering both slits gives us the pattern shown in Figure 22.2c for the probability of arrival. This pattern looks like the interference fringes obtained in the original Young's double-slit experiment with waves shown in

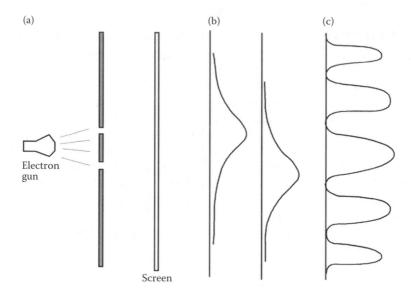

FIGURE 22.2 (a) Double-slit experiment with electrons. (b) Probability that an electron hits the screen at a particular point when only one slit is open. (c) Probability when both slits are open.

FIGURE 22.3 Photograph of a double-slit interference experiment with electrons. (Courtesy of Claus Jonsson.)

Figure 18.9! Figure 22.3 is a photograph of the interference pattern for electrons in a double-slit experiment. Observing the regions where the probability is minimum with both slits open, we notice that the number of counts with both slits open is smaller than the counts with only one slit open. This is due to destructive interference in those regions, a phenomenon characteristic of waves. The electrons are behaving like waves even though they arrive at the screen and make a flash, like particles. A wave does not hit the screen at a single point.

Could it be that the many electrons in the beam are somehow interacting with each other and "cooperating" so that when they arrive at the screen, an interference pattern is formed? To investigate this, we could reduce the intensity of the beam until we have just a single electron traveling at a time. We begin to notice the flashes on the screen and, as before, we record their position. At first, no pattern is noticeable. When enough flashes accumulate, we begin to see the interference pattern again, *as if each electron interferes with itself*!

How can an electron interfere with itself? Are the electrons, somehow, passing through both slits at once and then recombining to form the interference pattern? We could install small detectors near the slits to determine which slit the electron passes through. The detectors might shine some light on it so that we can observe the scattered light. The first thing we notice is that the electrons go through either one of the slits; no funny business of one electron going through both slits at once is observed. Everything seems normal, as we would expect particles to behave. As before, we record the flashes and when enough of them accumulate, we see that the interference pattern has disappeared. The probability that an electron hits the screen at a given location is what we would obtain for regular particles. *Observing the electrons destroys the interference pattern*. Determining the position of the electron destroys their ability to interfere.

It might be that the light we are using to observe the electrons disturbs them and changes their behavior. To get around this difficulty, we reduce the intensity of the light shining on the electrons and repeat the experiment. Reducing the intensity of the light reduces the energy of the photons. As we might recall, the energy of each photon, from Planck's formula, is given by hf, where f is the frequency of the light. If we have n photons in each flash, the total energy is $E = nhf$. We can reduce the intensity of the light shining on the electrons by sending very few photons so that n is small. When we use one photon for every electron that passes, no interference pattern is observed and we account for the position of each electron. Even more reduction in the number of photons results in some electrons not being seen. Not only do we lose information about the location of each electron, but we begin to observe a faint interference pattern on the screen, produced by the electrons that we failed to see. We need at least one photon for every electron if we want to know through which slit they pass.

Reducing the intensity of the light shining on the electrons by reducing the frequency increases the wavelength of the light which reduces the resolution of our detector; that is, its ability to determine the location of the electron. When the frequency is not too small and the wavelength small enough to let us see each electron, no interference pattern is produced. It might be that we are still disturbing the electrons. If we reduce the intensity of the light so as not to disturb the passing electrons, the wavelength becomes larger than the distance between the slits and we cannot tell which one the electron went through. And on top of that, the interference pattern returns!

Our attempt to observe the electrons without disturbing them has not worked. What Heisenberg's uncertainty principle tells us is that it cannot work. Our failure is not due to our lack of imagination, or to lack of better instrumentation; it will never work. It is an *intrinsic* limitation of nature. This is the way the universe works. It is not possible for us to determine through which slit the electron passes on its way to the screen. The only solution is that the electron goes through *both* slits at once and carries their imprint when it arrives at the screen. This concept is impossible for us to imagine and it is only through mathematics that physicists have been able to encapsulate this phenomenon.

Heisenberg derived a mathematical expression to explain this intrinsic limitation of nature. If the uncertainty in the measurement of the position of the electron along the vertical direction is Δq and the uncertainty in this component of the momentum is Δp, Heisenberg's uncertainty principle says that the product of these two uncertainties must never be *smaller* than a certain constant, which is equal to Planck's constant h divided by 2π that is,

$$\Delta p \Delta q \geq \frac{h}{2\pi}.$$

This constant $h/2\pi$ is equal to 1.055×10^{-34} J s and places a lower limit to the value of the product of the uncertainties. If our instruments allow us to measure the electron's position with an uncertainty of only one-millionth of a meter (10^{-6} m), for example, the uncertainty in the momentum must be no smaller than 1.055×10^{-28} kg m/s so that the product is at least $h/2\pi$. This makes the uncertainty in the electron's velocity equal to about 100 m/s.* If with better instruments we reduce the uncertainty of the position to one nanometer, 10^{-9} m, the uncertainty in the electron's velocity becomes one thousand times larger. The momentum has become so uncertain that we will not know whether, one second later, the electron will still be near the instruments or 100 km away!

We must emphasize here that the concept of uncertainty applies not to a single measurement but to the measurement of many electrons. Heisenberg's uncertainty principle is a statement about a statistical average over many measurements.

As can be seen from the above discussion, the name *particle* for electrons is an unfortunate one because it gives the idea that electrons are like dust particles or any other macroscopic particle. Electrons and other quantum "particles" are really *quantum entities* with properties given by the laws of quantum mechanics.

THE FRONTIERS OF PHYSICS: KNOWLEDGE AND CERTAINTY

For all its power and its implications for our knowledge of nature, the mathematical expression that Heisenberg developed for his uncertainty principle looks surprisingly simple. We can illustrate the method that Heisenberg used to obtain it as follows:

We can call the uncertainty in the position of the electron along the vertical direction Δq. The position of the electron when it hits the screen depends on the component of the electron's momentum along the vertical axis (q-axis); the uncertainty in this component of the momentum is Δp. If we use light of wavelength λ to observe each electron, then the uncertainty in the position is equal to λ, or $\Delta q = \lambda$. The momentum imparted to the particle is at the most equal to the momentum of the photon, which is given by de Broglie's

---

* Since momentum is the product of mass times velocity, and the mass of the electron is 9.11×10^{-31} kg. The uncertainty in the velocity is then the uncertainty in the momentum, 1.055×10^{-28} kg m/s, divided by is 9.11×10^{-31} kg, or 116 m/s.

relation. De Broglie's relation, from Chapter 21, is $\lambda = h/mv = h/\Delta p$, so that the uncertainty in the momentum is approximately $\Delta p \approx h/\lambda$. The product of the two uncertainties Δq and Δp is

$$\Delta q \, \Delta p \approx h.$$

What Heisenberg's uncertainty principle says is that the product of these two uncertainties must always be *greater* than Planck's constant h. Heisenberg's more detailed derivation gives the form of the uncertainty principle as

$$\Delta q \Delta o \geq \frac{h}{2\pi}.$$

PHYSICS IN OUR WORLD: ELECTRON MICROSCOPES

Soon after it was discovered that high speed electrons had wavelengths that were much shorter than the wavelengths of light, physicists showed that magnetic fields could act as lenses by causing electron waves to converge to a focus. In 1931, Max Knoll and Ernst Ruska built the first microscope that focused electron waves instead of light waves. The electrons were focused with electric or magnetic fields.

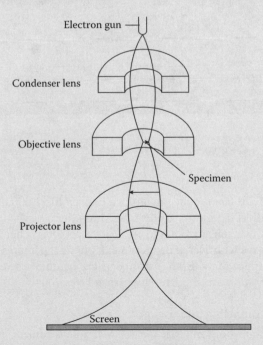

Diagram of a transmission electron microscope, TEM.

In the design of Knoll and Ruska which is based on an optical microscope and known as the *transmission electron microscope* (TEM), a hot tungsten wire emits electrons which are focused on the specimen by a magnet acting as a condenser "electron lens." An objective lens produces a magnified image which becomes the object for the projector lens. The image is further magnified and projected onto a fluorescent screen or a photographic plate.

A few years after the invention of the TEM, Knoll modified their earlier design so that the focused electron beam scanned the surface of the specimen. In the *scanning electron microscope* (SEM), as his invention is known, electrons from the surface of the specimen are knocked out by the incident electrons. Some of these electrons can be detected by an electrode. As the beam scans the surface, more electrons are knocked out from sharp edges than from a flat surface and a map of the object can be generated.

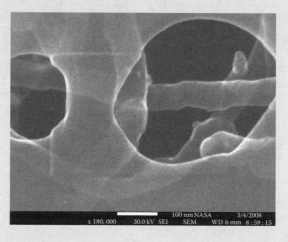

SEM image of carbon manotubes seen through a support grid. (Courtesy of P. Hintze, NASA Kennedy Space Center.)

A modern TEM can produce magnifications up to ×1,000,000 and resolve objects as small as 0.2 nm. A modern SEM, on the other hand, can produce magnifications that range from ×15 to about ×250,000. The SEM, however, produces an image with a large depth of field, giving the impression of a three-dimensional surface relief. Both instruments are based on the extremely short effective de Broglie wavelengths of high-speed electrons.

The scanning tunneling microscope

The scanning tunneling microscope (STM), an instrument capable of producing images with atomic resolution, was invented in 1981 by Gerd Binnig,

Heinrich Rohrer, and collaborators at the IBM Research Laboratory in Zurich, Switzerland. Their discovery won Binning and Rohrer the 1986 Nobel Prize in Physics.

The operation of the STM is based on the phenomenon of *tunneling*. Quantum mechanics tells us that a particle such as an electron has a finite probability of "tunneling" through a potential barrier that classically separates two regions. Electromagnetic theory tells us that when a voltage is applied between the tip of the instrument and the surface to be examined, there is a potential barrier between the tip and the surface. An electron with insufficient energy to surmount the barrier would not be able to move from the tip to the surface. However, quantum mechanics tells us that since the electron can be described by a wave function, it has a finite probability of crossing or tunneling through the barrier. This tunneling probability decreases very rapidly (exponentially) with the width of the potential barrier. As a consequence, observation of tunneling events is feasible only for relatively small potential barriers.

In the STM, a sharp metal tip is brought to within a nanometer of the surface, producing an overlap of the electronic wave functions. When a small voltage is applied, a tunneling current in the nanoampere range flows between the tip and the surface. The tip is scanned across the surface using piezoelectric crystals, which expand or contract in response to an applied voltage. A third crystal maintains the tip at a constant distance of a few tens of nanometers above the surface. The potential applied to this crystal determines the vertical position of the tip and thus gives the topography of the surface.

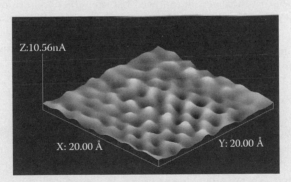

An image of graphite with the atoms arranged in an ordered array. An ångström (Å) is 0.1 nm.

THE NEW PHYSICS

What are we to make of all this? As the double-slit experiment with electrons showed, the act of observation changes the system being observed. Electrons and other quantum particles behave like particles and like waves, depending on the particular observation. If we set up an experiment to detect the wave nature of the electron, we find wave phenomena such as interference and diffraction. If we set up an experiment to detect the electron's particle nature, we detect particles. Which one is the correct interpretation? What is the true nature of the electron? According to Bohr, it is meaningless to ask what an electron *really* is. Physics does not tell us about what *is* but about what we can *describe* regarding the world. *Both* theoretical pictures, the wave picture and the particle picture, are equally valid in describing the electron and provide *complementary* descriptions of reality.

Heisenberg's uncertainty principle tells us that we cannot know both the position and the momentum of a particle simultaneously with infinite precision. The more precisely we demand our knowledge of position, for example, the less precision we are allowed in our knowledge of momentum. If we make Δq very small, Δp becomes large in such a way as to keep the product of the two uncertainties proportional to Planck's constant. We can choose to measure the position with infinite precision, but this means that we completely give up knowing anything about the particle's momentum; the uncertainty becomes infinite. Or we can determine precisely the particle's momentum but in this case, our knowledge of its position is undetermined; the particle can be anywhere. Both situations are complementary. Each quality—position and momentum—are complementary properties of the quantum particle.

An electron occupies a position in space only when an observation is made. Therefore, if, through observation, we determine that an electron is at a given location at one particular instant, and at another location at another particular instant, those two locations are the only ones that we can attribute to the electron. We cannot say what the electron did to move from the first location to the second, and we cannot speak of the *path* that the electron followed between the two locations, because the path does not exist.

Is this the way the world really is? Is the world this crazy place where the observer changes reality by the mere act of observation? Bohr was convinced that this was the case. His *Copenhagen interpretation* of quantum mechanics, as the ideas of Bohr, Heisenberg, and Born—uncertainty, complementarity, and the disturbance of the system by the observer—have been collectively called, is the interpretation accepted by the great majority of physicists today. The major dissenter was Albert Einstein. He never accepted that there could be chance and limitations imposed on the measurement of observed phenomena in nature. The uncertainty in the quantum world is not an intrinsic property of the universe but the result of our incomplete knowledge of it, he reasoned. The weather, for example, is unpredictable not because of any inherent uncertainty but because climate conditions involve too many parameters and too many particles which make it impossible with the tools available to us at the present to compute their behavior with the necessary detail for accurate prediction. If a large enough computer were available to take into consideration *all* these variables, accurate weather prediction would be possible. The unpredictability and uncertainty, then, are a result of these *hidden variables* that we have not included. In the case of quantum mechanics, the hidden variables have not been included because we do not know them yet. According to Einstein, quantum mechanics is not incorrect; it is simply *incomplete*. When this deeper level of hidden variables is discovered, a new correct and deterministic theory will be developed.

Measure momentum   Original particle   Measure
for particle 1   total momentum, $p = 0$ position
for particle 2

FIGURE 22.4 In the EPR experiment a particle at rest explodes into two fragments that move away from each other. We can measure the momentum of particle 1 and the position of particle 2. Because of conservation of momentum, it is possible to deduce the momentum of particle 2.

In 1935 Einstein wrote a paper with his colleagues Boris Podolsky and Nathan Rosen ("EPR") in which he proposed a thought experiment to show that it was possible for a particle to have both position and momentum at the same time and consequently, to show that quantum mechanics is incomplete. In the EPR experiment, a particle at rest splits into two fragments, 1 and 2 (Figure 22.4). The momentum of the original particle before splitting is zero since the particle is at rest. Since momentum is conserved, the total momentum of particles 1 plus 2 has to be zero also. When the particles are at a great distance from each other, the momentum of particle 1 is measured. This should not affect particle 2 in any way, since it is very far away. Heisenberg's uncertainty principle prohibits us from accurately measuring the position of particle 1 but, since the momentum of particle 2 has not been measured, we are free to determine its position with perfect accuracy. We now know particle 1's momentum, particle 2's position and the total momentum of both particles. The momentum of particle 2 is equal to the total momentum of both particles minus the momentum of particle 1, which we know. Although we did not measure particle 2's momentum, we are able to compute it with accuracy. Contrary to what quantum mechanics says, we have accurately determined the position and the momentum of particle 2. "I am therefore inclined to believe," Einstein wrote later, "that the description of [the] quantum mechanism ... has to be regarded as an incomplete and indirect description of reality, to be replaced at some later date by a more complete and direct one."

Bohr's response to the EPR argument was that the two particles, once in contact with each other, continue to be part of the system upon which one makes measurements. Therefore, Heisenberg's uncertainty principle prohibits us from knowing both the momentum and the position of particle 2 regardless of the technique used to measure these quantities. Einstein could not accept the idea that a measurement made at one location could affect a second particle at a distant location. There is no limitation as to how far apart the two particles have to be in this thought experiment. The measurement of the momentum of particle 1 could be made in a laboratory on earth and the measurement of particle 2's position could be done by a laboratory on the moon, when the particle gets there.

Why not resolve the controversy by direct experiment? It turns out that this kind of experiment is extremely difficult to perform. In 1965, the Irish physicist John Stewart Bell, working at CERN, the European accelerator center in Geneva, decided to study the EPR experiment and was able to state and prove a powerful theorem, known today as *Bell's theorem*, which deals with the correlations that could exist between the results of simultaneous measurements done on two separated particles.

Bell's theorem gives the theoretical limits on the correlations between these simultaneous measurements. The limits on these correlations were given by Bell in the form of an inequality, known appropriately enough as *Bell's inequality*. To give a simple illustration of the reasoning behind Bell's inequality, the amount of money in change that you have in your pocket at a particular moment cannot be greater than the amount in change plus the amount that you have in bills of different denominations at that same moment.

Bell's theorem opened the path for an experiment that would decide whether or not there are hidden variables in quantum mechanics. A real EPR experiment. In 1982, Alain Aspect in Paris performed such an experiment. The results were unequivocal. There are no hidden variables in quantum mechanics. Bohr was right and Einstein was wrong. The world is as strange as quantum mechanics has shown us it is. An electron is neither a particle nor a wave. It is not an entity out there, separated from the rest of the universe. An electron, a proton, a quantum particle, an atom—all of these "particles" are simply convenient ways that we have of thinking about what is only a set of mathematical relations that connect our observations. The universe is not a collection of objects; it is rather an inseparable net of *relations* in which no component, including the observer, has an independent reality.

THE FRONTIERS OF PHYSICS: QUANTUM TELEPORTATION

In Star Trek, when the members of the away team down on the surface of an alien world need to get out of a sticky situation in a hurry, they contact their ship and ask to be "beamed up." Until recently, teleportation, the idea of transporting matter instantaneously from one place to another, remained in the realm of science fiction.

Physicists had not paid much attention to teleportation because it seemed to violate the uncertainty principle of quantum mechanics, which forbids the simultaneous knowledge of all the information in an object. However, in 1993 a research team led by Charles H. Bennett of the IBM Thomas J. Watson Research Center in Yorktown Heights, NY, proposed a way to use the EPR experiment to transmit a quantum state instantaneously. It has been known for several decades that there are situations in which an atom emits two simultaneous photons which travel in opposite directions and remain forever correlated or "entangled." Quantum mechanics tells us that neither photon has a particular polarization until a measurement is performed on them. When such a measurement is done on one of the two photons and its polarization is determined, the second "entangled" photon immediately acquires the opposite polarization, regardless of how far away it is (at the other side of the laboratory or at the opposite end of the galaxy).

The EPR experiment does not allow the instantaneous transmission of information, however. The IBM team proposed a theoretical method to record, or scan, and "transmit" part of the information from a particle's quantum state to its entangled partner. Later, the second particle's state is modified using the scanned information to bring it to the original state of the first particle. Having been disturbed by the scanning process, the first particle is no longer in its original quantum state. The method provided a clever way to instantaneously teleport the quantum state of a particle to a second particle.

The work of the IBM team was theoretical. In December 1997, researchers at the University of Innsbruck in Austria reported the results of the first experiment in which a photon was teleported. A second group led by Francesco De Martini at the University of Rome "La Sapienza" also reported having teleported photon characteristics.

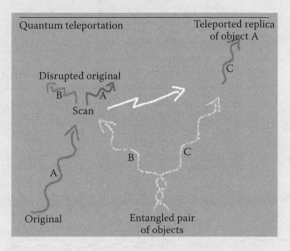

(Reprinted with permission from International Business Machines Corporation, copyright 2000.)

Scientists had been considering transporting electrons and even whole atoms and ions. In 2009, a team of researchers at the University of Maryland, College Park, demonstrated quantum teleportation with ytterbium ions that were located one meter apart. With teleportation of ions, information could be stored on trapped ions, shielded from the environment, thus providing a quantum memory for computers. Other ideas involve the use of quantum teleportation for transmission of information under noisy conditions, where messages would be degraded or for providing links between quantum computers. Physicists see that it would be difficult to go beyond those applications to the point where macroscopic objects could be instantly teleported. The problem is that once the transfer is made, the receiver's observation changes the transferred object's state. To interpret the result correctly, information about the process still needs to be sent to the receiver by conventional means (at speeds lower than the speed of light).

23 Nuclear Physics

BEYOND THE ATOM

Quantum mechanics provided the mathematical foundation for our understanding of the properties of atoms. With this foundation in place, progress came quickly. Some physicists used the new theory to develop a more complete picture of the electronic properties of atoms, while others went on to apply the theory to the study of the atomic nucleus.

We studied some of the basic properties of the nucleus in Chapter 8. Here we will examine nuclear transformations, nuclear energy, and the applications of nuclear physics to modern technology.

RADIOACTIVITY

As we saw in Chapter 8, atomic nuclei are composed of protons and neutrons bound by the strong force. Since the electric force has infinite range, diminishing gradually with distance, the positively charged protons feel the electrical repulsion of each and every one of the remaining protons in the nucleus. Since the nuclear force is a short range force, its influence dies out to almost nothing when the protons are separated by more than 10^{-14} m.

If a nucleus contains too many protons, the total electric repulsion can overwhelm the nuclear attraction and a fragment of the nucleus (an alpha particle) will fly out. This is known as *alpha decay*. (Alpha is the Greek letter α.) However, if the nucleus contains a large number of neutrons, they contribute to the nuclear force without being affected by the electric force, as they are neutral. This is why light nuclei usually have a similar number of protons and neutrons, such as $^4_2$He, $^7_3$Li, $^{12}_6$C, and $^{16}_8$O, whereas heavy nuclei require a great deal more neutrons to counterbalance the effect of the electrical repulsion. For example, $^{208}_{82}$Pb contains 126 neutrons and 82 protons, and $^{197}_{79}$Au has 118 neutrons and only 79 protons.

There is a limit to the number of neutrons that a nucleus can have and still be stable. The reason for this has to do with another type of force, the weak *nuclear force*, which is also a short-range force and is responsible for the instability of the neutron. On the average, a free neutron, outside the nucleus, can only last for about fifteen minutes before it disintegrates into a proton, an electron, and another particle called a neutrino. Inside the nucleus, however, this disintegration usually does not take place because of a principle of quantum mechanics, known as the *Pauli exclusion principle*, which states that no two protons can occupy the same quantum state. These quantum states are similar to the quantum states occupied by the electrons in the atom. Pauli's principle prevents a nuclear neutron from decaying into a proton because this new proton would have nowhere to go, all the available states being already occupied by other protons.

Neutrons also obey the Pauli exclusion principle. Thus, a nucleus with too many neutrons has to place those neutrons in higher energy states. In these states the neutrons have enough energy to place the protons into which they decay in unoccupied high energy levels. But where this happens the nucleus loses its identity. In other words, when a neutron inside a nucleus is able to decay through the weak nuclear force into a proton, an electron and a neutrino, the nucleus is transmuted into a different nucleus. Since the nucleus cannot contain free electrons, these are ejected. (According to Heisenberg's Uncertainty Principle, a particle with a mass as small as that of the electron and having the energies observed in beta decay cannot be confined to a volume as small as that occupied by the nucleus.) This phenomenon is called *beta decay*. (Beta is the Greek letter β.) This is the same phenomenon as that identified by Rutherford at the beginning of this century.

We now discuss these two types of radioactive decay, together with a third, known as *gamma decay*, in some detail. (Gamma is the Greek letter γ. Alpha, beta and gamma are the first three letters of the Greek alphabet.)

ALPHA DECAY

As we said earlier, if a nucleus contains too many protons it is electrically unstable and undergoes alpha decay. An example of a nucleus that alpha-decays is $^{238}_{92}\text{U}$. This decay is illustrated in Figure 23.1.

We can write this decay as follows:

$$^{238}_{92}\text{U} \rightarrow\, ^{234}_{90}\text{Th} + {}^{4}_{2}\text{He}.$$

We call the nucleus before the decay the *parent nucleus* ($^{238}_{92}\text{U}$ in our example) and the nucleus produced after the decay, the *daughter nucleus* ($^{234}_{90}\text{Th}$). Notice that the mass number of the parent nucleus is 238 and this is equal to the sum of the mass numbers of the daughter nucleus and the alpha particle ($^{4}_{2}\text{He}$), namely $234 + 4$. Since the mass number represents the number of nucleons in the nucleus, this is the same as saying that the number of nucleons in the parent nucleus must be equal to the total number of nucleons in the daughter plus those in the alpha particle. The

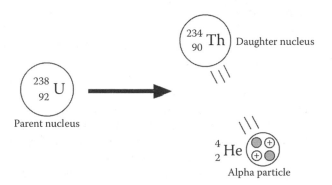

FIGURE 23.1   Alpha decay of $^{238}_{92}\text{U}$.

atomic numbers obey a similar rule. Since the atomic number represents the number of charges in the nucleus (number of protons), this second rule says that the number of charges in the parent nucleus must equal the total number of charges in the daughter and the alpha particle. These two rules hold for every radioactive decay and are based on conservation laws. These conservation laws are called *conservation of nucleon number* and *conservation of electric charge*.

BETA DECAY

A neutron-rich nucleus is also unstable and will convert spontaneously into another nucleus with an extra proton through the weak nuclear force. As we mentioned earlier, this "beta decay" happens because of the transformation of the neutron into a proton, an electron and a neutrino particle. The neutrino was proposed by the physicist Wolfgang Pauli as a solution to a problem that was observed when the energies of the nuclei involved in beta decay were measured. It seemed that in beta decay the principle of conservation of energy no longer held. Since physicists have always been reluctant to accept anything that violates this well-established principle, Pauli proposed in 1930 the existence of a small neutral particle that would take care of the observed discrepancy. The Italian physicist Enrico Fermi named it the *neutrino*, Italian for "little neutral one." Twenty years later Frederick Reines and Clyde L. Cowan detected the neutrino at the U.S. Atomic Energy Commission Savanna River Laboratory in Georgia.

A typical beta decay event is illustrated in Figure 23.2. In symbols, we write this decay as

$$^{14}_{6}C \rightarrow {}^{14}_{7}N + {}^{0}_{-1}e + \bar{\nu}.$$

Here the $\bar{\nu}$ symbol represents the neutrino particle (ν is the Greek letter nu). In this case it is an antineutrino, which is the antiparticle of the neutrino, identified by a bar over the symbol ν. An antiparticle is a particle of antimatter: antimatter consists of atoms with negatively charged nuclei and positive electrons, called *positrons*. (Antimatter will be studied in more detail in Chapter 22). Since the electron is not a

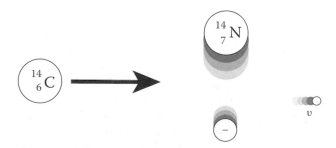

FIGURE 23.2 Beta decay of $^{14}_{6}C$. One of the 8 neutrons in $^{14}_{6}C$ is transformed into a proton, an electron, and an antineutrino. The electron and the antineutrino fly away. The daughter nucleus, $^{14}_{7}N$, has one more neutron and one fewer proton than the parent nucleus.

nucleon, its mass number is indicated as zero, and as it has the same electric charge as the proton but opposite sign, its atomic number appears as –1.

Emission of a positron (a positive electron or antielectron) together with a neutrino also takes place. An example of this type of beta decay is

$$\ce{^{12}_{7}N} \rightarrow \ce{^{12}_{6}C} + \ce{^{0}_{1}e} + \nu.$$

Note the + 1 charge on this antielectron.

Gamma Decay

In many cases the product of an alpha or beta decay process is a nucleus that is left in an excited state. When a nucleus is in an excited state, the nucleons can jump to a lower energy state and, in a process similar to the way atoms emit electromagnetic radiation, one or several photons are emitted from the nucleus. The energy of these photons is much higher than that of visible light. Photons so produced are called gamma rays.

In gamma decay there is no emission or absorption of nucleons. The only change is in the energy of the nucleus. Therefore, neither Z nor A change and the nucleus remains the same. We indicate a nucleus in an excited state by placing an asterisk next to its chemical symbol. One example of gamma decay is

$$\ce{^{12}_{6}C^{*}} \rightarrow \ce{^{12}_{6}C} + \gamma.$$

The products of radioactive decays (the alpha and beta particles and the gamma ray photons) can interact with the atoms and molecules in living cells or in electronic circuits. These interactions can lead to certain changes. A few of the changes that may occur in living cells may be beneficial and, if they are passed to offspring, these changes help to produce the process that we call evolution. Most of the changes in the cells of biological systems, however, are detrimental, causing damage to the organism. Just about every change in an electronic circuit produces some damage and can lead to serious disruption of the electronic systems in which they are installed.

Half-Life

In a radioactive sample there are vast numbers of nuclei. Every time a nucleus decays by alpha or beta emission, the number of radioactive nuclei in the sample decreases by one. The time required for half the nuclei in a given radioactive sample to decay is called the half-life T of the particular nuclide. In each subsequent half-life, half the remaining nuclei decay; thus after two half-lives, the remaining radioactive nuclei is one-fourth the original number. After three half-lives, the original number is reduced to one-eighth. For a given radioactive nuclear species, the half-life is the same regardless of the number of nuclei in the sample. The half-lives of several nuclei are listed in Table 23.1.

TABLE 23.1
Half-Lives of Selected Nuclei

Element	Half-Life	Element	Half-Life
Uranium-238	4.55×10^9 years	Cobalt-60	5.3 years
Uranium-234	2.48×10^5 years	Thorium-234	24.1 days
Carbon-14	5730 years	Iodine-131	8.1 days
Radium-226	1620 years	Radon-222	3.8 days
Strontium-90	28.9 years	Lead-214	26.8 minutes

Although the half-life of $^{14}_{6}C$, say, is known to be 5730 years, we cannot predict exactly when a particular $^{14}_{6}C$ nucleus will decay. The decay of a given nucleus is a random event. Since a macroscopic sample contains an enormous number of nuclei, we can use the laws of probability to determine how many nuclei (without specifying which ones) will decay in a given length of time. The probability that an unstable nucleus will decay spontaneously is the same at any instant, regardless of its past history, and it is the same for all nuclei of the same type.

NUCLEAR REACTIONS

All three radioactive decay processes are examples of nuclear transformations. These are not the only way a nucleus can be changed into one or more nuclei, however. If a fast moving nucleon collides with a nucleus, a nuclear transformation can occur. These *nuclear reactions* can take place in nature or they can be artificially produced.

The first nuclear reaction produced artificially was observed by Rutherford in 1919. He discovered that when an alpha particle passed through nitrogen gas, protons were emitted. He concluded correctly that nitrogen nuclei had been transformed into oxygen nuclei (Figure 23.3).

We can write this nuclear reaction in symbols, as

$$^{4}_{2}He + ^{14}_{7}N \rightarrow ^{17}_{8}O + ^{1}_{1}H,$$

which is commonly written in the abbreviated form, with p representing the proton, as

$$^{14}_{7}N(\alpha,p)^{17}_{8}O.$$

Thousands of nuclear reactions have been studied since Rutherford's time, most of them artificially produced with the aid of the particle accelerators invented during the 1930s. The products of these reactions are in many cases radioactive isotopes that have value in medicine and related fields where they are used as radioactive "labels" that show the path taken by a substance through the body (Figure 23.4).

Nuclear reactions also obey the two conservation rules introduced in our study of radioactive decay. Applying these conservation rules, that is, conservation of nucleon number and conservation of electric charge, to Rutherford's reaction, we can see that the number of nucleons is the same before and after the reaction takes place. To check this, notice that the sum of the mass numbers of the nuclei before the collision takes place, He and N, is $4 + 14 = 18$, which is equal to the sum of the mass numbers of the final nuclei, 17 for oxygen plus 1 for the ejected proton. Similarly, the conservation of charge gives 2 for helium plus 7 for nitrogen which equals the 8 for oxygen plus 1 for the proton.

Alpha particle   Nitrogen   Oxygen   Proton

FIGURE 23.3   An alpha particle strikes a nitrogen nucleus and is absorbed. The new nucleus ejects a proton, becoming an oxygen nucleus. This was the first nuclear reaction to be artificially produced.

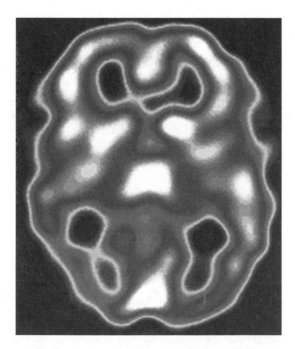

FIGURE 23.4   Image of the brain of a patient with a migraine is formed by detecting gamma rays from radioactive flourine-18 that has been injected into the patient. (Courtesy of Department of Nuclear Medicine, Charing Cross Hospital/Science Photo Library.)

Using these rules we could, for example, determine the product of the collision of a neutron ($_0^1$n) with $_{13}^{27}$Al, which ejects an alpha particle. Writing X for the unknown nucleus we have

$$_0^1\text{n} + _{13}^{27}\text{Al} \rightarrow _2^4\text{He} + \text{X}.$$

Conservation of nucleon number gives

$$1 + 27 = 4 + A_x$$

and conservation of charge,

$$0 + 13 = 2 + Z_x$$

where A_x and Z_x are the mass number and atomic number of our unknown nucleus, respectively. This tells us that the atomic number of X is 11, identifying the nucleus as sodium, Na, with a mass number equal to 24. This particular isotope of sodium, $_{11}^{24}$Na, is radioactive, and decays with a half-life of 15 hours through beta emission into $_{12}^{24}$Mg. This process is used in medicine to study the way sodium is transported across membranes.

NUCLEAR ENERGY: FISSION AND FUSION

NUCLEAR FISSION

During the summer of 1939, just before the start of World War II, the German scientists Otto Hahn and Fritz Strassmann published a paper where they stated that when uranium was bombarded with neutrons, it produced smaller nuclei which were about half the size of the original uranium nucleus. One of Hahn's previous collaborators, the physicist Lise Meitner and her nephew Otto Frisch, both of whom had fled Nazi Germany and were working at the Nobel Institute in Sweden, immediately realized that the uranium nucleus actually had split into two fragments. This new phenomenon was called *nuclear fission*.

In Figure 23.5 we have illustrated how nuclear fission works. An incoming neutron strikes a uranium-235 nucleus and is momentarily absorbed, turning the target nucleus into uranium-236. The absorbed neutron gives the uranium nucleus extra

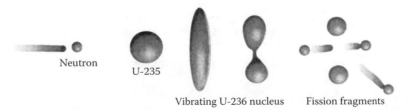

Neutron

U-235

Vibrating U-236 nucleus   Fission fragments

FIGURE 23.5 The steps in the nuclear fission of a uranium nucleus, as explained in the liquid drop model.

internal energy which causes it to split. Once the fragments are separated, the electrical repulsion between the fission fragments is stronger than the nuclear attraction and the two nuclei fly apart. Since a liquid drop splits into two parts in a similar fashion, physicists constructed a model based on this phenomenon, the *liquid-drop model of fission*, to explain how it works. If enough energy is provided to the drop, it will begin to take on increasingly larger elongated shapes, vibrating back and forth until it finally splits. In the uranium-235 nucleus, the neutron provides the energy needed to start the vibrations until the newly formed uranium-236 splits into fission fragments.

A typical fission reaction is

$$\frac{1}{0}n + \frac{235}{92}U \rightarrow \frac{236}{92}U^* \rightarrow \frac{139}{54}Xe + \frac{94}{38}Sr + 3\frac{1}{0}n.$$

If the neutrons released in this reaction could be used to initiate further reactions, a *self-sustaining chain reaction* could take place (Figure 23.6). Enrico Fermi and other physicists at the University of Chicago showed that this chain reaction was possible by constructing the first nuclear reactor in 1942.

In a uranium fission reaction like the one above, the mass of U-235 plus a neutron is much greater than that of the fission fragments. As we saw in Chapter 8, the curve of binding energy per nucleon versus mass number (Figure 23.7) shows that the values of the binding energy for light and for heavy nuclei are smaller than those for medium nuclei. Therefore, these nuclei are not as tightly bound as medium-weight nuclei (iron is the most stable nucleus). Splitting a heavy nucleus, like U-235, to produce medium-weight nuclei, which are more stable, releases energy. We can calculate this energy by determining the mass deficiency. Before the reaction takes place, the total mass is the mass of the approaching neutron plus the mass of the U-235 nucleus, or

$$m(\tfrac{1}{0}n) + m(\tfrac{235}{92}U) = 1.008665 \text{ amu} + 235.043925 \text{ amu}$$

$$\cong 236.0526 \text{ amu}$$

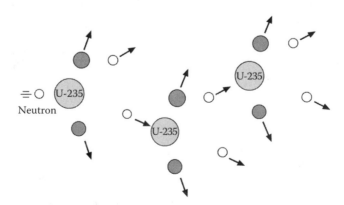

FIGURE 23.6 Chain reaction. After the first reaction is initiated by one neutron, the neutrons released by this reaction can strike uranium nuclei causing additional fissions which in turn produce more neutrons. The process multiplies very rapidly.

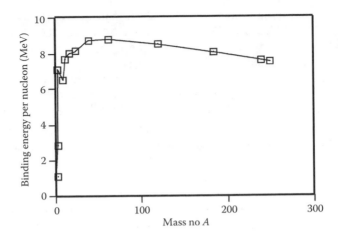

FIGURE 23.7   Binding energy per nucleon plotted versus atomic mass number. This curve shows that medium nuclei are more tightly bound than light or heavy nuclei.

and the mass of the fission fragments is

$$m(^{139}_{54}\text{Xe}) + m(^{94}_{38}\text{Sr}) + 3m(^{1}_{0}\text{n}) = [138.91843 + 93.91547 + 3(1.008665)]\text{amu}$$

$$= 235.8599 \text{ amu}.$$

The mass deficiency is thus $(236.0526 - 235.8599)$ amu $= 0.1927$ amu. That is, the fission fragments have less mass than the U-235 nucleus and the incoming neutron. The energy equivalent of this mass difference is

$$E = 0.1927\text{amu} \times \frac{931 \text{ MeV}}{1 \text{ amu}}$$

$$= 179 \text{ MeV}.$$

This is the energy released in each fission event. At the nuclear level, this is an enormous amount of energy. At the macroscopic level, 179 MeV correspond to about 3×10^{-11} J, which is a very tiny amount of energy. In a macroscopic sample of uranium, however, there are billions of nuclei, so that if a chain reaction is triggered, a huge amount of energy can be released. This is the principle behind the nuclear bomb, a nuclear chain reaction without control.

For a nuclear explosion to occur, there must be enough fissionable uranium to sustain the chain reaction. Natural uranium consists of several isotopes of uranium, the most abundant of which is U-238 (99.3%). Unlike uranium-235, U-238 is not a good prospect for a chain reaction because when neutrons collide with it they are either scattered or captured. Since the natural concentration of U-235 is only 0.7%, it is necessary to increase it to over 90% in order to obtain a chain reaction. The minimum mass required to produce a chain reaction is called the *critical mass* and

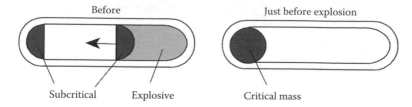

FIGURE 23.8 One design for a nuclear bomb. An explosive propels the two pieces together to form the critical mass. The bomb explodes in about one microsecond. This design was used in Little Boy, the bomb dropped on Hiroshima in 1945.

it depends not only on the amount of uranium-235 but on the design of the bomb. An amount of U-235 as small as 1 kilogram is a significant quantity for a good design.

A critical mass of U-235 will explode on its own, as any stray neutron can initiate a chain reaction. For this reason nuclear bombs are designed so that the uranium cannot form a critical mass until the precise moment when the bomb is to explode. Little Boy, the bomb that exploded on Hiroshima on August 6, 1945, was designed so that the uranium was kept in two *subcritical* masses at the two ends of a cigar-shaped container (Figure 23.8). An explosive propelled the two pieces together at a speed of several kilometers per second to form a critical mass. After the critical mass is formed the entire explosion takes place in about one microsecond. 99.9% of the energy is released during the last tenth of a microsecond of the explosion.

In a nuclear reactor (Figure 23.9), a low concentration of U-235, usually about 3% ensures that most of the neutrons released by one fission reaction are not absorbed by other U-235 nuclei. But even at this low concentration, the fission rate would eventually increase. To prevent this increase and to control the rate of reactions, cadmium rods that absorb neutrons are inserted inside the uranium core (Figure 23.10). In the reactor, the splitting of U-235 nuclei in a controlled environment produces energy that is used to heat water and obtain steam that powers a turbine to produce electricity.

In the 1970s scientists found evidence of a natural nuclear reactor that existed almost two billion years ago during the Precambrian era in Gabon, West Africa. Examining samples of uranium from a mine, they discovered that a chain reaction had taken place there for almost a million years with water trapped in the sandstone absorbing some of the neutrons, like the cadmium rods in modern reactors.

Nuclear Fusion

Another source of nuclear energy comes from *nuclear fusion*, the process that fuels the sun and the stars (Figure 23.11). Nuclear fusion occurs when two light nuclei are fused together to form a heavier nucleus (Figure 23.12). As we know, the mass of every stable nucleus is less than the mass of its protons and neutrons taken together. This is seen better by reexamining the binding energy per nucleon curve of Figure 23.7. Since light nuclei are less tightly bound than medium-weight nuclei, joining light nuclei produces a more massive nucleus which is more stable. Thus if two light nuclei were to come together to form a new nucleus, energy would be released.

FIGURE 23.9   Core of a nuclear reactor. (Courtesy of NASA.)

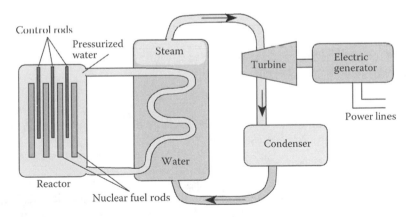

FIGURE 23.10   Schematic diagram of a nuclear reactor.

Uncontrolled fusion reactions are responsible for the fury of hydrogen bombs. Physicists, however, have been attempting to achieve *controlled* fusion reactions for many years and, although great advances have been achieved, complete success is not yet at hand. The fusion of hydrogen could be an almost unlimited source of

FIGURE 23.11   (See color insert following page 268.) Nuclear fission powers the sun. (Courtesy of NASA.)

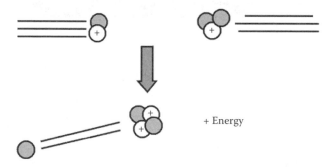

FIGURE 23.12   Nuclear fusion: a nucleus of deuterium and a tritium nucleus come together and form He-4 with the release of one neutron and energy.

energy as hydrogen is so plentiful. The main difficulty lies in bringing the hydrogen nuclei close enough for the attractive nuclear force to overcome the electrical repulsion. Two techniques have been proposed for a practical fusion reactor: magnetic confinement and laser fusion. The magnetic confinement scheme uses magnets to keep the protons confined to a small doughnut-shaped region. The *tokamak* reactor is one example of this technique. The Princeton Tokamak Fusion Test Reactor, which operated at the Princeton Plasma Physics Laboratory between 1982 and 1997, was able to confine protons for 1/6 of a second. The second technique, the *inertial confinement method*, uses lasers to bring together protons so that fusion can take place.

At the Lawrence Livermore laboratory in California, lasers bombard a deuterium–tritium pellet with enormous amounts of energy so that the outer layers of the pellet burn away, causing the rest of the pellet to implode and compress the fuel to densities high enough to trigger nuclear fusion.

PIONEERS OF PHYSICS: ENRICO FERMI

After Lise Meitner and Otto Frisch proved experimentally that the uranium nucleus did split into two fragments, Niels Bohr and the American physicist John Wheeler developed the liquid drop model to explain the newly discovered phenomenon. One important feature of this model was the release of neutrons as the uranium nucleus split, which made a chain reaction possible.

After War World II started in 1939, the possibility that the Nazis could use this discovery to build a powerful bomb prompted the United States to start an urgent research program in nuclear physics. In December 1941, Enrico Fermi was called to Chicago to direct the "uranium project." On December 2, 1942, Fermi's team achieved the first controlled chain reaction on an "atomic pile," the first nuclear fission reactor.

Fermi had just been awarded the Nobel Prize in physics for his work with neutrons as projectiles to produce nuclear reactions. It had occurred to Fermi to bombard uranium with neutrons to produce a new artificial element of higher mass number. Although these transuranium elements were later discovered in nature, they were not known at the time. To Fermi's consternation, his discoveries were publicized by the Italian press as a Fascist triumph.

Fermi had received his doctoral degree magna cum laude from the University of Pisa in 1922, a few months before Mussolini seized power. He did postdoctoral work in Germany and returned to Italy a few years later to become professor of physics at the University of Rome. By the time of his trip to Stockholm to accept the Nobel Prize, life had become very difficult for the anti-Fascist Fermi. His wife was Jewish and Italy had just passed several anti-Jewish laws. The controlled Italian press was also increasingly critical of Fermi's refusal to wear the Fascist uniform at the Nobel ceremonies or to give the Fascist salute. The Fermi family decided to sail for the United States from Stockholm.

After the war, Fermi became professor of physics at the University of Chicago, where he taught until his early death at the age 53 from stomach cancer. The element with atomic mass 100, discovered a year later, was named fermium in his honor.

APPLICATIONS OF NUCLEAR PHYSICS

Nuclear reactions have practical applications in archeology, medicine, and industry. We shall review here some of the most important applications.

RADIOACTIVE DATING

The known half-lives of certain radioactive elements can be used as a clock for determining the age of some objects. This process is known as radioactive dating. Carbon-14, a radioactive isotope of carbon with a half-life of 5730 years, is used to date archeological findings.

The upper atmosphere of the earth is continuously being bombarded by cosmic rays. These are charged particles, electrons and nuclei of hydrogen and helium together with some heavier nuclei, moving at speeds close to the speed of light in the interstellar medium. When cosmic rays collide with the atoms in the upper atmosphere, neutrons are liberated; these in turn collide with nitrogen atoms to produce carbon-14. The reaction is

$$\textstyle{}^{1}_{0}n + {}^{14}_{7}N \rightarrow {}^{14}_{6}C + {}^{1}_{1}H$$

Even though carbon-14 is constantly being produced in the earth's atmosphere, only about 1 in 10^9 carbon atoms is of this radioactive isotope. All living plants absorb carbon dioxide from the air, and chemically do not distinguish between radioactive and nonradioactive carbon. Since animals and humans eat plants, all living things take in carbon-14 in the same proportion. When a plant or animal dies no new C-14 is ingested, and the carbon-14 present at death decays, with a half-life of 5730 years.

Carbon-14 decays through beta emission according to the reaction

$$\textstyle{}^{14}_{6}C \rightarrow {}^{14}_{7}N + {}^{0}_{-1}e + \bar{\nu}.$$

Therefore, when the plant or animal dies, the ratio of C-14 to ordinary C-12 begins to decrease as the C-14 decays, and from this decrease the elapsed time since the death took place can be determined. If, for example, the ratio of C-14 to C-12 in a wooden artifact unearthed in an archeological site is one-fourth of the ratio for living trees, we can infer that the tree from which the artifact was made died two half-lives (11,460 years) ago.

Carbon-14 dating is used to determine the age of fossils and archeological artifacts made from organic materials up to 40,000 years (about seven half-lives). Beyond that there is not enough C-14 remaining in the objects for an accurate age determination. Other isotopes with longer half-lives, such as U-238, with a half-life of 4.5×10^9 years, can be used for the determination of the age of rocks, for example.

Using these radioactive clocks, scientists have been able to determine that hominids have been around for 3.5 million years, that life on earth appeared some three to four billion years ago, and that the earth itself was formed some four and a half billion years ago.

BIOLOGICAL EFFECTS OF RADIOACTIVITY

Radioactive isotopes are used as "labels" that reveal the passage of a substance through the body (Figure 23.13). They are also used in radiation treatment, to destroy cancerous growths from the body where surgery is either inadvisable or impossible.

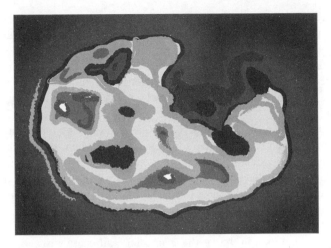

FIGURE 23.13   (See color insert following page 268.) Radioisotope image of a cross section of a human brain, showing an area damaged by a stroke (darkened area). The bright regions indicate normal blood flow. (From Barron Storey/National Geographic Image Collection. With permission.)

However, radiation not only destroys malignant tissue; it also destroys normal living cells by altering or damaging the structure of key molecules.

The radioactive decay of nuclei produces different kinds of ionizing radiation with different energies, reaching up to several million electron-volts per photon. The term "ionizing radiation" means that this radiation interacts with matter, stripping electrons off of atoms, thus forming ions. In addition to the alpha and beta particles and the gamma rays already discussed (all produced in nuclear processes), ionizing radiation also includes photons of lower energies which are produced in atomic processes.

In recent years, radioactive isotopes have been used to trace chemicals in several reactions as they move through an organism. If an atom is replaced in a chemical reaction by its radioactive counterpart, the compound is "tagged" and can be traced with the aid of a detector. Radioactive isotopes can be effectively detected by means of a scanning device. One such device is the Positron Emission Tomography (PET) scanner, a machine that produces cross-sectional views of the body, and requires only low levels of radiation. Radioactive nuclides that decay through emission of positrons (antielectrons) are introduced into the body. One such decay is

$$^{12}_{7}N \rightarrow {}^{12}_{6}C + {}^{0}_{1}e + \nu.$$

As we shall study in more detail in Chapter 24, when these positrons encounter regular electrons, they annihilate each other, emitting gamma rays. These gamma rays are emitted in pairs, emerging in opposite directions. Gamma ray detectors that recognize photons emitted simultaneously but in opposite directions are used to triangulate several such events so that the exact position where they were produced can be determined.

As we have said, ionizing radiation, which can be used to diagnose, treat, and sometimes cure malignant growths in humans, can also be harmful. Radiation damage in organisms is due to ionization effects in cells that result in structural changes. When such damage affects the DNA molecules of the reproductive organs, it is called genetic damage. This may lead to genetic mutations which, in their great majority, are harmful. Radiation damage which affects other cells in the body is called somatic damage.

It is, of course, desirable to protect ourselves from ionizing radiation. However, it is sometimes necessary to risk some radiation exposure when benefits can be derived from it, as in certain medical applications when other methods are not feasible. It is, nevertheless, impossible to avoid all radiation exposure because cosmic rays and radioactive rocks are natural sources of radiation that deliver a dose every year to each person equivalent to the dose received from medical x-rays by the average person in one year.

Uranium is ubiquitous in the earth's crust. The radioactive isotope uranium-238 alpha-decays into thorium-234—itself a radioactive isotope which beta-decays into another radioactive isotope. After several decays, the stable isotope lead-206 is produced and the sequence of nuclides that results is called a decay series. One of the products of this *decay series* is radium-226 which alpha decays into radon-222. The radioactive isotope radon-222 is an inert gas, invisible and odorless, which may seep into a house to build up a hazardous concentration. When radon decays, with a half-life of 3.82 days, it emits an alpha particle of 5.5 MeV energy in the reaction

$$^{222}\text{Rn} \rightarrow {}^{218}\text{Po} + {}^{4}\text{He}$$

and this is followed by a series of other alpha and beta particles produced in the subsequent decays of the remaining nuclides in the decay series. The total energy released in these processes is 21.5 MeV. When a person breathes in radon and other nuclides in the decay series—some of which are solid and might lodge in the lung—these emit radioactive alpha and beta particles that can cause biological damage.

Although the average house contains radon at a concentration of about one-fourth what the environmental protection agency (EPA) considers as the threshold of concern, recent tests have determined that nearly one third of the residences tested by the EPA exceeded that threshold. Fairly simple devices can be used to record the concentration of radon in homes. They fall into two categories: alpha-track devices and activated-charcoal detectors. Alpha-track devices, developed by General Electric's Research and Development Center as part of the lunar exploration program during the 1970s, consist of a small sheet of polycarbonate plastic kept in a plastic bottle (Figure 23.14a). When an alpha particle strikes the plastic sheet, it causes a microscopic pockmark which can be identified by a laboratory with the appropriate equipment. In a typical test, fewer than 100 pocks are produced in a span of three months; for this reason, the alpha-track detectors are best for long-term measurements. For short-term measurements, the activated-charcoal detectors are better. These are flat canisters of activated-charcoal granules which trap radon gas (Figure 23.14b). After a short period of exposure to the air in the house, they are sent back to a laboratory for analysis.

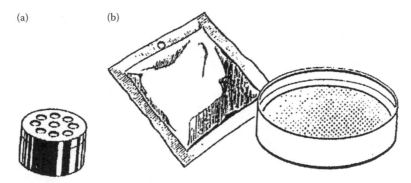

FIGURE 23.14 (a) Commercially available alpha-track detector to measure radon concentration in homes. (b) Activated-charcoal detector.

PHYSICS IN OUR WORLD: PROTON BEAMS FOR CANCER THERAPY

The first hospital-based proton accelerator for cancer treatment was installed in 1990 at the Loma Linda University Medical Center. The 50-ton machine, built by the Fermi National Accelerator Laboratory, strips electrons from hydrogen atoms and accelerates the remaining protons to great speeds.

The patient is partly enclosed in a tube at the center of a three-story-high, 90-ton rotating gantry, which directs the proton beam to its target for about a minute. Like the electron beam in a television set or computer monitor, the proton beam at Loma Linda scans a thin slice of tumor line by line. By varying the number of protons in the beam and their energy, the beam can be distributed directly in successive layers inside the tumor.

Unlike x-rays, which transmit some of their ionizing radiation over a few centimeters and are not completely stopped by the tissue, proton beams are stopped completely by the tumor, where they deposit most of their energy, leaving surrounding tissue untouched. The malignant tissue is destroyed as the beam splits apart DNA molecules.

At Loma Linda, the accelerator is used to treat prostate, lung, and brain cancers. Other proton accelerators have been installed in a few other U.S. hospitals, but the Loma Linda University Medical Center still treats more patients than any other center in the world.

24 Elementary Particles

ANTIMATTER

Paul Adrien Maurice Dirac was born in Bristol, England, on August 8, 1902, one of the three children of Charles Dirac, a Swiss émigré, and Florence Holten. Dirac was brought up to be bilingual in French and English. "My father made the rule that I should only talk to him in French. He thought it would be good for me to learn French in that way. Since I found that I couldn't express myself in French, it was better for me to stay silent than to talk in English. So I became very silent at that time—that started very early." Dirac remained very reserved throughout his life. "My father always encouraged me towards mathematics... He did not appreciate the need for social contacts. The result was that I didn't speak to anybody unless spoken to," he was to confess later. "I was very much an introvert, and I spent my time thinking about problems in nature."

At the age of 16, Dirac entered the University of Bristol where, afraid that there would be no jobs in mathematics, he studied engineering, graduating in 1921. Ironically, he could not find a job with his engineering degree and decided to stay on to study pure and applied mathematics. In 1923, he enrolled at Cambridge where he received his PhD in May of 1926, with a thesis entitled "Quantum Mechanics." In 1932, at the age of 30, he became Lucasian Professor of Mathematics at Cambridge, the chair once held by Newton.

In 1927 Dirac attended the Solvay Conference, a periodic meeting of prominent physicists paid for by the wealthy Belgian industrial chemist Ernest Solvay, where he met Einstein for the first time. During a conversation with Bohr at the conference, Bohr asked Dirac what he was working on. "I am trying to get a relativistic theory of the electron," answered Dirac. Bohr claimed that the problem had already been solved. Dirac did not agree.

Dirac knew that quantum mechanics and relativity theory were both correct. Quantum mechanics, however, could not be applied to quantum particles moving at relativistic speeds. Since the wave properties of the electron had already been confirmed, Dirac decided to develop a relativistic quantum theory of the electron using the concept of *quantum field*. According to this approach, every particle type has its own field. The field of the photon is the electromagnetic field. The electron also belongs to a field, now called the *Dirac field*. These fields interact with each other according to laws expressed in certain quantum field equations.

Dirac's theory—the first *quantum field theory*—is known today as *relativistic quantum field theory*. It is Maxwell's electromagnetism applied to relativistic quantum particles. In his theory, Dirac generalized the Schrödinger equation to include electrons moving at relativistic speeds with a new equation now called the *Dirac equation*. This equation had remarkable and profound consequences. "The relativistic wave equation of the electron [the Dirac equation] ranks among the highest achievements of twentieth-century science," says Abraham Pais of Rockefeller University.

The Dirac equation predicted the observed properties of the electron. For example, it predicted that an electron had an intrinsic spin angular momentum. The effects of this spin of the electron had been observed experimentally, although Schrödinger's equation did not account for it.

The most remarkable prediction of the Dirac equation had to do with its two solutions for the energy of an electron, one positive and the other negative. What to do with the negative energy solutions? Nobody knew what electrons with negative energy could be. "I was reconciled to the fact that the negative energy states could not be excluded from the mathematical theory," wrote Dirac, "and so I thought, let us try and find a physical explanation for them." The physical explanation that Dirac eventually gave was that the negative energy solution described a new kind of particle with the same mass as the electron but with *positive* electric charge—opposite to that of the electron. He initially called it the *antielectron*. That was a bold prediction. "I didn't see any chance of making further progress," he wrote later. "I thought it was rather sick." To his great triumph, antielectrons or *positrons*, as they were later called, were discovered four years later in cosmic rays by Carl D. Anderson of the California Institute of Technology.

Physicists soon realized that Dirac's theory predicted the existence of other *antiparticles*, in addition to the positron. For every quantum particle there was an antiparticle. These antiparticles constitute a new kind of matter, *antimatter*, identical to ordinary matter except that if the particle has electric charge, its antiparticle has the opposite charge. When a particle encounters its antiparticle, they annihilate each other, disappearing in a burst of photons in a process called *pair annihilation* (Figure 24.1a). The inverse process, in which high energy photons create a

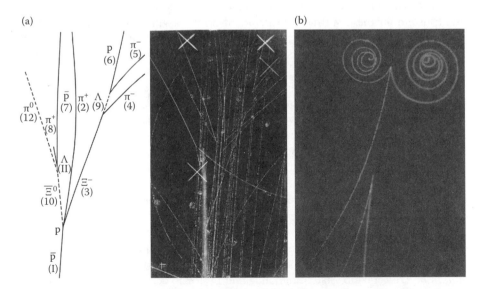

FIGURE 24.1 (a) Pair annihilation and (b) pair production. (Courtesy of Brookhaven National Laboratory and Lawrence Berkeley Laboratory.)

positron–electron pair, is also possible (Figure 24.1b). This process is called *pair production*. Recall that, according to the special theory of relativity, an electron has a rest energy m_0c^2. Thus, photons of energy hf equal to at least $2m_0c^2$ can produce an electron and a positron.

THE FUNDAMENTAL FORCES

Before we continue delving more deeply into the structure of matter, let us summarize what we have learned so far about the known interactions in nature. All the variety of phenomena in the universe, from the explosion of a supernova in another galaxy to the falling of a leaf during an autumn afternoon on earth, from the appearance of a dark spot on the atmosphere of a distant star, the explosion of a volcano on Io or the collapse of a distant red giant to the whirr of the wings of a hummingbird, all of these events can be ultimately explained by only four fundamental forces. Everything that happens anywhere in the universe is ultimately controlled by the operation of these four forces. Understanding the properties of these forces—gravity, electromagnetism, and the strong and weak nuclear forces—is, perhaps, the most important task in physics today.

- *Gravity*, the first one of the four forces to be discovered, was formulated by Newton in his universal law of gravitation. Gravity controls a wide range of phenomena, from the falling of an apple or the motions of ocean tides to the expansion of the universe. Newton's universal law of gravitation is a simple inverse-square law; that is, the gravitational force is a long-range force that decreases in strength in proportion to the square of the distance between the two interacting bodies. Newton's theory, however, fails to explain *why* gravity exists. Einstein's general theory of relativity presented gravity not as a force but as a product of geometry, the geometry of space-time. In doing so, Einstein placed gravity farther away from the rest of physics.
- The *electromagnetic interaction* is described by Maxwell's equations. These equations concisely summarize all known electric and magnetic phenomena. The electromagnetic force is responsible for the binding of atoms and molecules; it binds electrons to nuclei and atoms and binds atoms together into molecules, holding them together in solids. Like Newton's universal law of gravity, the electromagnetic force is an inverse-square force.
- The *strong force*, one of the two nuclear forces, holds the nucleons together in a nucleus. It is a short-range force, becoming negligible at distances greater that 10^{-15} m. The strong force acts on protons and neutrons but not on electrons, neutrinos, or photons. The strong force is 137 times stronger than the electromagnetic force.
- The *weak force* is also a short-range nuclear force. It is responsible for radioactive beta-decay processes, such as the transformation within the nucleus of a neutron into a proton or a proton into a neutron. The weak force controls many of the reactions that produce energy in the sun and the stars. The weak force is of the order of a hundred thousand times weaker than the strong force.

TABLE 24.1

Properties of the Fundamental Forces

Force	Range	Relative Strength
Strong	10^{-15} m	1
Electromagnetic	infinite	7.3×10^{-3}
Weak	10^{-17} m	10^{-5}
Gravitational	infinite	6×10^{-39}

Table 24.1 summarizes the properties of the fundamental forces. As we can see, gravity is the weakest of the forces. The gravitational force of attraction between two electrons, for example, is 10^{36} times weaker than their electrostatic repulsion. The gravitational force between large bodies becomes important because it involves a tremendous number of atoms, and because these atoms are neutral, so that their electrostatic interaction cancels out. The two nuclear forces are short-range and play no role in the interaction between large bodies at normal distances.

EXCHANGE FORCES

What is a force? How does a force work? How does an electron know that there is another electron nearby or the earth that the sun is 150 million kilometers away? How is a force transmitted? Newton never attempted to explain how his universal law of gravitation worked. Gravity was accepted as an action at a distance phenomenon. Faraday introduced the concept of field to explain how electric and magnetic forces were transmitted. Einstein, in his general theory of relativity, said that gravity was geometry. According to Einstein, the presence of an object disturbs the space-time around it, altering its geometry, so that when a second object enters this space, it experiences the distortions.

In 1928 Dirac developed his relativistic quantum theory of the electron by combining Maxwell's electrodynamics, relativity, and quantum mechanics. Dirac's theory, a quantum field theory, was further developed by Richard Feynman, Sin-Itiro Tomonaga, and Julian Schwinger into a complete theory, known today as *quantum electrodynamics* or QED. According to electrodynamics, when an electron accelerates, it radiates energy in the form of an electromagnetic wave. From quantum mechanics, we know that this electromagnetic wave consists of photons. Thus the acceleration of an electron results in the emission of one or more photons. When an electron approaches another electron, the mutual electrostatic repulsion that results is interpreted in QED as due to an interaction with the photons emitted by the accelerating electrons. According to QED, the approaching electron emits a photon which is then absorbed by the second electron. We can illustrate this process with the diagram shown in Figure 24.2. An electron is represented by a straight line and the photon they exchange by a wiggly line. These diagrams were first used by Richard Feynman as symbolic representations of equations and are known today as *Feynman diagrams*.

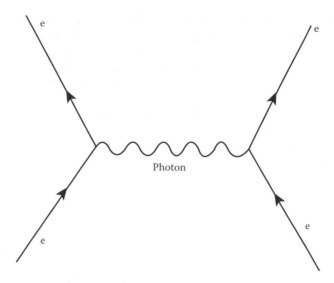

FIGURE 24.2   Feynman diagram of the interaction between two electrons. The two electrons exchange virtual photons which carry energy and momentum. This exchange causes the electromagnetic repulsion between the two electrons.

Where do the photons exchanged between two interacting electrons come from? From nowhere. They are allowed by the uncertainty principle. The electron creates a photon which is absorbed by the interacting electron before the change in energy associated with its creation can be detected. If the uncertainty in our knowledge of the energy of the electron is ΔE and Δt the uncertainty in our knowledge of the time during which the electron has that energy, Heisenberg's uncertainty principle states that

$$\Delta E \times \Delta t \geq \frac{h}{2\pi}.$$

This expression means that during the interval Δt a photon of energy as large as ΔE is created. If the second electron absorbs this photon before the time Δt is up, no experiment can detect any missing energy. Particles that exist for only the time allowed by the uncertainty principle are called *virtual* particles. A virtual photon exists only for the brief, fleeting moment permitted by the uncertainty principle. Real photons exist forever, provided they do not interact with other particles.

Quantum field theory provides an answer to the questions that opened this section. All the forces of nature can be explained as being the result of the exchange of some virtual particle.

PIONS

Is it possible to understand the strong interaction between nucleons in terms of the exchange of virtual particles? The Japanese physicist Hideki Yukawa made such a suggestion in 1933. "By confronting this difficult problem I committed myself to long days of suffering," writes Yukawa in his autobiography. "Is the new nuclear force a primary one?... It seemed likely that [it] was a third fundamental force unrelated to gravitation and electromagnetism... Perhaps the nuclear force could find expression as a field... If one visualizes the force field as a game of 'catch' between protons and neutrons, the crux of the problem would be the nature of the 'ball' or particle."

Heisenberg had had a similar idea before and had proposed that the "ball" was the electron. At first, Yukawa thought also that the electron was the intermediary particle. Soon, he realized that this was not possible. "Let me not look for the particle that belongs to the nuclear force among the known particles... If I focus on the characteristics of the nuclear force field, then the characteristics of the particle I seek will become apparent," he wrote. Then came the stroke of genius; "The crucial point came to me in October [1934]. The nuclear force is effective at extremely small distances. My new insight was that this distance and the mass of the new particle are inversely related to each other."

If a virtual particle of mass m can only exist for a time Δt allowed by the uncertainty principle and we assume that this particle can travel at the speed close to c, the maximum range R of the particle (that is, the maximum distance that the particle can travel during its existence) is

$$R = c\,\Delta t.$$

The uncertainty in energy equals the mass energy of the virtual particle

$$\Delta E = mc^2.$$

The uncertainty principle $\Delta E\,\Delta t \geq h/2\pi$ gives us

$$\Delta t = \frac{h}{2\pi \Delta E} = \frac{h}{2\pi mc^2}.$$

The maximum range of the virtual particle is

$$R = \frac{h}{2\pi mc}.$$

The maximum range is then inversely proportional to the mass. If the mass is zero, the range becomes infinite. A massive virtual particle has a finite range; the more massive the particle is, the shorter the range. The electromagnetic interaction had an infinite range and was mediated by a massless particle, the photon. The nuclear force, a very short-range force, had to be mediated by a massive particle. Knowing that nuclear forces occur over less than 10^{-14} m, Yukawa calculated the mass of the new particle to be 140 MeV/c^2. Since its mass fell between the mass of the electron and the mass of the proton, the new particle was called *meson* (from the Greek meso, "middle").

Three years later, in 1937, a team of American scientists found a particle with a mass of about 100 MeV/c^2 and it was immediately assumed that it was the meson

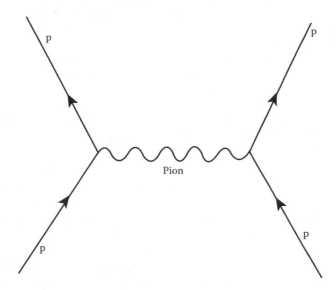

FIGURE 24.3   Feynman diagram for the strong interaction between two protons via the exchange of a virtual pion.

predicted by Yukawa. Experiments carried out in 1945, however, showed that this new particle interacted weakly with protons and neutrons and consequently could not be the mediator of the nuclear force. The following year, the real Yukawa particle, the particle exchanged in the nuclear interaction, was found in cosmic rays and was named the π *meson* or *pion*. (The earlier particle, later called the *muon*, turned out to be a completely different beast that participates only in weak and electromagnetic interactions and has nothing to do with Yukawa's particle.) There are three kinds of pions: π^+; π^-, which is the antiparticle of the π^+; and π^0, which is its own antiparticle. Figure 24.3 illustrates the strong interaction between two protons via the exchange of a virtual pion.

PARTICLE CLASSIFICATION: HADRONS AND LEPTONS

In the subsequent years, experiments with cosmic rays and with particle accelerators yielded a plethora of new particles. In addition to the protons, neutrons, pions, muons, neutrinos, and photons; particles with names like sigma, delta, lambda, tau meson, and kaon were added to the arsenal of high-energy physicists. What to make of all these particles?

To make sense of the hundreds of particles that have been discovered, a classification scheme based on the interaction in which they participate has been devised. Particles that participate in the strong interaction, the interaction that holds the nucleus together, are called *hadrons* (from the Greek *hadros*, meaning *strong*). Hadrons with half-integer spin in units of $h = 2\pi$ (such as the proton or the neutron) are called *baryons*, whereas hadrons with integer spin (such as the pion) are called *mesons*. Since all baryons, except the proton itself, ultimately decay into protons and other particles, they are all more massive than the proton (baryon means *heavy* in Greek).

TABLE 24.2

Lepton Classification

Generation	Charged Lepton	Neutrino	Charged Antilepton	Antineutrino
First	Electron, e^-	ν_e	Positron, e^+	$\bar{\nu}_e$
Second	Muon, μ^-	ν_μ	Antimuon, $\bar{\mu}^+$	$\bar{\nu}_\mu$
Third	Tau, τ^-	ν_τ	Antitau, $\bar{\tau}^+$	$\bar{\nu}_\tau$

Particles that do not interact via the strong force are called *leptons*, from the Greek *leptos*, meaning light. The electron, the neutrino, the muon, and the tau particle are all leptons. To the best of anyone's knowledge, the muon is exactly the same as the electron except that it is 200 times more massive. The same can be said for the tau; in this case, it is 3500 times heavier than the electron. All leptons have spin 1/2 and all are truly elementary particles, without any internal structure. As far as we know, leptons are point particles with no dimensions. The most precise experiments have determined that the electron is no larger than 10^{-17} cm in diameter. This is one thousand-millionth the size of an atom and one ten-thousandth the size of a proton.

In a weak interaction involving the electron, like the decay of the neutron, one kind of neutrino is involved, the *electron neutrino*. In the decay of the pion, however, another kind of neutrino, the *muon neutrino*, is involved. When this neutrino strikes a target, it produces muons and not electrons. There is a third type of neutrino emitted in conjunction with the tau particle, the *tau neutrino*. There is then a grouping of leptons into pairs or *generations*, the electron with the electron neutrino, the muon with the muon neutrino, and the tau particle with the tau neutrino. Each one of these leptons has its corresponding antiparticle. This classification is summarized in Table 24.2.

CONSERVATION LAWS

Interactions between elementary particles take place according to a set of conservation laws. One such law is the principle of *conservation of energy*. In any process between particles, the mass energy of the decay products must equal the mass energy of the original particles. It was the principle of conservation of energy that led Wolfgang Pauli in 1932 to predict the existence of the neutrino from the observation of an apparent violation of this law in beta-decay processes. The neutrino was found in 1950.

Conservation of electric charge is another very important law. It says that the sum of all the charges of the original particles must equal the sum of the charges of the particles produced in the interaction. For example, in the reaction between two nucleons that produces pions:

$$p + n \rightarrow p + p + \pi^-$$

electric charge: $1 + 0 \rightarrow 1 + 1 - 1$

the sum of the charges of the original nucleons is 1 and the sum of the charges of the products is $1 + 1 - 1 = 1$.

Conservation of energy and conservation of charge—two laws we had encountered before—are, as far as we know, universal laws. We mentioned before that all baryons finally decay into protons. This is a consequence of another important conservation law, the *law of conservation of baryons*. In any interaction, *the creation of a baryon must be accompanied by the creation of an antibaryon*. If we assign a *baryon number $B = +1$* to all baryons, $B = -1$ to all antibaryons, and $B = 0$ to all other particles, this law can also be stated as follows:

> The law of conservation of baryons: In any interaction, the total baryon number must remain constant.

Since the proton is the lightest baryon, all other baryons must ultimately decay to a proton, which can decay no further without changing baryon number.

Leptons also obey similar conservation laws. Leptons, as we recall, participate in weak interactions. A weak decay process is the decay of the neutron into a proton, an electron and an electron-neutrino. Another weak decay is that of the pion, which decays into a muon and a muon-neutrino. In all of these processes, the creation of a lepton always takes place with the creation of an antilepton. If the leptons (electron, muon, and tau) are assigned a *lepton number* $+ 1$ and the antileptons a lepton *number* -1, this principle can be stated as follows:

> The principle of conservation of leptons: In all interactions between particles, the total lepton number for each variety (electron, muon, and tau) must remain constant.

As we indicated before, each charged lepton has its own associated neutrino, so that a pion cannot decay into a muon and an electron-neutrino. The electron and the electron-neutrino are then assigned an electronic lepton number $L_e = + 1$. The positron and the positron-neutrino are assigned an electronic lepton number $L_e = -1$. All other leptons and in fact all other particles are assigned an electronic lepton number $L_e = 0$. The muon and the muon-neutrino, on the other hand, are assigned a muonic lepton number $L_\mu = + 1$; the antimuon and its associated neutrino have $L_\mu = -1$; similarly for the tau and the tau-neutrino. Table 24.3 summarizes the lepton quantum numbers of the six leptons and their antiparticles.

STRANGE PARTICLES

There are some important differences between hadrons and leptons. Hadrons have a definite extension in space whereas leptons behave like point particles. In addition, there are only six leptons (and their corresponding six antiparticles), while the hadrons, it is believed today, are infinite in number. Moreover, all leptons obey the same law of conservation of lepton number. For hadrons, only the baryons have a conservation law; there is no meson number conservation law.

To complicate matters, new particles were discovered in the early 1950s. These were produced via the strong interaction, making them hadrons; but they decayed in strange

TABLE 24.3

Lepton Quantum Numbers

Lepton	L_e	L_μ	L_τ
e^-	+1	0	0
ν_e	+1	0	0
μ^-	0	+1	0
ν_μ	0	+1	0
τ^-	0	0	+1
ν_τ	0	0	+1
e^+	−1	0	0
$\bar{\nu}_e$	−1	0	0
$\bar{\mu}^+$	0	−1	0
$\bar{\nu}_\mu$	0	−1	0
$\bar{\tau}^+$	0	0	−1
$\bar{\nu}_\tau$	0	0	−1

ways, as if they were leptons decaying through the weak interaction. These particles, the K mesons or kaons, and the Λ (Greek capital lambda) and Ξ (Greek capital xi) baryons, were called *strange particles* owing to their strange behavior. In 1952, Abraham Pais of the Institute for Advanced Studies in Princeton introduced the idea of *associated production*, according to which, the strange particles were always produced in pairs.

In 1956 Murray Gell-Mann in the United States and Kazuhiko Nishijima in Japan independently suggested that Pais's associated production phenomenon was the result of the conservation of a new property, which they called *strangeness*. It turned out that strangeness, with a new quantum number S, must be conserved in strong interactions. The strange particles can only be produced in pairs of overall zero strangeness.

QUARKS

In 1961, Gell-Mann, and independently Yuval Neeman in Israel, noticed a pattern in the hadrons which was based on a mathematical symmetry which they called SU(3). The mathematical tool that Gell-Mann and Neeman used to discover this symmetry is called *group theory*, a technique that had been formulated by a 20-year-old French mathematician, Evariste Galois, the night before his death in 1832. Galois had become involved in a duel challenge over a woman and, anticipating his death, spent his last night writing out his ideas on group theory. Galois was killed but his ideas survived, and Gell-Mann and Neeman used them in their theory. According to the symmetric group SU(3), each hadron is a member of a specific family of hadrons with 1, 8, 10, and 27 members each. Gell-Mann called this scheme the *eightfold way*. The proton and the neutron belong to the same family of eight hadrons, called the *baryon octet*. The pion also belongs to a family with eight members called the *meson octet*.

The eightfold way was very successful in the prediction of unknown particles. The most spectacular prediction was that of the Ω^- (Greek capital omega minus) particle, the tenth member of a family of hadrons called a decuplet. The mass and the quantum properties of the Ω^- had been predicted by Gell-Mann in 1962. In December 1963 the Ω^- was found by a team of physicists at Brookhaven National Laboratory with exactly the properties predicted by the eightfold way.

The questions that scientists were asking then were: Why did the eightfold way work? Why were the hadrons grouped into families of 1, 8, 10, and 27 members each? It was again Gell-Mann with yet another physicist working independently, George Zweig, who provided the answers. Zweig, who like Gell-Mann was a professor of physics at Caltech, developed his ideas while on sabbatical leave at CERN, the European research center in Geneva. They suggested that the families in the eightfold way could all be generated from a basic triplet which corresponded to three new fundamental particles that made up all the hadrons. Hadrons, according to the new theory, were not elementary particles but were composed of members of a fundamental triplet that Gell-Mann called quarks. A proton, for example, was supposed to be composed of three quarks. Gell-Mann borrowed the name quark from a passage in James Joyce's novel Finnegans Wake that reads:

> *Three quarks for Muster Mark!*
> *Sure he hasn't got much bark*
> *And sure any he has it's all beside the mark.*

The "three quarks" may refer to the three children of Mr. Mark (or Mr. Finn) who occasionally represent him.

The three quarks in the original theory are point particles like the electron and with the same spin ½. According to this theory, all hadrons are made of combinations of two or three quarks, called the *up* or *u* quark, *down* or *d* quark and *strange* or *s* quark. The three varieties of quark are known as *flavors*. The term comes from the initial whimsical use of the names chocolate, vanilla, and strawberry for the three quarks. Although these names did not stick, the term "flavor" for the different types of quark did.

Quarks have unusual quantum numbers (Table 24.4). The most unusual is the fractional electric charge. The baryon number is also fractional. All baryons, like the proton and neutron, are combinations of three quarks, and their antiparticles are combinations of three antiquarks. With three quarks to a baryon, the baryon number of a quark has to be (1/3). Mesons, on the other hand, are combinations of a quark and an antiquark, giving a baryon number of zero, as it should be, since they are not baryons. The rules of SU(3) govern the combination of quarks into baryons and mesons to produce the correct baryon, charge, and other quantum numbers. According to these rules, the proton is composed of two *u* quarks and one *d* quark, as shown in Figure 24.4a. As shown in Table 24.4, the *u* quarks have a charge of + (2/3) and the *d* quark has a charge of −(2/3); the total charge of the proton is then + (2/3) + (2/3) − (1/3) = 1, in units of the fundamental charge *e*, which is the charge of the proton. The pion is composed of a *u* quark and an anti-*d* quark. Its total charge is + (2/3) + (1/3) = 1.

TABLE 24.4

Quantum Numbers of the Original Three Quarks and Their Antiparticles

Quark	Charge (e)	Baryon Number	Strangeness	Spin
u	$+\dfrac{1}{3}$	$\dfrac{1}{3}$	0	$\dfrac{1}{2}$
d	$-\dfrac{1}{3}$	$\dfrac{1}{3}$	0	$\dfrac{1}{2}$
s	$-\dfrac{1}{3}$	$\dfrac{1}{3}$	-1	$\dfrac{1}{2}$
$\bar{u}$	$-\dfrac{2}{3}$	$-\dfrac{1}{3}$	0	$\dfrac{1}{2}$
$\bar{d}$	$+\dfrac{1}{3}$	$-\dfrac{1}{3}$	0	$\dfrac{1}{2}$
$\bar{s}$	$+\dfrac{1}{3}$	$-\dfrac{1}{3}$	$+1$	$\dfrac{1}{2}$

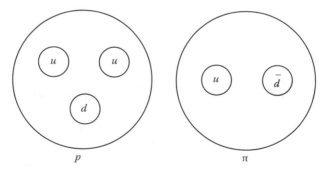

FIGURE 24.4 (a) Quark structure of the proton and (b) the pion.

PIONEERS OF PHYSICS: GELL-MANN'S QUARK

James Joyce's last novel, *Finnegans Wake*, is a massive, dauntingly obscure work in which technique is more important than content. The story centers on a Dublin pub owner named H. C. Earwicker who dreams throughout the entire novel. Through his dreams Earwicker reenacts myths and great historical events in a haphazard progression of reflection, complete with paradoxes, repetitions, and sudden changes of focus. The novel is also full of puns, deliberate misspellings, and curious linguistic turns, beginning with the missing apostrophe in the book's title.

When Gell-Mann was struggling with his eightfold way, he was trying to explain to a colleague why some of the conclusions of his calculations seemed crazy because they implied the existence of fractional electric charge. Thinking more about it the following day, Gell-Mann began to accept the strange conclusions. Since these fractional charges were so peculiar, he used an odd term for them: *quork*.

Several months later, when he was ready to publish his idea, he found himself reading some passages in *Finnegans Wake* ("you know how you read *Finnegans Wake*," he said in an interview), when he came across "Three quarks for Muster Mark." That was it! Three *quarks* make a proton or a neutron! Although Joyce's word rhymes with bark, it was close enough to his *quork*. *Quark* it was from then on.

PARTICLES WITH CHARM

In 1973, the theoretical physicist Sheldon Glashow of Harvard University and his collaborators developed a more complete theory of quarks based on mathematical symmetries which led them to propose the existence of a fourth quark flavor which Glashow named the charm or c quark. The introduction of this fourth quark was done on a theoretical basis alone, for no hadrons had been found that required its existence. In November of 1974 two experimental teams, one led by Samuel Ting at Brookhaven National Laboratory and the other led by Burton Richter at Stanford Linear Accelerator simultaneously announced their independent discovery of a new particle that behaved like a meson but was heavier than the proton. Ting, who is of Chinese descent, named the particle the J because his name, in Chinese, is written with a character similar to this letter. The Stanford group, however, had named the particle the Ψ (Greek capital psi). Since the discovery was made simultaneously by the two groups, the particle is known today as the J/Ψ. The behavior of this particle could only be explained in terms of the c quark predicted by Glashow. Sometime later, other charged mesons were discovered that confirmed the existence of the c quark. Ting and Richter won the Nobel Prize in physics in 1976 for their work.

In the early 1970s, there were four quark flavors and four leptons. In 1975, experiments conducted at Stanford University led to the discovery of the fifth lepton, the tau, and this discovery persuaded physicists to consider the possibility of a fifth and even sixth quark. These quarks were named bottom or b quark and top or t quark. In 1977, Leon Lederman and his group at the Fermi National Laboratory reported the discovery of a very massive meson, the Υ, which consists of a $b-\bar{b}$ pair, thus confirming the existence of the b quark. Evidence for the t quark was found at Fermilab in 1995. Table 24.5 lists the quantum numbers of the six quarks.

Just as the six leptons were classified into three generations, the six quarks are also classified in three generations each containing two associated quarks or *flavor doublets* (Table 24.6). The three generations of quarks and leptons are shown in Table 24.7.

TABLE 24.5
Quark Quantum Numbers

Quark	Charge (e)	Baryon Number	Strangeness	Charm	Bottom	Top
u	$+\dfrac{2}{3}$	$\dfrac{1}{3}$	0	0	0	0
d	$-\dfrac{1}{3}$	$\dfrac{1}{3}$	0	0	0	0
s	$-\dfrac{1}{3}$	$\dfrac{1}{3}$	−1	0	0	0
c	$+\dfrac{2}{3}$	$\dfrac{1}{3}$	0	1	0	0
B	$-\dfrac{1}{3}$	$\dfrac{1}{3}$	0	0	−1	0
t	$+\dfrac{2}{3}$	$\dfrac{1}{3}$	0	0	0	1

TABLE 24.6
Quark Classification

Generation		Quark
First	u	d
Second	s	c
Third	b	t

TABLE 24.7
Quarks and Leptons

Charge / Symbol / Name			
$+\dfrac{2}{3}$	$-\dfrac{1}{3}$	−1	0
u Up	d Down	e Electron	ν_e Electron neutrino
$+\dfrac{2}{3}$	$-\dfrac{1}{3}$	−1	0
c Charm	s Strange	μ Muon	ν_μ Muon neutrino
$+\dfrac{2}{3}$	$-\dfrac{1}{3}$	−1	0
t Top	b Bottom	τ Tau	ν_τ Tau neutrino

(Courtesy of Thaves.)

THREE GENERATIONS

The classification of the quarks and leptons into the three generations of Table 24.7 tells us that these building blocks of matter form a pattern. Recognition of this pattern is about all we can say about the existence of the three families. We have reached the limit of our current understanding of nature and we cannot tell what this classification means or even whether it is correct. As far as we know, the first generation is all that is needed to construct the universe. Yet, nature provides us with two additional generations of quarks and leptons with the same properties as those of the first, only more massive. Why do they exist? No one knows yet. Perhaps one day, this classification will automatically flow from a new symmetry.

We are not completely lost, however. There is experimental evidence that the top quark is the last one and that there are no additional generations beyond these three. Moreover, the enormous mass of the top quark, 178 GeV/c^2, suggests that it is intimately connected with the mechanism that explains where and how particles obtain their masses.

25 Unifying Physics

SYMMETRY

Gottfried Wilhelm Leibniz, the seventeenth-century German philosopher and mathematician, formulated the principle of the identity of the indiscernibles, which states that if it is not possible to establish a difference between two objects, they are identical. Since it is not possible to distinguish between two identical objects, an interchange in their positions has no effect in the physical state of the two objects or of the system they belong to. This interchange is an example of what a mathematician calls *symmetry*.

When a system remains unchanged after some operation is performed on it (such as the exchange of two identical particles) we say that the system is *invariant* under that particular operation. Another example of a symmetric operation is the rotation of crystals and other objects that have symmetry. If we rotate a snowflake through an angle of 60°, the new orientation is indistinguishable from the original orientation. Rotating a cube through an angle of 90° around any one of its three axes gives us back the original orientation. A sphere can be rotated through any angle around any axis and the new position is indistinguishable from the original. We say that the sphere is invariant under any rotation. Figure 25.1 illustrates these symmetries.

After a symmetric operation, such as the exchange of two identical electrons in an atom, the mathematical equation that describes the system must remain invariant. To deal with these symmetries, physicists use a mathematical technique called group theory, which we mentioned in connection with Gell-Mann's Eightfold Way. Group theory was developed by Galois in France in 1831. Toward the end of the nineteenth century, the Norwegian mathematician Sophus Lie classified all possible groups of a particular type into seven classes. The Lie group O(3), for example, describes the symmetry of a billiard ball; the ball looks exactly the same after it is rotated through any angle.

The connection between symmetry and the laws of physics stems from the work of the brilliant German mathematician Amalie Emmy Nöther. Born in the university town of Erlangen in 1882, Nöther attended the all-male University of Erlangen where her father was a professor of mathematics, after obtaining special permission to enroll. In 1907 she received her PhD degree summa cum laude with a dissertation on invariants. After graduation, she taught without salary at a mathematical school and occasionally substituted for her father at the university. In 1915, she joined a research team working on Einstein's general relativity at Göttingen but was not accepted at the university even with an unpaid contract. The eminent mathematician David Hilbert, who supported her appointment and even had her lecture in his classes, angrily said, "I do not see that the sex of the candidate is an argument against her admission as [an instructor]. After all, we are a university and not a bathing establishment." She finally won admission as a university instructor in 1919.

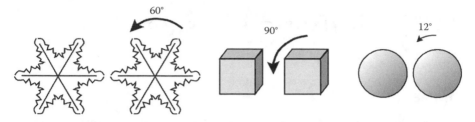

FIGURE 25.1   A snowflake is unchanged when rotated 60°. A rotation of 90° around an axis keeps the cube unchanged. A sphere can be rotated through any angle and remain unchanged.

By 1930 Nöther had become the center of mathematical research at Göttingen. In 1933, she fled Nazi Germany for the United States and became a visiting professor of mathematics at Bryn Mawr College and a member of the research faculty at the Institute for Advanced Study in Princeton. After her death in 1935, Einstein wrote of her work in *The New York Times*:

> Pure mathematics is, in its way, the poetry of logical ideas. One seeks the most general ideas of operation which will bring together in simple, logical and unified form the largest possible circle of formal relationships. In this effort toward logical beauty spiritual formulas are discovered necessary for the deeper penetration into the laws of nature.

The theorem that made Nöther famous among physicists was developed for the paper that she presented before the faculty at Göttingen as part of her application for the position, as was customary in German universities. *Nöther's theorem* can be stated as follows:

> For every continuous symmetry in the laws of physics there exists a corresponding conservation law.

A continuous symmetry is one in which the corresponding transformation can be varied continuously, as is the case with rotation. The angle of rotation can be continuously changed. This symmetry leads to the conservation of angular momentum.

GLOBAL AND LOCAL SYMMETRIES

Nöther's elegant theorem prompted physicists to re-examine the conservation laws from the perspective of symmetry. In 1918, the year that Nöther introduced her theorem, the physicist Hermann Weyl was attempting to formulate a theory that would combine electromagnetism and the general theory of relativity by showing that both theories were linked to a symmetry of space. He proposed a theory that remained invariant under arbitrary space dilations or contractions in which the standards of length and time changed at every point in space-time. This reminded him of the gauge steel blocks used by railroad engineers as standards of length in their determination of the distance between tracks and so he referred to the invariance in his theory as *gauge invariance.*

The gauge invariance in Weyl's theory was a *local* one because the standards of length and time changed at every point. In a *global* gauge invariance, on the other hand, the symmetric transformation must not change at every point but must be the same everywhere at once. An example of global symmetry is the charge symmetry of electromagnetism. Suppose that after we carefully measure the interactions between all pairs of charges in some region of space the sign of each individual charge is changed. Measuring the forces between all pairs again produces exactly the same result. The forces are unchanged by a change in the sign of *all* particles in the region of interest (Figure 25.2).

As another example of global symmetry, suppose that one evening some phenomenon causes the dimensions of everything on the earth to be reduced by one half, including the earth itself. When you wake up the next morning, everything will appear to be normal. There is no way to determine that a change took place because everything changed in the same proportion; all the measuring devices were reduced in the same proportion and the phenomenon is undetected. Precision measurements of the diameter of the moon, the orbits of the planets and the frequency of the light reaching us from the sun and other stars would of course make possible the detection of this hypothetical phenomenon. Therefore, to be precise, we would have to say that the change has to take place in the entire universe at once.

Clearly, if only some objects change size, the phenomenon is easy to detect. How can we then have local symmetries? That is, how can we produce *different* changes at different points or to different objects and still keep the system unchanged? By finding another part of the system that produces a *compensating* change. If you are photographing an approaching water skier and want the size of the image to remain constant in your viewfinder, you can zoom out to compensate. If you are skillful enough, you may succeed in maintaining the size unchanged. The zooming effect of your lens compensates for the apparent change in size of the approaching skier.

Suppose that we measure the electric field at a particular point due to a stationary electric charge located a certain distance away. If the entire laboratory is then raised to a potential of, say, 2000 volts, a determination of the electric field should yield the same value. The reason is that the electric field is determined only by differences in the electric potential, not by the absolute value of the potential. An analogous situation occurs when we measure the energy required to lift a 10-kg box from the floor up to a height of one meter. If we first perform this experiment in a laboratory at sea level and later in a laboratory in the mountains, at an altitude of 2000 meters, the amount of energy will be the same. The energy required to lift the box depends only on the *difference* in height. The symmetries involved here are global symmetries. In the case of the electric field experiment, the entire laboratory was raised to the same potential. Neither electricity nor Newtonian gravitation contains local symmetries.

FIGURE 25.2 Changing the sign of all the individual charges produces no effect on the forces between them.

What happens if the electric charge in our experiment is not static? Maxwell's theory of electromagnetism, we recall, also deals with moving charges. A charge in motion relative to the laboratory generates a magnetic field. Just as there is an electric potential associated with the electric field, a *magnetic potential*, associated with the magnetic field, can be defined. The interplay between the two potentials allows us to establish the local symmetry. Any arbitrary *local* change in the electric potential can be compensated by a local change in the magnetic potential. This means that we are free to set our own reference potential at every point in space in any way we want. Regardless of how we vary the one potential throughout space, the other potential can be adjusted to cancel out the differences. A charged particle moving through this region of space will experience no change in the electromagnetic force acting on it as a consequence of the changes in each individual potential.

Seen this way, the electromagnetic field is not merely a force field which happens to exist in nature; it is actually a manifestation of a simple local gauge symmetry. Although Newtonian gravity exhibits only global symmetry, Einstein's general relativity is a theory based on a local gauge symmetry, as well. We can say that the force of gravity is also a manifestation of a simple local gauge symmetry. As we shall see soon, all four forces of nature can be generated from local gauge symmetries.

THE ELECTROWEAK UNIFICATION

Maxwell unified electricity and magnetism without any knowledge of group theory, local gauge symmetries or Nöther's theorem. His only guide was the symmetry observed in electric and magnetic phenomena and in the equations that described them. From a modern perspective, we can see that although neither electricity nor magnetism alone exhibits local symmetry, the unified theory of electromagnetism does exhibit this more powerful local gauge symmetry.

Nearly a hundred years later, two physicists at Brookhaven National Laboratory, C. N. Yang and Robert L. Mills, took the first steps toward the next unification. Yang, the son of a mathematics professor, immigrated to the United States from his native China in 1945 at the age of 23 and entered the University of Chicago as a graduate student in physics. At Chicago, Yang began thinking about a way to generalize Maxwell's theory of electromagnetism to see if there was some sort of connection between it and the theory of the weak interaction. Electromagnetism, Yang knew, possessed a kind of symmetry known as U(1). That is, electromagnetism was invariant under U(1) gauge transformations.

Werner Heisenberg had shown in 1932 that if a proton is changed into a neutron everywhere in the universe, the strong force between nucleons remains unchanged. The reason for this invariance is that, as we saw in the previous chapter, the strong force can be thought of as due to the exchange of a meson, and this exchange takes place regardless of whether the particles are protons or neutrons. The symmetry in Heisenberg's theory is known as global SU(2) gauge symmetry. Could it be possible to construct a theory with *local* SU(2) symmetry? In 1954, Yang and Mills succeeded in constructing such a theory. To convert the global symmetry into a local one, they introduced a set of new fields that described a family of spin-1 particles. Two of these fields were identified with the electric and magnetic fields, and they described the photon. The remaining fields described *charged* photons, one positive and the other negative.

Charged photons had never been observed in nature. However, physicists knew that the weak interaction, like the electromagnetic interaction, was mediated by spin-1 particles. Could the new charged photons be the mediators of the weak interaction? The fact that they were charged made the particles possible candidates for the positions of mediators. The only problem was that the weak interaction was known to be short range, requiring a particle with mass, and the new charged photons of Yang and Mills were massless. The way out of this difficulty came from three physicists who were to succeed where others had failed; in the unification of electromagnetism with one of the nuclear forces.

Abdus Salam from Imperial College, London, and Sheldon Glashow from Harvard University were working independently on gauge theories and became aware of each other's work when Glashow, just after completing his PhD, gave a lecture on his work in London. Although the two were later to share the Nobel Prize for their work in the unification of the weak and electromagnetic interactions, the encounter in London did not result in collaboration. Salam, who was in the audience when Glashow presented some of his results in his early attempt at a gauge theory linking the weak and electromagnetic forces, did not believe that "this slip of a boy" was correct. "Naturally," said Salam later, "I wanted to show he was wrong, which he was. As a consequence, I never read anything else by Glashow, which of course turned out to be a mistake."

Salam was born and raised in what is now Pakistan and entered Punjab University with the highest marks ever recorded in the entrance examination. In 1946 he traveled to England where he obtained two separate undergraduate degrees in mathematics and in physics at Cambridge University. After obtaining his PhD in theoretical physics, Salam returned to Pakistan to teach at the university. The work that he had done for his doctoral thesis, however, had made him well known among physicists and when a position opened at the University of Edinburgh in 1954, it was immediately offered to him.

In a paper on gauge theories, Salam advanced the idea that systems that are not symmetrical can be described by symmetrical equations. This idea, now called *spontaneous symmetry breaking*, had actually been proposed in another context by Heisenberg in 1928. Heisenberg used the temperature dependence of ferromagnetic materials to illustrate symmetry breaking. A magnet has a north and a south pole formed by the alignment of millions of atoms, as we saw in Chapter 14. A magnet, then, has a very specific orientation in space; the force that it exerts on an object made of iron located near its north pole is different from the force on the same object when the magnet is rotated 90°, for example (Figure 25.3a). The interaction with the magnet is not invariant under rotations in three dimensions. If, however, we heat up the magnet above a certain temperature, the alignment of the atoms is lost due to the increased random motion caused by the added energy. Now the magnet can be rotated in any direction without any change in the interaction with the object (Figure 25.3b). In the language of groups, we say that the interaction is invariant under the symmetry group O(3); that is, under rotations in three dimensions. When the temperature falls below the critical temperature, this symmetry is broken. Below

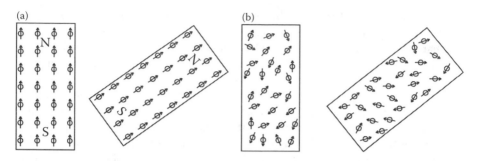

FIGURE 25.3 (a) Iron atoms in a magnet at normal temperatures. Rotating this magnet produces a different magnetic field configuration. (b) Heating up the magnet above a certain critical temperature destroys the alignment of the iron atoms. Rotation of the magnet now produces no change in the space around it.

this temperature, the symmetry is *hidden.* Adding energy to the magnet allows us to observe this symmetry.

In 1967, Steven Weinberg from Harvard University published a paper in which he showed that the massless charged photons predicted by Yang and Mills could be considered to be the mediators of the weak interaction when spontaneous symmetry breaking was taken into account. Abdus Salam in London was to make the same discovery a few months later. To achieve the spontaneous symmetry breaking, Weinberg and Salam made use of an idea developed by F. Englert and Robert H. Brout of the University of Brussels and by Peter Higgs of the University of Edinburgh. According to this mechanism, known today as the *Higgs mechanism,* a new field—the *Higgs field*—is introduced to the theory which has its lowest energy state with broken symmetry. We can illustrate the Higgs field with the situation presented in Figure 25.4. A top balanced on a smooth surface with its axis oriented in the vertical direction has an obvious symmetry; that is, the top remains unchanged when it is rotated around its vertical axis. However, this position is unstable; any small perturbation will cause the top to topple, losing energy. A rotation around the vertical axis now does not keep the top unchanged; the symmetry is broken. The position with broken symmetry has lower energy than the unstable equilibrium with rotational symmetry. Likewise, the behavior of the Higgs mechanism is such that the state in which the Higgs field has its lowest energy is one of broken symmetry.

According to this idea, the Higgs field couples with the Yang–Mills fields and in the process it gives mass to the Yang–Mills photons. Salam refers to this mechanism in a picturesque way: the massless Yang–Mills particles "eat" the Higgs particles in order to gain weight, and the eaten particles become "ghosts." Weinberg and Salam were able to construct a theory of the weak and electromagnetic interactions based on the Yang–Mills theory and the Higgs mechanism. According to their model, at very high energies the mediators of the interaction are massless. At low energies, due to the spontaneous symmetry-breaking Higgs mechanism, these particles acquire masses. At the outset, the four fields in the Weinberg–Salam model have infinite range and are therefore mediated by massless particles. The Higgs mechanism

(a) (b)

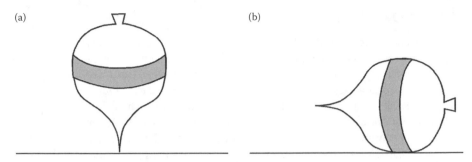

FIGURE 25.4 (a) Balancing a top on its tip so that its long axis is vertical produces an unstable equilibrium; the top easily falls. This position has rotational symmetry. (b) When the top falls, losing energy, the symmetry is broken.

that causes the system to fall to the lower, stable energy state selects the broken symmetry solution and gives masses to three of the four particles. These particles are known today as the W^+, the W^-, and the Z^0. The fourth particle remains massless; this particle is the photon, mediator of the electromagnetic interaction.

The photon is the mediator of the electric and magnetic forces. We can thus consider electricity and magnetism to be two different manifestations of the same force because the mediator of both interactions is the photon. The electromagnetic interaction and the weak interaction are both gauge theories whose forces are mediated by the exchange of the same family of particles. These interactions appear very different to us because our observations are done at low energies, and at these energies, due to the spontaneous symmetry breaking mechanism, the mediators of the interactions are different. The mediator of the electromagnetic interaction, the photon, is massless and the interaction has infinite range. The mediators of the weak interaction, the W^+ and W^- and the Z^0 particles, are very massive and the interaction is short range. At extremely high energies, the theory predicted, the symmetry is restored, all the mediators become massless photons, and the two forces become one: the *electroweak force*.

The electroweak theory of Weinberg and Salam was very beautiful. Originally formulated in the late 1960s, it was cleaned up of certain mathematical difficulties by a very bright graduate student at the University of Utrecht, Gerald 't Hooft, and by his thesis advisor, Martinus Veltman, in 1971. The only difficulty remaining with the theory was that no W or Z particles had ever been observed. In 1983, an extremely elaborate experimental setup conducted by two teams of several hundred physicists led by Carlo Rubbia at CERN, the European center for nuclear research, discovered the W and Z particles and triumphantly confirmed the predictions of the electroweak theory. In 1979, Weinberg, Salam, and Glashow shared the Nobel Prize in physics for the formulation of their theory. In 1984, Carlo Rubbia and Simon van der Meer, the Dutch physicist who discovered the brilliant experimental technique, shared the Nobel Prize in physics for the detection of the W and Z particles. Veltman and 't Hooft won the prize in 1999.

THE COLOR FORCE

The enormous success of the electroweak theory encouraged physicists to seek further unification of the fundamental forces of nature. The next obvious step was to try to integrate the theory of the strong interaction with the electroweak theory. The strong interaction arises from the interactions between quarks. Quarks, as we have seen, have fractional electric charge, and the different combinations of quarks to form hadrons obey the law of conservation of charge so that, for example, the two u quarks with charge + 2/3 and the d quark with charge −1/3 that form a proton add up to a total charge of +1, the charge of the proton in units of the fundamental charge e (the charge of the electron). The way that these charges add up, however, does not explain why quarks combine to form the different hadrons observed.

One way to explain this interaction is to imagine that quarks possess a new kind of charge, in addition to their electric charge. For mesons, which consist of a quark and an antiquark, this charge acts in a fashion analogous to the electric charge that binds the electrons to the nucleus to form an atom; that is, opposite charges attract. For protons or neutrons which consist of combinations of three quarks, the situation is more complicated and we need to consider the attraction of *three* charges. This new kind of charge is called *color charge*, although it has nothing whatsoever to do with the colors of the visible spectrum. The three color charges are called *red, green,* and *blue,* like the primary colors, because the mathematical rules obeyed by these three charges are similar to the way the three primary colors combine to form white. Each quark ($u, d, s, c, b,$ and t) has the three possible color charges red, green, and blue. Quarks can combine in groups of two or three to form a particle only when the resulting color is white, in a way similar to the formation of neutral atoms by an equal number of positively charged protons in the nucleus and negatively charged orbiting electrons. Thus, a proton, which we know consists of two u quarks and one d quark has a red u quark, a green d quark, and a blue u quark. Like the primary colors, the red, green, and blue quarks combine to produce a *white* or color-neutral proton (Figure 25.5).

What about mesons, which are composed of a quark and an antiquark? It turns out that antiquarks have the *anticolors: cyan,* the anticolor of red; *magenta,* the anticolor of green; and *yellow,* the anticolor of blue. As pigments, these colors absorb the primary colors and are said to be *subtractive colors* or *complementaries.* Cyan is a greenish-blue color that absorbs red; magenta absorbs green; and yellow absorbs blue. When we add a subtractive color to its corresponding primary color we get white. Likewise, when a quark with red color charge is combined with an antiquark with cyan color charge, the result is white, as is the case with the pion (Figure 25.6).

The theory that explains the color interactions between quarks is called *quantum chromodynamics* or QCD, and is modeled after quantum electrodynamics or QED, the most successful theory ever devised. Although from our discussion above it might seem that QCD is a fairly straightforward and simple theory, the equations are much more complicated than in any other theory and the calculations extremely complicated and tedious. QCD is a gauge theory and it is this fact what allows its unification with the electroweak theory. The gauge symmetry associated with QCD is the invariance in the local transformations of color. Since the invariance is local,

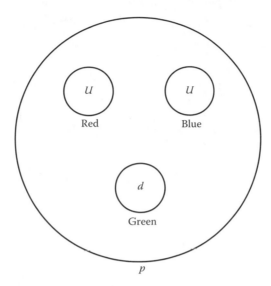

FIGURE 25.5 The combination of a red, a green, and a blue quark to form a proton gives a color-neutral or white particle.

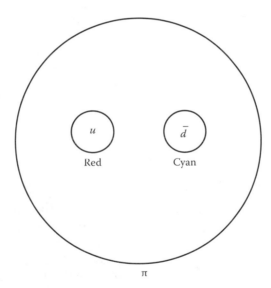

FIGURE 25.6 A pion is formed with a red *u* quark and a cyan *d* antiquark. The particle has no color charge, since red and cyan combine to produce white.

the color of one quark can change while the other quarks remain unmodified. This change would give color to the particle of which the quark is a constituent. As with the electroweak theory, the way to balance this change is with the introduction of new fields. The messenger particles of these fields are called *gluons*, of which there are eight, one for each compensating field. The gluons play the same role in quantum

" GLUONS, ... DID I CREATE GLUONS ? "

(Courtesy of Leo Cullum.)

chromodynamics as the photon in electrodynamics and the W^+, W^- and Z^0 particles in the weak interaction; they are the messenger particles that convey the force.

Gluons carry one color and one anticolor, although not necessarily of a corresponding pair. A gluon can carry, for example, blue-antigreen or red-antiblue. When a quark changes color a colored gluon is emitted, which is then absorbed by another quark. Since the absorbed gluon carries color, the second quark shifts its color in exactly the right way to compensate for the color change that took place in the first quark, keeping the hadron formed by these quarks white. This is the gauge symmetry of QCD. Although the quark colors vary from point to point, all hadrons remain white due to the continual compensation carried by the gluon fields.

AN ATTEMPT AT A THIRD UNIFICATION

With quantum chromodynamics and the electroweak theory based on the gauge symmetry principle, the next problem was whether these two theories could be unified into a single one. Would it be possible to build a theory treating quarks and leptons on an equal footing? In 1973 Sheldon Glashow and Howard Georgi of Harvard University proposed such a theory. In their *Grand Unified Theory* or GUT, the electroweak and strong forces are merely two aspects of the same grand force. Symmetry breaking differentiates the two forces into separate fields.

As before, the local gauge symmetry required in the grand unified theory is compensated by the introduction of new fields. In the GUT of Glashow and Georgi, the first and simplest of the theories, twenty-four compensating force fields are required. Twelve of the messenger particles for these fields are to be identified with the known quanta of the electroweak and strong interactions: the photon, the two Ws, and the Z, and the eight gluons of the strong force. The remaining twelve particles are new and are responsible for the conversion of quarks into leptons and vice versa. This conversion cannot be seen even at the high energies of the new Large Hadron Collider at CERN.

If the quarks inside a proton change, the proton would not be stable but could decay with a half-life of 10^{30} years. The physicist Maurice Goldhaber once said that the lifetime of a proton must be greater than the life of the universe or we would all die of cancer. He reasoned that since there are about 10^{29} protons in a human body, the decay of a couple of dozen protons in our body would damage several thousand molecules, likely causing cancer at an early age. Although this half-life is longer than the age of the universe, we would not have to wait that long to see a proton decay. We could have 10^{30} protons available for our test if we fill a large tank with water. In a year, on average, one of the protons in the hydrogen and oxygen atoms forming the water molecules could decay. In the 1980s and 1990s, several such experiments were set up throughout the world; none detected a single decay. The failure of these experiments to detect proton decay has made it very difficult for physicists to accept the GUT proposed by Glashow and Georgi.

Similar efforts have been proposed to develop GUT versions, some without the problem of proton decay. However, more recent developments to go beyond these theories are showing more promise, as we can see in the next chapter.

THE STANDARD MODEL

The particle theories that led to the successful unification of the non-gravitational forces described in this chapter—the electroweak theory and QCD—are collectively known as the *Standard Model* of particle physics. The Standard Model is a complete description of the particles and forces that make up the universe, and provides, for the first time in history, an accurate picture of the way the universe works. The model requires a relatively small number of particles: the quarks, the leptons, and the force-carriers. All the fundamental matter particles, the quarks and leptons, have half-integer spin and are collectively known as fermions, in honor of Enrico Fermi, who introduced the statistical description of their behavior. All the force carriers, the photon, gluon, and the W and Z particles, have integer spin and are collectively known as *bosons*, in honor of Sarayenda Bose who, along with Einstein, worked out the statistical quantum mechanical description of their behavior. A list of the Standard Model particles appears in Table 25.1.

The Standard Model is a comprehensive theory that has withstood the extremely accurate experimental tests performed during the last decades. In spite of its success, the Standard Model does not give us all the answers. As we saw in Chapter 24, the particles in the Standard Model are classified into three generations. However, as far as we know, only the first generation is needed to construct the universe. The theory does not explain why there are two additional generations or even what the grouping of particles in generations actually means.

Technically, the Standard Model is a consistent theory of all particles and all non-gravitational forces only if the Higgs particle exists. The current Standard Model with the Higgs field included works flawlessly, leaving little doubt that its description

TABLE 25.1

Standard Model of Matter Particles and Force Carriers

		Fermions			*Bosons*	
Matter Particles	**Quarks**	u up	c charm	t top	γ photon	**Force Carriers**
		d down	s strange	b bottom	g gluon	
	Leptons	ν_e electron neutrino	ν_μ muon neutrino	ν_τ tau neutrino	Z Z boson	
		e electron	μ muon	τ tau	W W boson	
	Generation	I	II	III		

of the way nature works is correct. However, the Higgs particle has not been found. There is experimental evidence that the mass of this particle is not much greater than about 100 GeV/c². Although, these energies are well within reach of the Tevatron Collider at Fermilab, the particle's signature is not easy to differentiate from that of other more common processes. New experiments on the new Large Hadron Collider offer hope of discovering the elusive Higgs particle.

26 Superstrings

SUPERSYMMETRY

The Standard Model—perhaps the most complete theory ever developed—gives us a full description of the known building blocks of matter, the particles, and forces that make up the universe. It describes, with uncanny accuracy, the way these particles and forces shape our world. But the Standard Model is not yet the final answer. Although, it successfully unifies the electromagnetic and weak forces into the single electroweak force, it still hasn't been able to take the next step, the grand unification of the three non-gravitational forces. The brave attempt at a grand unified theory by Glashow and Georgi has not stood up to newer and more accurate observations.

For a while, it appeared as if the Standard Model had reached a dead end. When physicists started looking to other possibilities, they found hope in *supersymmetry*, a theory that had been proposed back in 1971, before Glashow and Georgi had started to think about grand unification. Supersymmetry introduces a powerful new gauge symmetry that unites quarks and leptons—the matter particles—with the messenger particles that carry the forces. This supersymmetry brings together all quantum particles as components of a single master superfield. Since all matter particles are made up of quarks and leptons, which are fermions by virtue of having half-integer spins, and all messenger particles are bosons with integer spins, by uniting quarks and leptons with the messenger particles, supersymmetry actually connects fermions and bosons.

Supersymmetry proposes to add new dimensions of space that behave like fermions of spin ½. When a fermion, like a quark of spin ½ is pushed into one of the new fermionic dimensions, it becomes a *squark* of spin 0, or a boson. Or when a boson of spin 1, like a photon, is thrown into the new dimension, it becomes a *photino*, a fermion of spin ½. If the theory is correct and nature is supersymmetric, all particles come in pairs, called *superpartners*, whose spins are half a unit apart.

The supersymmetric Standard Model, the extension of the Standard Model that includes supersymmetry, unifies the three non-gravitational forces, but it does so at the expense of doubling the Standard Model particles. There is no experimental evidence yet of the existence of any of these superpartners. It is likely that in the present state of the universe, the superpartners have very different properties and are perhaps much more massive, requiring energies much larger than we have at our disposition for their detection. In this case, supersymmetry would be a broken symmetry. The new Large Hadron Collider at CERN, the European Center for Nuclear Research, should provide some answers in the near future.

SUPERSTRINGS

Even if the superpartners are found, the supersymmetric Standard Model will not be the final theory. It promises to unify all the forces of nature into a single superforce. All the forces, except for one: gravity.

Before the two nuclear forces were known, Einstein started a lifelong and fruitless effort to unify electromagnetism with gravity. As we saw in the previous chapter, the path to unification was different and gravity has been the last one to be considered. Einstein's general theory of relativity, a theory of gravity, is a gauge theory. As a gauge theory, it should fit with the other forces in a complete description of the world. Gravity is not, however, a quantum field theory; and therein lays the main obstacle toward unification with the other forces.

The best candidate for a supersymmetric theory that could fulfill Einstein's dream for the unification of all forces is *superstring theory*. Superstring theory unifies the Standard Model and gravity into one single theory. One particular important development physicists discovered was that the mathematical structure of superstring theories could be greatly simplified if more than the four space-time dimensions are considered. The most popular supersymmetric theory, for example, was formulated in 11 space-time dimensions: ten space dimensions and one time dimension. Only four of the 11 space-time dimensions are observable; the remaining seven are rolled up to a very small size.

The idea of considering higher dimensions in an attempt at unification is not new. In 1919, the German mathematician Theodor Kaluza generalized Einstein's gravitational field equations to a five-dimensional space-time. In this representation, the extra fifth dimension produced a set of equations that turned out to be Maxwell's equations for the electromagnetic field. Electromagnetism, in the context of Kaluza's theory, is nothing more than a form of gravity, the gravity of the unseen additional dimension. In 1926, the Swedish physicist Oscar Klein extended and cleaned up Kaluza's theory and calculated the radius of the extra fifth dimension to be about 10–30 cm. We can visualize the existence of an extra fifth dimension that cannot be observed by considering, for example, a pipeline, like the Alaska pipeline that brings crude petroleum to the lower 48 states. When seen from a large distance, the pipeline appears to be a wiggly line, with only one dimension. Of course when we look at it from a closer distance, the other two dimensions become apparent and what looked like a point of the line is actually a circle (Figure 26.1). Although, a nice idea at the time, the Kaluza–Klein theory turned out to be extremely restrictive and no one has been able to apply it to the other forces.

The Kaluza–Klein theory was almost forgotten until the early 1960s when Gabriele Veneziano of CERN proposed a mathematical model to explain the existence of many very short-lived hadrons that had been observed in experiments at CERN and at other high-energy physics laboratories throughout the world. Closer investigation of this new mathematical model by Veneziano and other physicists revealed that this new theory described the quantized motion of subatomic *strings*. During the late 1970s and early 1980s, Michael Green of Queen Mary College in London and John Schwarz of the California Institute of Technology advanced

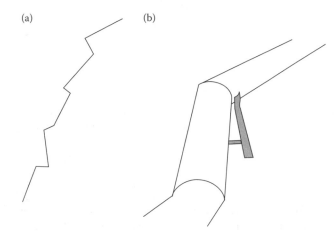

(a)   (b)

FIGURE 26.1   (a) A pipeline seen from afar looks like a broken line. (b) When seen from a closer distance, its real dimensions become apparent.

Veneziano's idea and solved some mathematical inconsistencies. In 1984 Green and Schwarz showed that all mathematical inconsistencies disappeared if string theory incorporated supersymmetry. The new theory is now known as *superstring theory*.

According to superstring theory, all elementary particles are represented by strings no longer than 10^{-35} m, which make them some 10^{20} times smaller than a proton. These strings can vibrate like a guitar string and each vibrational mode corresponds to a particle. The frequency of vibration of a string determines the energy and therefore the mass of the particle, according to $E = mc^2$. The vibrational mode of the string determines the particle's identity and all of its properties, like electric charge or spin. There are open strings and closed strings. Open strings have *endpoints*. Quantities such as electric or color charge are attached to the endpoints of open strings. These open strings can interact with other open strings by joining their endpoints to form another string (Figure 26.2). Open strings can also split into two strings. An open string can become a closed string by joining its endpoints.

During the decade that followed the work of Green and Schwarz, a number of physicists began to pay a great deal of attention to this new theory that promised to unify all of physics in one single sweep. Soon, they realized that supersymmetry could be incorporated into the theory in five distinct ways, and five slightly different superstring versions appeared. They are called *Type I*, *Type IIA*, *Type IIB*, *Heterotic-O*, and *Heterotic-E*. The Type I theory is the only one to include both open and closed strings, as in Figure 26.2. Things were not looking good; one could not really have five different versions of the super theory that was supposed to unify physics. The hope was that eventually, four versions would be shown to be faulty, leaving the remaining version to reign supreme as the theory of all of physics.

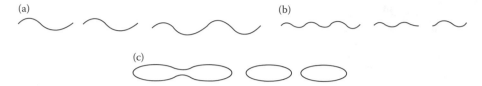

FIGURE 26.2   (a) Two open strings can combine to form a third string. (b) An open string can split into two new strings. (c) Closed strings can also combine and split.

M-THEORY

In 1995, Edward Witten of the Institute for Advanced Studies in Princeton showed that the five different superstring theories were actually five versions of the same single theory. Witten provisionally called his new approach M-theory and the name stuck.

Shortly after Witten's discovery, Joseph Polchinski of the University of California at Santa Barbara showed that M-theory includes objects of higher dimensions than the original two-dimensional strings. As he and his collaborators examined the behavior of open strings, they realized that the motion of the endpoints of these strings, which are free to move as the string vibrates, describe two-dimensional surfaces, a type of membrane (Figure 26.3). Their calculations showed that these surfaces or *branes*, as they became known, were actually a new type of mathematical object within M-theory. They also showed that higher dimensional p-branes were also possible, where they used the letter p to denote the brane dimension, which can take values up to ten.

What is important to understand is that these branes are more than mathematical constructs; they posses their own physical properties, just as strings do. An important example is the possibility that charges can reside on them. These charges can be the electric charges that give rise to electromagnetic fields or the Yang–Mills charged force carriers. Closed strings, on the other hand, having no open ends, cannot be attached to branes and must exist in the full 11-dimensional space. String theory associates closed strings with the graviton, the carrier of the gravitational force. Thus, M-theory shows that gravity acts throughout all space, both inside and outside brane space.

According to M-theory, photons, the uncharged carriers of the electromagnetic force, are open string vibrations that reside on three-branes. Photons, then, are confined to move only within our three-dimensional space. That would mean that the three-brane where our three spatial dimensions reside is transparent to photons, and would be invisible to us. We could not detect directly the existence of the brane with our naked eyes or with any instrument that uses any region of the electromagnetic spectrum. Therefore, radio, infrared, ultraviolet, or x-ray telescopes cannot be used to see the brane. The electromagnetic force is confined to the three-brane of our three-dimensional space; the higher dimensions of the ten-brane universe are invisible to us.

What about other detecting techniques based on any of the other three forces? The short answer is no. The matter particles, the electrons, and the quarks are

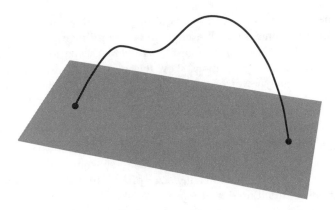

FIGURE 26.3 The endpoints of open strings move on two-dimensional surfaces called *two-branes*. Higher dimensional *p-branes* also exist, according to calculations based on M-theory.

vibrations of open strings and reside on the three-brane. The same is true for the messenger particles. The higher-dimensional universe is undetectable to us through any mechanism based on these three forces. The fourth force, the force of gravity, on the other hand, is not confined to the branes and acts in the entire multidimensional space. If M-theory is correct, the higher dimensions of the universe could one day be detected through the gravitational force.

If M-theory is correct, the fundamental constituents of matter may not be one-dimensional strings but multi-dimensional branes. In the approximations required during the earlier development of string theory, the additional dimensions were curled up on a much smaller scale. What appeared as strings in those early versions were actually branes with extra dimensions wrapped around the strings. Like the pipeline seen from afar (Figure 26.1), the branes appeared as strings as seen through the distorting lens of the early approximate theories.

In 1997, Juan Maldacena—then a graduate student at Princeton and now a Harvard University professor—used M-theory to propose that our four-dimensional universe may be the boundary of a five-dimensional universe. As in the Kaluza–Klein idea, where electromagnetism was proposed to be a form of gravity in a fifth dimension, the four-dimensional universe of our experience is not the one that exists. In Maldacena's conjecture, our four-dimensional universe may be imagined as the holographic projection of a five-dimensional universe. A hologram, as we saw in Chapter 18, is a two-dimensional image that contains all the information of a three-dimensional image. It is produced by the interference of two light beams, a direct beam from the light source that serves as reference and a reflected beam from the object. As in the hologram, our three-dimensional universe contains all the information of the five-dimensional universe.

Other researchers have considered the problem of our four-dimensional universe embedded in a five-dimensional world. In the four dimensions of our experience, the three spatial dimensions reside on a three brane. All the electrons and quarks that make up our world—the Standard Model particles—reside on this three-brane while gravity, as the curvature of space, roams through the fifth dimension. However, if gravity fills the fifth dimension, its strength in our four-dimensional world is diminished. Gravity is the weakest of the four forces and this scenario would explain why it is so. However, simple calculations show that in five dimensions, gravity would decrease with the four power of the distance, rather than with the square of the distance, as we know it does. In other words, gravity's strength will be diluted too much.

The physicists Arkani-Hamed, Dimopoulos, and Dvali looked at this problem and suggested that if the fifth dimension was much larger than what was previously thought possible, the strength of the gravitational force would be maintained to what is actually observed. Rather than having sizes of 10^{-17} m, their calculations showed that the additional curled up dimensions could be as large as a millimeter. Recent measurements have failed to show evidence of new dimensions at those large sizes but have not ruled out dimensions as large as a tenth of a millimeter. New more precise experiments being planned with the new particle accelerators currently being completed may give us an answer to this problem.

Lisa Randall of Harvard University is not waiting for these experiments. She and her colleague Raman Sundrum believe that the fifth dimension is only moderately large. In their model, the fifth dimension extends between two branes (Figure 26.4) with all the Standard Model particles, including the Higgs boson, confined to one of the branes. As in the earlier models, gravity acts throughout the fifth dimension but its strength varies exponentially with distance from one of the branes. The reason for

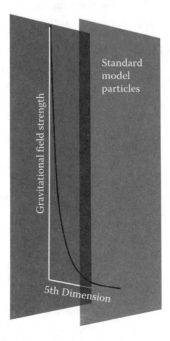

FIGURE 26.4   The warped five-dimensional space-time geometry proposed by Randall and Sundrum.

the variation in the strength of the gravitational force throughout the fifth dimension is that, according to general relativity, the presence of the branes themselves should curve the five-dimensional space-time of their model.

THE ORIGIN OF THE UNIVERSE

The recent developments in our understanding of matter have allowed us to pierce a few of the veils that cloud the age-old question of the origin of the universe. As we said in Chapter 11, matter and energy during the first few moments after the universe began were under conditions of very high pressure and very high energy. In our cold universe today, we recognize four different interactions: the electromagnetic force, the weak nuclear force, the strong nuclear force, and the gravitational force.

The strength of the electromagnetic force is determined by the electron. The force between two electrons, however, depends on how fast they are moving. Each electron is surrounded by a sea of virtual electron–positron pairs that owe their existence to Heisenberg's uncertainty principle (Figure 26.5). These particles appear out of nothing and disappear almost immediately after being created. The only requirement is that they recombine and annihilate in time to satisfy the uncertainty principle. As we saw in Chapter 22, Heisenberg's uncertainty principle states that

$$\Delta E \Delta t \geq \frac{h}{2\pi}.$$

This uncertainty relation means that during the interval Δt an electron–positron pair of total energy ΔE can be created out of nothing; if the pair recombines and annihilates itself before the time Δt is up, no experiment can detect any missing energy.

The electron–positron pairs that surround an electron at any given time are oriented so that the virtual positrons are attracted toward the electron and the negatively charged virtual electrons are repelled by it. These particles are continuously appearing and disappearing, creating a cloud of virtual pairs around the electron. This cloud shields the electron from other incoming electrons, reducing the strength of their interaction. For an electron that stays far away, the positive and negative virtual

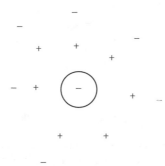

FIGURE 26.5 Cloud of virtual electron–positron pairs surrounding an electron.

charges surrounding the first electron will cancel out and the interaction is the regular Coulomb interaction that we studied in Chapter 12. However, if this electron approaches the first electron with high enough energy, it can penetrate this cloud, leaving some of the virtual electrons behind. What lies ahead is an electron surrounded by a shield of virtual positrons which reduce the effect of the repulsion. If the energy of the approaching electron is high enough, it can penetrate far into the cloud, leaving these positrons behind and seeing the *bare* electron, thus feeling the full strength of the electrical repulsion. The electric force, then, seems stronger to electrons with high energy or equivalently to those at high temperatures.

Quarks are also surrounded by a cloud of virtual quark–antiquark pairs with the color charge polarized in a similar fashion as with the electrons. There is a screening of the color charge due to this shielding effect. The situation here is more complicated, however, because the gluons, carriers of the color charge, also carry color. There is a second cloud of colored gluons surrounding the quark which carry color charge of the *same* type (Figure 26.6). If, for example, the original quark carries red color charge, the cloud of virtual quark–antiquark pairs will carry red and anti-red color charges arranged so that the anti-red is closer to the red quark. The second cloud of virtual gluons carries red color charge and this produces an *anti-screening* effect, opposite to the screening effect of the quark–antiquark pair, which *weakens* the force between interacting quarks. There are two competing phenomena taking place here. One is the screening of the virtual gluons which has the same effect as that between interacting electrons. The other is the anti-screening of the gluons just mentioned. When a high-energy quark approaches another quark and gets past the screening, the anti-screening of the gluons makes the interaction weaker, exactly the opposite of what takes place between interacting electrons. This property of quarks is called *asymptotic freedom*, conveying the idea that if the energy were infinite, the interaction would become weaker and weaker until it would finally vanish; the quarks would feel no forces and behave as free particles.

The weak interaction is also mediated by massive particles, the W^+, W^-, and Z particles. A similar effect takes place here. The weak interaction becomes weaker as the energy of the interaction increases.

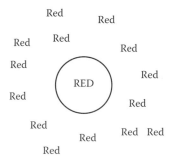

FIGURE 26.6 A quark is surrounded by a cloud of quark–antiquark pairs and by an overlapping cloud of virtual gluons that carry the same type of color charge as the original quark. These two clouds produce competing effects and the overall result is a cloud with a net color charge of the same type as that of the quark.

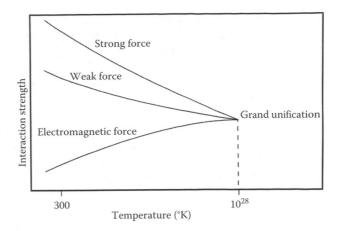

FIGURE 26.7 Evolution of the forces of nature. Due to symmetry breaking, the forces have very different strengths. In the past, when the universe was very hot and dense, the electromagnetic, weak, and strong forces were one and the same.

In our cold universe, due to symmetry breaking, the four interactions have different strengths. The strong and weak interactions are stronger than the electromagnetic interaction, and gravity is the weakest. However, as we have just seen, at very high energies the electromagnetic interaction becomes stronger and the strong and weak interactions weaker. Calculations in the supersymmetric Standard Model show that, at some point very high in energy, these three interactions have equal strength (Figure 26.7). Theoretical calculations place this energy at about 10^{15} GeV. These enormous energies only existed for a brief moment after the creation of the universe. In our laboratories, we have achieved energies of about 1000 GeV.

At those very high energies, then, all the forces of nature except gravity were unified. For a few very tiny fractions of a second only gravity and this unified force existed. At energies even higher than that, when the universe was less than 10^{-43} second old, gravity was unified with the other forces and only one *superforce* existed.

THE FIRST MOMENTS OF THE UNIVERSE

The picture that modern physics paints of the evolution of the universe is that of complete simplicity from the moment of the Big Bang up to 10^{-43} second, when gravity became a separate force. This incredibly small interval of time is called the *Planck time*. Physicists are making great progress toward the understanding of this small period in the life of the universe. The main difficulty lies with the fact that, during that extremely brief time, the universe was a microscopic object, subject to the laws of quantum mechanics, which means that the theory of gravity must be combined with quantum mechanics before we can achieve a complete understanding. At the present, we lack a theory of quantum gravity.

At the Planck time, the universe was confined to a region 10^{-33} cm across, a distance known as the *Planck length*. During those first instances, the universe was in a symmetric state described by the *inflation field*. This inflation field is described by the *inflationary theory* of the early universe developed by Alan Guth of MIT, Andre Linde of Stanford University, Andreas Albrecht of the University of California at Davis, and Paul Steinhardt of Stanford University. According to their theory, the inflation field started a period of extremely rapid expansion that lasted for 10^{-35} second, during which the universe doubled in size every 10^{-37} second, expanding the diameter of the universe 10^{100} times.

10^{-35} second after the Big Bang, the inflationary energy decayed into a hot plasma of radiation and elementary particles at a temperature of 10^{28} K. At this time, the color force becomes distinct, and the grand unification symmetry was broken. With the period of rapid inflation over, the Big Bang expansion that still drives our universe today takes over.

During this period in the life of the universe only three forces existed: gravity, the color force, and the electroweak force. At 10^{-11} second the universe had cooled down to 10^{15} K and the weak and electromagnetic forces separate by the process of symmetry breaking. These conditions are now possible to reproduce in the laboratory. From now on, the four forces familiar to us today control everything in the universe. Figure 26.8 illustrates this symmetry-breaking process. The universe consists of quarks, antiquarks, electrons, positrons, and all the matter particles of the Standard Model together with all the force carrier bosons. All these particles are very close together, colliding with each other very frequently. The photons, moving at the speed of light, change directions at each interaction with the matter particles which move at subluminal speeds. As a result, matter particles and photons remain together to form a hot plasma in thermal equilibrium.

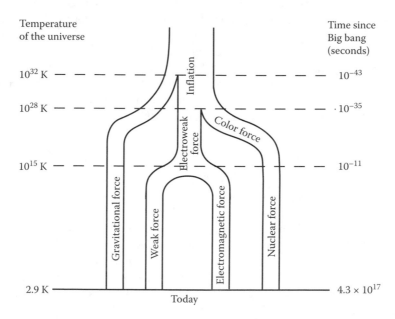

FIGURE 26.8 The universe has evolved from total symmetry through a process of symmetry-breaking where the four known forces of nature are separated.

When the universe was about 10^{-5} second old (one hundredth of a millisecond) and had cooled down to 10^{12} K, the quarks became confined into protons and neutrons. However, at that temperature, the energy was still too high for these protons and neutrons to form nuclei.

One second after the Big Bang, there were eight protons for every neutron, the present abundance ratio. One minute after the instant of creation, the universe had cooled down due to its expansion, so that nuclei could be formed without immediately being shredded into pieces again. This process lasted for about ten additional minutes and the universe behaved like a giant fusion reactor producing helium from the fusion of hydrogen nuclei. Calculations show that at the end of that era the ratio of hydrogen to helium was three to one, exactly the proportion observed today.

Nothing much happened for the next 300,000 years. Then the universe entered a new phase: the age of the atom. Up until then, the universe had been too hot for electrons and nuclei to combine and form atoms. The universe was filled with a plasma of charged particles—electrons and nuclei—interacting with photons. But now the universe has cooled down to about 3000 K, and one by one, electrons become attached to nuclei to form neutral atoms. When all the electrons are bound to nuclei to form neutral matter, the photons begin to uncouple. Three hundred thousand years after its birth, the universe became transparent to light. Photons embarked now on an independent existence and continued to cool off as the universe expanded. We see them today as the 2.7 K background radiation that permeates the universe.

After the uncoupling of the photons from charged matter, the universe entered the present phase: the era of structure. After 200 million years, matter began to coalesce into the first galaxies and gas began to accrete into stars. One billion years after the Big Bang, the large galaxies formed and star formation peaked. Most of those first-generation stars did not live for very long and ended their existences in powerful explosions called supernovas (Figure 11.6). These spewed out carbon, oxygen, silicon, and iron. New stars were born out of the matter of the explosion mixed with the surrounding gas in the galaxies. Millions of years later, the protostar nebula from which our sun and the planet Earth were formed contained the ashes of the early supernova explosions.

THE FRONTIERS OF PHYSICS: THE COSMIC BACKGROUND EXPLORER

In April 1992 scientists presented in Washington, DC the results of what the renowned physicist Stephen Hawking called "the discovery of the century, if not of all time." The physicists showed the extremely precise measurements taken with NASA's Cosmic Background Explorer (COBE) satellite which allowed them to detect the very remnants of creation.

Instruments aboard the COBE satellite measured small fluctuations in the cosmic microwave background radiation. These fluctuations are believed to represent gravitational ripples that could have seeded galaxy formation. After its launch in late 1989, COBE's instruments had measured the spectrum of the microwave background radiation. COBE's microwave spectrum reproduced perfectly a black body radiation curve of 2.73 K, in accordance with the basic Big Bang theory. However, a perfectly smooth radiation curve would not have allowed the existence of tiny concentrations of matter that must have been formed in the early universe. This unevenness in the early universe would eventually evolve into the galaxies and clusters of galaxies observed today.

The COBE more recent measurements showed minuscule variations of 30 millionths of a kelvin from the 2.73 K background temperature. "It's the missing link," said Berkeley cosmologist Joseph Silk. The very small fluctuations detected, however, do not seem to be enough to account for the rapid (on a cosmic scale) formation of galaxies observed. Although scientists are very happy to have finally detected the predicted fluctuations, they are hard at work attempting to devise other mechanisms that could help explain the formation of the cosmos.

Appendix A: Powers of Ten

In physics we deal with quantities that range from the very small to the immensely large. The distance to the nearest galaxy in kilometers, for example, would require many zeros if we were to write it in the conventional way. The mass of an electron in kilograms, a very small number, would require writing 26 zeros after the decimal point. We can avoid these difficulties if we use powers of ten to write these very large and very small numbers.

The product

$$10 \times 10 \times 10 \times 10 \times 10 \times 10$$

where the factor 10 occurs 6 times, can be written as

$$10^6$$

Suppose that we now have the product

$$10 \times 10 \times 10 \times 10 \times 10 \times 10 \times 10 \times 10 \times 10 = 10^9$$

where the factor 10 occurs 9 times. We can write this last product as

$$(10 \times 10 \times 10 \times 10 \times 10 \times 10) \times (10 \times 10 \times 10)$$

or

$$10^6 \times 10^3$$

We can see from the above example that

$$10^6 \times 10^3 = 10^{(6+3)} = 10^9$$

and, in general, that

$$10^m \times 10^n = 10^{(m+n)}.$$

To multiply powers of ten, then, we *add* the exponents.

Suppose that now we want to obtain the result of

$$\frac{10^6}{10^3}$$

which we can write as

$$\frac{10 \times 10 \times 10 \times 10 \times 10 \times 10}{10 \times 10 \times 10} = 10 \times 10 \times 10 = 10^3$$

We can then see that

$$\frac{10^6}{10^3} = 10^3$$

and, in general,

$$\frac{10^m}{10^n} = 10^{(m-n)}.$$

To divide powers of ten, we *subtract* the exponents of the numerator and denominator.

If the exponents m and n are the same (equal to 4, for example), the expression

$$\frac{10^4}{10^4} = 10^{(4-4)} = 10^0.$$

Since $10^4/10^4$ is equal to 1, then

$$10^0 = 1.$$

Another important case occurs when the exponent of the numerator is equal to 0:

$$\frac{10^0}{10^4} = 10^{(0-4)} = 10^{-4}$$

since 10^0 equals 1, then

$$\frac{1}{10^4} = 10^{-4}$$

or, in general,

$$\frac{1}{10^n} = 10^{-n}.$$

For example,

$$\frac{1}{10^7} = 10^{-7} \quad \text{and} \quad \frac{1}{10^{-12}} = 10^{12}$$

What happens when the exponent is a fraction? Consider for example $10^{1/2}$ (or $10^{0.5}$). If we multiply 10 raised to the one half by itself we get 10, since we know already that to multiply powers of ten we add the exponents. That is,

$$10^{1/2} \times 10^{1/2} = 10$$

Since $10^{1/2}$ multiplied by itself gives 10, $10^{1/2}$ must be the square root of 10, or

$$10^{1/2} = \sqrt{10} = 3.16$$

If we now multiply $10^{1/3}$ by itself three times, we also get 10, since $1/3 + 1/3 + 1/3 = 1$:

$$10^{1/3} \times 10^{1/3} \times 10^{1/3} = 10$$

We can see that $10^{1/3}$ is the cube root of 10, or

$$10^{1/3} = \sqrt[3]{10} = 2.15.$$

Appendix B: The Elements

Element	Symbol	Atomic number, Z
Actinium	Ac	89
Aluminum	Al	13
Americium	Am	95
Antimony	Sb	51
Argon	Ar	18
Arsenic	As	33
Astatine	At	85
Barium	Ba	56
Berkelium	Bk	97
Beryllium	Be	4
Bismuth	Bi	83
Boron	B	5
Bromine	Br	35
Cadmium	Cd	48
Calcium	Ca	20
Californium	Cf	98
Carbon	C	6
Cerium	Ce	58
Cesium	Cs	55
Chlorine	Cl	17
Chromium	Cr	24
Cobalt	Co	27
Copper	Cu	29
Curium	Cm	96
Dysprosium	Dy	66
Einsteinium	Es	99
Erbium	Er	68
Europium	Eu	63
Fermium	Fm	100
Fluorine	F	9
Francium	Fr	87
Gadolinium	Gd	64
Gallium	Ga	31
Germanium	Ge	32
Gold	Au	79
Hafnium	Hf	72
Helium	He	2
Holmium	Ho	67
Hydrogen	H	1

Element	Symbol	Atomic number, Z
Indium	In	49
Iodine	I	53
Iridium	Ir	77
Iron	Fe	26
Krypton	Kr	36
Lanthanum	La	57
Lawrencium	Lw	103
Lead	Pb	82
Lithium	Li	3
Lutetium	Lu	71
Magnesium	Mg	12
Manganese	Mn	25
Mendelevium	Md	101
Mercury	Hg	80
Molybdenum	Mo	42
Neodymium	Nd	60
Neon	Ne	10
Neptunium	Np	93
Nickel	Ni	28
Niobium	Nb	41
Nitrogen	N	7
Nobelium	No	102
Osmium	Os	76
Oxygen	O	8
Palladium	Pd	46
Phosphorus	P	15
Platinum	Pt	78
Plutonium	Pu	94
Polonium	Po	84
Potassium	K	19
Praseodymium	Pr	59
Promethium	Pm	61
Protactinium	Pa	91
Radium	Ra	88
Radon	Rn	86
Rhenium	Re	75
Rhodium	Rh	45
Rubidium	Rb	37
Ruthenium	Ru	44
Samarium	Sm	62
Scandium	Sc	21
Selenium	Se	34
Silicon	Si	14
Silver	Ag	47
Sodium	Na	11
Strontium	Sr	38

Element	Symbol	Atomic number, Z
Sulfur	S	16
Tantalum	Ta	73
Technetium	Tc	43
Tellurium	Te	52
Terbium	Tb	65
Thallium	Tl	81
Thorium	Th	90
Thulium	Tm	69
Tin	Sn	50
Titanium	Ti	22
Tungsten	W	74
Uranium	U	92
Vanadium	V	23
Xenon	Xe	54
Ytterbium	Yb	70
Yttrium	Y	39
Zinc	Zn	30
Zirconium	Zr	40

Appendix C: Nobel Prize Winners in Physics

1901	Wilhelm Konrad Rontgen	1845–1923	For the discovery of x-rays.
1902	Hendrik Antoon Lorentz	1853–1928	For their work on the influence of
	Pieter Zeeman	1865–1943	magnetism on radiation.
1903	Antoine Henri Becquerel	1852–1908	For his discovery of radioactivity.
	Pierre Curie	1859–1906	For their joint research on nuclear
	Marie Sklowdowska-Curie	1867–1934	radiation phenomena.
1904	Lord Rayleigh	1842–1919	For his research on the densities of the gases
	(John William Strutt)		and for his discovery of argon.
1905	Philipp Eduard Anton von Lenard	1862–1947	For his work on cathode rays.
1906	Joseph John Thomson	1856–1940	For his research on the conduction of electricity by gases.
1907	Albert Abraham Michelson	1852–1931	For his optical instruments and for measuring the speed of light.
1908	Gabriel Lippmann	1845–1921	For his method of reproducing colors photographically based on the interference techniques.
1909	Guglielmo Marconi	1874–1937	For their development of wireless
	Carl Ferdinand Braun	1850–1918	telegraphy.
1910	Johannes Diderik van der Waals	1837–1932	For his research on the equation of state for gases and liquids.
1911	Wilhelm Wien	1864–1928	For his work on heat radiation.
1912	Nils Gustaf Dalen	1869–1937	For his invention of automatic regulators for use in lighthouses.
1913	Heike Kamerlingh Onnes	1853–1926	For his work on the properties of matter at low temperatures and for liquefying helium.
1914	Max von Laue	1879–1960	For his discovery of the diffraction of x-rays in crystals.
1915	William Henry Bragg	1862–1942	For their analysis of crystal structure using
	William Lawrence Bragg	1890–1971	x-rays.
1917	Charles Glover Barkla	1877–1944	For his study of atoms by x-ray scattering.
1918	Max Planck	1858–1947	For his discovery of quanta of energy.
1919	Johannes Stark	1874–1957	For his discovery of the splitting of spectral lines in electric fields.
1920	Charles-Edouard Guillaume	1861–1938	For his discovery of invar, a nickel-steel alloy.
1921	Albert Einstein	1879–1955	For his explanation of the photoelectric effect.
1922	Niels Bohr	1885–1962	For his model of the atom.

1923	Robert Andrews Millikan	1868–1953	For his measurement of the charge of the electron and for his experimental work on the photoelectric effect.
1924	Kark Manne Georg Siegbahn	1888–1979	For his research in x-ray spectroscopy.
1925	James Franck	1882–1964	For their research on electron-atom
	Gustav Hertz	1887–1975	Collisions.
1926	Jean Baptiste Perrin	1870–1942	For his work on the discontinuous structure of matter and for measuring the size of atoms.
1927	Arthur Holly Compton	1892–1962	For his discovery of the Compton effect.
	Charles Thomson Rees Wilson	1869–1959	For inventing the cloud chamber which makes visible the paths of charged particles.
1928	Owen Willans Richardson	1879–1959	For his discovery of the thermionic effect.
1929	Prince Louis-Victor de Broglie	1892–1987	For his discovery of the wave nature of electrons.
1930	Sir Chandrasekhara Venkata Raman	1888–1970	For his work on light scattering.
1932	Werner Heisenberg	1901–1976	For the development of quantum mechanics.
1933	Erwin Schrodinger	1887–1962	For the development of wave mechanics.
	Paul Adrien Maurice Dirac	1902–1984	For the development of relativistic quantum mechanics.
1935	James Chadwick	1891–1974	For the discovery of the neutron.
1936	Victor Franz Hess	1883–1964	For the discovery of cosmic radiation.
	Carl David Anderson	1904–1984	For the discovery of the positron.
1937	Clinton Joseph Davisson	1881–1958	For their experimental discovery of the
	George Paget Thomson	1892–1975	diffraction of electrons by crystals, confirming de Brogie's hypothesis.
1938	Enrico Fermi	1901–1954	For producing new radioactive elements by means of neutron irradiation.
1939	Ernest Orlando Lawrence	1901–1958	For the invention of the cyclotron.
1943	Otto Stern	1888–1969	For his discovery of the magnetic moment of the proton.
1944	Isidor Isaac Rabi	1898–1988	For his discovery of the nuclear magnetic resonance method that records the magnetic properties of nuclei.
1945	Wolfgang Pauli	1900–1958	For the discovery of the exclusion principle.
1946	Percy Williams Bridgman	1882–1961	For his work in the field of high-pressure physics.
1947	Sir Edward Victor Appleton	1892–1965	For his study of the physics of the upper atmosphere.
1948	Patrick Maynard Stuart Blackett	1897–1974	For his discoveries in nuclear physics with cloud-chamber photographs of cosmic rays.
1949	Hideki Yukawa	1907–1981	For his prediction of the existence of mesons.
1950	Cecil Frank Powell	1903–1969	For his photographic method of studying nuclear processes and his discoveries of new mesons.
1951	Sir John Douglas Cockcroft	1897–1967	For their work on the transmutation of atomic
	Ernest Thomas Sinton Walton	1903–1995	nuclei using a particle accelerator.
1952	Felix Bloch	1905–1983	For their discovery of nuclear magnetic
	Edward Mills Purcell	1912–1997	resonance in liquids and gases.

1953	Frits Zernike	1888–1966	For his invention of the phase-contrast microscope.
1954	Max Born	1882–1970	For his interpretation of the wave function as probability.
	Walther Bothe	1891–1957	For his coincidence method for studying subatomic particles.
1955	Willis Eugene Lamb	1913–2008	For his discoveries concerning the fine structure of the hydrogen spectrum.
	Polykarp Kusch	1911–1993	For his precision determination of the magnetic moment of the electron.
1956	William Shockley	1910–1989	For their development of the
	John Bardeen	1908–1991	transistor.
	Walter Houser Brattain	1902–1987	
1957	Chen Ning Yang	1922–	For their prediction that parity is not
	Tsung Dao Lee	1926–	conserved in beta decay.
1958	Pavel Aleksejevic Cerenkov	1904–1990	For the discovery and interpretation of
	Il' ja Michajlovic Frank	1908–1990	Cerenkov radiation.
	Igor' Evgen' evic Tamm	1895–1971	
1959	Emilio Gino Segre	1905–1989	For their discovery of the antiproton.
	Owen Chamberlain	1920–2006	
1960	Donald Arthur Glaser	1926–	For the development of the bubble chamber.
1961	Robert Hofstadter	1915–1990	For his discovery of the internal structure of the nucleons.
	Rudolf Ludwig Mössbauer	1929–	For his discovery of the Mössbauer effect regarding recoilless emission of γ-rays.
1962	Lev Davidovic Landau	1908–1968	For his theoretical work on the superfluidity of liquid helium.
1963	Eugene P. Wigner	1902–1995	For his discovery and application of symmetry principles to elementary particle theory.
	Maria Goeppert Mayer	1906–1972	For their work concerning the shell structure
	J. Hans D. Jensen	1907–1973	of the nucleus.
1964	Charles H. Townes	1915–	For fundamental work in quantum electronics,
	Nikolai G. Basov	1922–2001	which has led to the construction of
	Alexander M. Prochorov	1916–2002	oscillators and amplifiers based on the maser-laser principle.
1965	Sin-itiro Tomonaga	1906–1979	For their development of quantum
	Julian Schwinger	1918–1994	electrodynamics.
	Richard P. Feynman	1918–1988	
1966	Alfred Kastler	1902–1984	For the development of optical methods for studying energy levels in atoms.
1967	Hans Albrecht Bethe	1906–2005	For discoveries concerning the energy production in stars.
1968	Luis W. Alvarez	1911–1988	For the discovery of resonance states of elementary particles.
1969	Murray Gell-Mann	1929–	For his theoretical work regarding the classification of elementary particles.
1970	Hannes Alvén	1908–1995	For discoveries in magneto-hydrodynamics.
	Louis Neel	1904–2000	For his discoveries concerning antiferromagnetism and ferrimagnetism.

1971	Dennis Gabor	1900–1979	For his development of the principles of holography.
1972	John Bardeen	1908–1991	For their development of a theory of superconductivity.
	Leon N. Cooper	1931–	
	J. Robert Schrieffer	1931–	
1973	Leo Esaki	1925–	For the discovery of tunneling in semiconductors.
	Ivar Giaever	1929–	For the discovery of tunneling in superconductors.
	Brian D. Josephson	1940–	For his theoretical work on the properties of currents through a tunnel barrier.
1974	Antony Hewish	1924–	For the discovery of pulsars.
	Sir Martin Ryle	1918–1984	For his work in radio interferometry.
1975	Aage Bohr	1922–	For their work on the structure of the atomic nucleus.
	Ben Mottelson	1926–	
	James Rainwater	1917–1986	
1976	Burton Richter	1931–	For their independent discovery of J and psi particles.
	Samuel Chao Chung Ting	1936–	
1977	Philip Warren Anderson	1923–	For their theoretical investigations of the electronic structure of magnetic and disordered systems.
	Nevill Francis Mott	1905–1996	
	John Hasbrouck Van Vleck	1899–1980	
1978	Peter L. Kapitza	1894–1984	For his fundamental work in low-temperature physics.
	Arno A. Penzias	1926–	For the discovery of cosmic microwave background radiation.
	Robert Woodrow Wilson	1936–	
1979	Sheldon Lee Glashow	1932–	For their unified theory of the weak and electromagnetic forces and their prediction of the weak neutral current.
	Abdus Salam	1926–1996	
	Steven Weinberg	1933–	
1980	James W. Cronin	1931–	For the discovery of parity violations in the decay of neutral K mesons.
	Val L. Fitch	1923–	
1981	Nicolaas Bloemergen	1920–	For their development of laser spectroscopy.
	Arthur Leonard Schawlow	1921–1999	
	Kai M. Siegbahn	1918–2007	For the development of high-resolution electron spectroscopy.
1982	Kenneth Geddes Wilson	1936–	For his work regarding phase transitions.
1983	Subrehmanyan Chandrasekhar	1910–1995	For his work on the structure and evolution of stars.
	William A. Fowler	1911–1995	For his work on the formation of the chemical elements.
1984	Carlo Rubbia	1934–	For their discovery of the W and Z particles, the carriers of the weak interaction.
	Simon van der Meer	1925–	
1985	Klaus von Klitzing	1943–	For his discovery of the quantized Hall effect.
1986	Ernst Ruska	1906–1988	For the invention of the electron microscope.
	Gerd Binnig	1947–	For the invention of the scanning-tunneling electron microscope.
	Heinrich Rohrer	1933–	
1987	Karl Alex Muller	1927–	For their discovery of high temperature superconductors.
	J. George Bednorz	1950–	

1988	Leon Lederman	1922–	For their production of neutrino
	Melvin Schwartz	1932–2006	beams and their discovery of the
	Jack Steinberger	1921–	mu neutrino.
1989	Norman Ramsay	1915–	For the development of atomic resonance spectroscopy.
	Hans Dehmelt	1922–	For their development of techniques for
	Wolfgang Paul	1913–1993	trapping single atoms.
1990	Jerome Friedman	1930–	For experiments that revealed the existence
	Henry Kendall	1926–1999	of quarks.
	Richard Taylor	1929–	
1991	Pierre-Gilles de Gennes	1932–2007	For discovering that methods developed for studying order phenomena in simple systems can be generalized to polymers and liquid crystals.
1992	Georges Charpak	1924–	For his development of elementary particle detectors.
1993	Russel Hulse	1950–	For discovering evidence for gravitational
	Joseph Taylor	1941–	waves.
1994	Bertram N. Brockhouse	1918–2003	For contributions to the development of
	Clifford G. Shull	1915–2001	neutron scattering techniques in condensed matter studies.
1995	Martin L. Perl	1927–	For experimental contributions to lepton
	Frederick Reines	1918–1998	physics.
1996	David M. Lee	1931–	For their discovery of superfluidity in
	Douglas D. Osheroff	1945–	helium-3.
	Robert C. Richardson	1937–	
1997	Steven Chu	1948–	For the development of methods to cool and
	Claude Cohen-Tannoudji	1938–	trap atoms with laser light.
	William D. Phillips	1948–	
1998	Robert B. Laughlin	1950–	For their discovery of a new form of
	Horst L. Stomer	1949–	quantum field with fractionally
	Daniel C. Tsui	1939–	charged excitations.
1999	Gerardus 'T Hooft	1946–	For elucidating the quantum structure of
	Martinus J.G. Veltman	1931–	electroweak interactions in physics.
2000	Zhores I. Alferov	1930–	For basic work on information and
	Herbert Kroemer	1928–	communication technology and specifically for developing semiconductor heterostructures used in high-speed and optoelectronics.
	Jack S. Kilby	1923–2005	For basic work on information and communication technology and specifically for his part in the invention of the integrated circuit.
2001	Eric A. Cornell	1961–	For the achievement of Bose-Einstein
	Wolfgang Ketterie	1957–	condensation in dilute gases of alkali atoms,
	Carl E. Wieman	1951–	and for early fundamental studies of the properties of the condensates.

2002	Raymond Davis Jr.	1914–2006	For pioneering contributions to astrophysics,
	Masatoshi Koshiba	1926–	in particular for the detection of cosmic neutrinos.
	Riccardo Giacconi	1931–	For pioneering contributions to astrophysics, which have led to the discovery of cosmic x-ray sources.
2003	Alexei A. Abrikosov	1928–	For pioneering contributions to the
	Vitaly L. Ginzburg	1916–	theory of superconductors and superfluids.
	Anthony J. Legget	1938–	
2004	David J. Gross	1941–	For the discovery of asymptotic freedom
	H. David Politzer	1949–	in the theory of the strong interaction.
	Frank Wilczek	1951–	
2005	Roy J. Glauber	1925–	For his contribution to the quantum theory of optical coherence.
	John L. Hall	1934–	For their contributions to the development of
	Theodor W. Hänsch	1941–	laser-based precision spectroscopy, including the optical frequency comb technique.
2006	John C. Mather	1946–	For their discovery of the blackbody form and
	George F. Smoot	1945–	anisotropy of the cosmic microwave background radiation.
2007	Albert Fert	1938–	For the discovery of Giant
	Peter Grünberg	1939–	Magnetoresistance.
2008	Yoichiro Nambu	1921–	For the discovery of the mechanism of spontaneous broken symmetry in subatomic physics.
	Makoto Kobayashi	1944–	For the discovery of the origin of the broken
	Toshihide Maskawa	1940–	symmetry which predicts the existence of at least three families of quarks in nature.

Appendix D: Physics Timeline

ANCIENT GREEKS

425 BC   Democritus proposes that all matter is made of small indivisible particles that he calls "atoms."

280 BC   Aristarchus of Samos determines the relative distances of the sun and the moon from the earth. He also determines the relative sizes of the sun, the moon, and the earth. These considerations lead him to propose that the earth revolves around the sun.

240 BC   Archimedes discovers his principle of buoyancy (Archimedes' Principle).

235 BC   Eratosthenes develops a method to measure the circumference of the earth.

130 BC   Hipparchus estimates the size of the moon from the parallax of an eclipse.

130 AD   Ptolemy develops his theory of the motion of the heavenly bodies. According to his theory, the earth is at the center of the universe, and the sun and known planets revolve around it.

PRE-GALILEAN PHYSICS

1269   Petrus de Maricourt conducts experiments with magnets and magnetic compasses.

1514   Nicolaus Copernicus develops his heliocentric theory. He publishes it in 1543, a few days before his death.

CLASSICAL PHYSICS

1592   Galileo Galilei invents the thermometer.

1600   William Gilbert publishes *De Magnete* which starts the modern treatment of magnetism. He also shows that the earth is a magnet.

1604   Galileo Galilei proves that falling bodies are accelerated toward the ground at a constant rate. He also shows that the distance for a falling object increases as the square of the time.

1609   Johannes Kepler publishes his first and second laws of planetary motion in a book entitled *Astronomia Nova*.

1609   Galileo Galilei builds a telescope after hearing of its invention.

1613   Galileo Galilei introduces his principle of inertia.

1619   Johannes Kepler publishes his third law of planetary motion.

1621   Willebrord van Roijen Snell introduces the law of refraction.

1638   Galileo introduces the concept of the relativity of motion in his *Two New Sciences*.

1651   Blaise Pascal shows that pressure applied at one point in a liquid is transmitted unchanged to all points in the liquid (Pascal's Principle).

1662   Robert Boyle, while experimenting with gases, shows that if a fixed amount of a gas is kept at a constant temperature, the pressure and the volume of the gas follow a simple mathematical relationship.

1665–   Isaac Newton begins his work on the motion of bodies. He also completes his theory of
1666   colors, develops the main ideas on the calculus, and his law of gravitation.

1668   Isaac Newton designs and builds a reflecting telescope.

1672	Isaac Newton, in a letter to the Royal Society, describes his experiments explaining the nature of color. This letter became Newton's first published scientific paper.
1676	Robert Hooke proposes his law relating the elongation of a spring to the force applied to produce that elongation.
1714	Gabriel Fahrenheit introduces the mercury thermometer and his new scale of temperature.
1738	Daniel Bernoulli develops the foundations of hydrodynamics.
1742	Anders Celsius proposes a new temperature scale.
1838	Friedrich Bessel first observes the parallax of a star with the aid of a telescope.
1747	Benjamin Franklin conducts experiments that show that one type of electrification could be neutralized by the other type. This indicated to him that the two types of electricity were not just different; they were *opposites* and calls one type positive and the other negative.
1848	William Thomson, Lord Kelvin, devises what is now known as the absolute temperature scale or Kelvin scale.
1766	Joseph Priestley proposes that the force between electric charges follows an inverse square law.
1777	Charles de Coulomb invents a torsion balance to measure the force between electrically charged objects (Coulomb's law).
1787	Jacques-Alexander Charles discovers the relationship between change in the volume of a gas and its temperature. He fails to publish his discovery.
1798	Henry Cavendish adapts the torsion balance invented by Coulomb to measure the gravitational constant.
1798	Benjamin Thompson, Count Rumford, introduces the idea that heat is a form of motion.
1800	Alessandro Volta invents the battery.
1802	Thomas Young, in a landmark experiment, demonstrates that light is a wave phenomenon.
1802	Gian Domenico Romagnosi proposes in a newspaper article that an electric current affects a magnetic current. His discovery is largely ignored. Oersted, a better known scientist, would discover the same phenomenon in 1819.
1804	Joseph Louis Gay-Lussac, without knowledge of Charles' work of 1787, discovers the relationship between the expansion of a gas at constant pressure and the temperature. This discovery is known as Gay-Lussac's law.
1808	John Dalton develops his atomic theory.
1814	Joseph von Fraunhöfer invents the spectroscope, and with it he observes the absorption lines in the sun's spectrum two years later.
1819	Hans Christian Oersted discovers that an electric current deflects a magnetic compass. His discovery, published in a scientific journal, gets noticed.
1820	André Ampère gives mathematical form to Oersted's discovery. In modern language, Ampère's law is stated as follows: an electric current creates a magnetic field.
1820	Biot and Savart propose a force law between an electric current and a magnetic field.
1822	André Ampère shows that two wires carrying electric currents attract each other.
1827	Georg Ohm shows that current and voltage are related by a very simple relationship, known today as Ohm's law.
1831	Michael Faraday showed experimentally that a changing magnetic field produces an electric current (Faraday's law).
1842	Christian Doppler proposes his Doppler Effect for sound and light waves.
1843	James Joule measures the electrical equivalent of heat.
1846	Gustav Kirchhoff proposes his rules of electrical circuits (Kirchhoff's Rules).
1850	Rudolf Gottlieb, known as Clausius, states the second law of thermodynamics.
1851	Armand Fizeau measures the velocity of light in a moving medium.
1868	James Clerk Maxwell proposes the electromagnetic nature of light and suggests that electromagnetic waves exist and are observed as light.

1869	Dmitri Mendeleyev proposes his periodic table of the chemical elements.
1873	Johannes van der Waals develops his theory of intermolecular forces in fluids.
1887	Heinrich Hertz generates electromagnetic waves in his laboratory.

MODERN PHYSICS

1887	Albert Michelson and E.W. Morley, in a landmark experiment, determine the absence of the ether, a substance postulated to fill all space.
1895	Wilhelm Roentgen discovers x-rays.
1890	James Prescott Joule measures the mechanical equivalent of heat.
1897	J. J. Thomson determines the charge to mass ratio of the electron.
1898	Pierre and Marie Curie discover the radioactive elements radium and polonium.
1898	Ernest Rutherford discovers alpha and beta radiation.
1900	Max Planck introduces the concept of quanta in black body radiation and Planck's constant.
1905	Albert Einstein explains Brownian motion.
1905	Albert Einstein explains the photoelectric effect.
1905	Albert Einstein publishes his special theory of relativity.
1905	Albert Einstein postulates the equivalence of mass and energy.
1906	Albert Einstein proposes quantum explanation of the specific heat laws for solids.
1909	Robert Millikan measures the charge on the electron.
1911	Heike Kamerlingh Onnes introduces his theory of superconductivity.
1911	Ernest Rutherford discovers the nucleus of the atom.
1913	Niels Bohr proposes his quantum theory of atomic orbits.
1915	Albert Einstein publishes his general theory of relativity.
1916	Karl Schwarzschild calculates the critical radius of curvature of space-time around a collapsing star at which light cannot escape.
1917	Albert Einstein presents his theory of stimulated emission, the foundation for the laser.
1918	Emmy Nöther proposes the mathematical relationships between symmetry and conservation laws of physics.
1923	Louis de Broglie predicts the wave nature of particles.
1925	Werner Heisenberg develops matrix mechanics, the first quantum mechanical theory.
1926	Erwin Schrödinger develops wave mechanics, an alternate quantum mechanical theory.
1926	Werner Heisenberg proposes the uncertainty principle.
1927	Niels Bohr proposes the principle of complementarity.
1927	Niels Bohr develops the Copenhagen interpretation of quantum mechanics.
1930	Ernest Orlando Lawrence and M. Stanley Livingston invent the cyclotron.
1932	James Chadwick identifies the neutron.
1932	Werner Heisenberg proposes that the nucleus of an atom is composed of protons and neutrons.
1942	Enrico Fermi obtains the first self-sustaining fission reaction.
1948	Shin'ichiro Tomonaga, Julian Schwinger, and Richard Feynman develop quantum electrodynamics or QED.
1948	John Bardeen, Walter Brattain, and William Shockley invent the transistor.
1953	Charles Townes invents the maser.
1954	C. N. Yang and Robert L. Mills propose a non-abelian gauge theory.
1956	Murray Gell-Mann and Kazuhiko Nishijima introduce the strangeness quantum number.
1957	John Bardeen, Leon Cooper, and John R. Schrieffer propose their BCS theory of superconductivity.

1961	Sheldon Glashow introduces the neutral intermediate vector boson of electroweak interactions.
1961	Murray Gell-Mann and Yuval Ne'eman independently discover the SU(3) octet symmetry of hadrons.
1964	Peter Higgs, Robert Brout, and François Englert introduce the Higgs mechanism of symmetry breaking.
1964	Murray Gell-Mann and George Zweig independently propose the quark theory of hadrons.
1965	John Stewart Bell states and proves a powerful theorem (Bell's Theorem), which gives the theoretical limits on the correlations between the results of simultaneous measurements done on two separated particles. The limits on these correlations are given by Bell in the form of an inequality.
1965	Arno Penzias and Robert Wilson measure cosmic background radiation.
1967	Steven Weinberg and Abdus Salam independently propose the electroweak unification which is based on significant contributions by Sheldon Glashow. The three would later share the Nobel Prize in physics for their theory.
1974	Howard Georgi and Sheldon Glashow propose the SU(5) as a Grand Unified Theory and predict decay of the proton.
1977	A Fermilab team detects the bottom quark.
1981	Michael Green and John Schwarz propose what becomes known as Type I superstring theory.
1981	Gerd Binnig and Heinrich Rohrer invent the scanning tunneling microscope.
1982	Alain Aspect performs an experiment that is considered to confirm the non-local aspects of quantum mechanics.
1983	Carlo Rubbia leads a team that detects the W and Z bosons at CERN.
1990	Measurements of NASA's Cosmic Microwave Background Explorer (COBE) satellite reproduced a blackbody radiation curve of 2.73 K with an accuracy of 1%, as predicted by the Big Bang theory.
1992	NASA's COBE satellite measured minute variations of 30 millionths of a kelvin from the 2.73 K microwave background temperature. These fluctuations are believed to have given rise to the galaxies and other large structures observed in the universe today.
1994	A Fermilab team detects the top quark.
1995	Eric Cornell and Carl Wieman produce a Bose-Einstein condensate by trapping a cloud of 2000 metallic atoms cooled to less than a millionth of a kelvin.
1997	Quantum teleportation is experimentally demonstrated by physicists at the University of Innsbruck by transferring the properties of a photon to another photon at another location.
1998	Saul Perlmutter of the Lawrence Berkeley National Laboratory and Brian Schmidt of the Australian National University discovered, through careful observations of supernova explosions, that for the past 5 billion years, the expansion of the universe has accelerated.
2002	Adam Riess of NASA's Space Telescope Science Institute used the Advanced Camera for Surveys on the Hubble Space Telescope to confirm the 1998 observations of Perlmutter and Schmidt that the expansion of the universe has accelerated for the past 5 billion years.
2003	NASA's Wilkinson Microwave Anisotropy Probe (WMAP) discovers that the geometry of the universe is Euclidean and that the universe is flat.
2003	NASA's WMAP satellite measures the age of the universe to be 13.7 billion years.
2003	NASA's WMAP satellite measurement of the temperature fluctuations of the early universe agrees with the predictions of the inflationary theory of the evolution of the universe.
2007	Electron tunneling in atoms in real time is observed with the use of attosecond light pulses.

2007   NASA's Gravity Probe B orbiting observatory measures the geodesic effect, the warping of space-time in the vicinity of and caused by the earth, with a precision of 1%, confirming a prediction of Einstein's theory of general relativity.

2008   Unusual combinations of quarks observed. A Fermilab experiment discovers a particle containing two strange quarks and one bottom quark. The Belle detector in Japan observes a meson-like particle with four quarks instead of the usual two.

Appendix E: Discover What You Have Learned

The ideas and concepts described in this book should provide you with the ability to answer the questions in this appendix. These questions are conceptual and are designed to be answered without having to remember specific facts.

CHAPTER 1

1. A theory makes several hundred predictions, all of them correct. Is the theory correct?
2. If, for some particular reason, all the swans that you have seen in your life have been white, are you justified in saying that *all* swans are white?
3. If hundreds of predictions made by a certain theory are corroborated while one particular prediction, after many tests, fails, does the theory have to be abandoned? Explain.

CHAPTER 2

1. Is it possible to add three vectors and obtain a zero resultant?
2. Is it possible for two vectors of the same magnitude to have a resultant also of the same magnitude?
3. Is it possible for two vectors of different magnitudes to have a zero resultant?
4. Two vectors have the same magnitude. Is it possible for the vector sum to have a smaller magnitude than each of the two vectors?
5. Is it possible for an object to have zero velocity and still have a nonzero value of acceleration?
6. Can a body have nonzero velocity and zero acceleration?
7. Is it possible for an object to have a decreasing acceleration while increasing its speed?
8. What is the acceleration of a ball that you drop from a balcony?
9. What is the acceleration of a ball that you throw straight down from a balcony?
10. If you throw a ball upwards, what is the speed of the ball at the top of its motion?
11. You throw a rock horizontally from a bridge over a pond and the rock hits the water in one second. How long would the rock have taken to reach the water if you had instead dropped it from the bridge?

12. A young child throws a rock horizontally from a bridge onto a pond. A college student also throws a rock horizontally from the same bridge with a greater speed. Neglecting air resistance, which rock hits the water first?

13. If you throw a ball straight up while standing on a moving train, where is the ball going to fall—ahead, behind, or on top of you? Neglect air resistance. (Remember that before you throw it, the ball has the same horizontal velocity, with respect to the ground, as that of the train.)

CHAPTER 3

1. Which has greater inertia, a train moving at 160 km/h or the same train moving at 100 km/h?
2. Which has greater inertia, a pickup truck or a sports car?
3. Would it be easier to throw a soccer ball horizontally on the moon or on the earth? How about vertically?
4. The bus you are traveling on suddenly stops, and you let go of a notebook and a pencil that you had in your hands. Which object would keep moving with the larger speed? Why?
5. An astronaut in deep space gives a slight push to a small box. What is the net force acting on the box after the astronaut releases it? How does the box move after it is released?
6. How can an astronaut in the space shuttle tell the difference between two identical and sealed boxes, one containing an instrument and the other one empty, without opening them?
7. Is it possible for a shuttle astronaut in orbit to weigh themselves—that is, to determine their mass?
8. A bicyclist pedals uphill, maintaining a constant velocity. Is there a net force on the bicycle?
9. If you push a very heavy crate and fail to move it, is this a violation of Newton's second law? Explain.
10. Consider an object of mass m out in space, far away from all other objects. Would the force required to set this object into motion be zero?
11. Could you pitch a ball faster in outer space, where the ball has no weight?
12. The drivers of a small car and a full-size sedan race each other when the light turns green at an intersection. If they approach the next intersection at the same speed, which car experiences the greatest force? Why?
13. What force returns the basketball you are dribbling back at you?
14. A child pulls a wagon with a certain force. What is the reaction to this force?
15. Do the forces (action and reaction) in the previous question balance each other out? Is the net force on the wagon then zero? Explain.
16. A paper clip hangs from a small magnet that a boy is playing with. The magnet exerts an attractive force of about one-hundredth of a newton on the paper clip. What is the reaction to this force?
17. The earth revolves around the sun due to the force of attraction that the sun exerts on the earth. Does the earth exert a force on the sun? Explain.

18. The earth exerts a force on the moon which keeps it in orbit about the earth. The moon exerts an equal and opposite force on the earth. Why does the earth not revolve around the moon?
19. If you push a shopping cart along the floor in a supermarket, is the force that it exerts on you smaller or greater than the force that you apply to it? Consider the action of the force of friction in this case.
20. A 170-lb student and his 110-lb girlfriend are ice skating together. If they hold hands and pull each other closer, which one exerts the greater force? Which one experiences the greater acceleration?
21. The earth pulls on the moon with a force of 1.98×10^{22} N. What is the force with which the moon pulls on the earth? Do these forces cancel each other out?

CHAPTER 4

1. Can you do work on an object and not move it?
2. A woman is pushing a stroller in a park at a constant velocity. Is work done on the stroller?
3. Can an object have negative potential energy?
4. Does an unbalanced force acting on a body always change its kinetic energy?
5. If you do work on an object as you move it along a level, frictionless surface, does the object accelerate?
6. An object is moving with a steadily increasing speed. Is work being done on the object?
7. You apply a force to an object and the object does not move. Did you do work on the object?
8. Engineers move a 30-story building a couple of meters to give way for a new avenue. Does gravity do work on the building as it is raised a few centimeters off the ground on hydraulic jacks? Does it do work as the building is being moved horizontally?

CHAPTER 5

1. At what point in the trajectory of a falling walnut is its kinetic energy equal to its potential energy?
2. If you toss your keys vertically up and catch them as they come back down, is the speed with which they arrive at your hand the same as the initial speed with which they left your hand? Why?
3. A person throws a ball vertically up from a balcony and lets it fall to the ground below. Is the speed with which the ball arrives at the ground greater than the speed with which it was thrown? Explain.
4. You need to move a chair from the basement up to the ground floor. Would you do more or less work if you take one minute instead of two to bring up the chair? In which case would the power be larger?

5. If one device generates twice the power of another, does it do twice as much work? Explain.
6. Can a small family car have the same momentum as a large bus?
7. Two billiard balls are moving with the same speed along different directions on a table. Do they have the same momentum?
8. Two objects have the same momentum. Do they necessarily have the same velocity?
9. The moon revolves around the earth once every 27.32 days in an almost perfect circular orbit. Is the momentum of the moon constant?
10. You are standing on very slick ice. Without digging your skates in the ice, can you throw a heavy rock horizontally across the ice while you remain stationary?
11. The propelling rockets of an untethered astronaut working on a space station malfunction. How can the astronaut propel himself to return to the station?
12. What does a rocket push against when firing its engines in outer space?

CHAPTER 6

1. The parents of a very young child are afraid that the child might get scared the first time he rides the merry-go-round at the amusement park. What should they do if they want to start him slowly?
2. Which has a greater moment of inertia, a ring or a disc of the same diameter and mass?
3. If you are trying to remove a screw with a medium-sized screwdriver without much luck, could you use an adjustable wrench to help you in the process?
4. Can a large force applied to an object produce no torque?
5. Do you apply a torque when you throw a Frisbee?
6. Is a torque required to keep the earth rotating around its axis?
7. Would you switch to a large-diameter gear or a small-diameter gear if you wanted to go fast on your bicycle?
8. Can an object that is not rotating have a moment of inertia?
9. Is the moment of inertia of an object a fixed quantity (like its mass,) or does it depend how it rotates?
10. Does the moment of inertia of an object depend on how fast it rotates?
11. Why does the moment of inertia of a spinning ice skater decrease when she draws in her arms closer to her body?
12. Would the increase in angular velocity be larger if the spinning ice skater were holding small weights in her arms as she draws them in?
13. If you were asked to start the ice skater spinning, would it require more effort when she has her arms extended or close to her body?
14. Astrophysicists tell us that in about five billion years the sun will begin to expand until it becomes a red giant with a size of about 100 times its present size. Since the sun is rotating, what will happen to its angular velocity in its red giant stage?

15. About which axis would the moment of inertia of a pencil be largest?
16. Are the moments of inertia about all axes perpendicular to each face of a cube the same? What about any arbitrary axis?
17. If you swing a ball attached to a string, why does the pull on the hand increase when the length of the string holding the ball is lengthened?
18. Does the centripetal acceleration of a geosynchronous satellite depend on its mass?
19. Are all geosynchronous satellites placed at the same height above the earth?
20. Should a synchronous satellite around another planet be placed at the same distance from the center of the planet as a geosynchronous satellite is placed from the center of the earth?
21. Is the acceleration due to gravity at the top of the mountain different from the value at sea level?
22. An astronaut in the Space Shuttle is given two apparently identical spheres made, however, of different materials. How can they determine which sphere has the larger mass?

CHAPTER 7

1. Do Rutherford's experiments imply a model of the atom where the electrons move in circular orbits around the nucleus? Would a *static* model, with the electrons stationed at some distance from the nucleus, be consistent with Rutherford's results?
2. Do Rutherford's experiments imply a model of the atom where the electrons move in circular orbits around the nucleus? Would a *static* model, with the electrons stationed at some distance from the nucleus, be consistent with Rutherford's results?
3. Could you explain, on the basis of the principle of conservation of energy, why the Rutherford's atom with orbiting electrons is unstable?
4. What is the difference between the quanta or photons that make up a weak beam of red light and the photons that make up a stronger beam of the same red light?
5. An electron is in the ground state of an atom. Can it emit energy?

CHAPTER 8

1. Two nuclei have the same mass number. Do these nuclei have to be isotopes of the same element?
2. Why do we say that the number of protons in a nucleus determines the element?
3. Compare the nuclear force between two protons to the nuclear force between one proton and one neutron.
4. Neutrons have no electric charge and protons have positive electric charge. Since like charges repel each other, why doesn't the nucleus fly apart?

5. The nucleus of an isotope of calcium has 40 nucleons. How much larger would the volume of the nucleus of Bromine be, if this nucleus has 80 nucleons?

CHAPTER 9

1. Since their weights don't change, why does crawling help rescue personnel avoid falling through thin ice?
2. Does it make any difference whether the area of the top of the balance you stand on to weigh yourself is large or small?
3. It is often difficult to squeeze out the frozen juice from a can. However, if you punch a hole in the closed end of the can with a bottle opener, the frozen juice slides out easily. Why?
4. If you pull a filled straw from a drink while keeping your finger over the top end, the liquid will not spill out of the straw. Why?
5. A hydraulic lift acts as a force magnifier. Does this violate the principle of conservation of energy?
6. You place a fish bowl without the fish on a balance and then place the fish in the water. Does the balance reading change if the fish does not touch the bottom while you take the reading?
7. Why does a helium-filled balloon rise in the air?
8. Can a helium-filled balloon rise all the way up into space?
9. A person lying on an inflatable mattress in the middle of a pool decides to take a swim and jumps in the water. Does the water level rise or stay the same?
10. A popular question in puzzle and game books asks what happens to the water in a glass filled to the brim when an ice cube floating in the water melts. Give your answer using Archimedes' principle.
11. There is a famous problem in physics, asked by scientists like George Gamow and J. Robert Oppenheimer (who, incidentally, initially answered it incorrectly), about throwing a rock in the water from a boat floating in a swimming pool. What happens to the level of the water in the pool after the rock is thrown into the water?
12. Is the buoyant force on a submerged submarine different at different depths?
13. When it rains, water beads up on a well-waxed automobile. Why?
14. Sand at the beach is wet above the water line. Explain how this can happen.
15. Explain why the canvas top of a convertible bulges up at high speeds.
16. Should your blood pressure be lower when you are lying down?

CHAPTER 10

1. If you stir a liquid at room temperature, does the temperature of the liquid change? Explain.
2. Can we speak of the temperature of a single molecule?

3. If two different liquids are at the same temperature, are their average kinetic energies per molecule the same?
4. Two 1-liter bottles are filled to the brim with water. It is determined that their thermal energies are the same. Are they at the same temperature?
5. Why is wet sand on the beach cooler than dry sand during a hot sunny day?
6. Why are ice skates so narrow? It would seem that a wider skate would make it easier to stand.
7. In warmer areas, farmers sometimes spray water on plants to protect them from freezing when they suspect that the temperature might fall slightly below the freezing point of water. Explain how this works.
8. Explain how panting helps dogs cool off when they are hot.
9. Why is it much worse to get burned by steam at 100°C than by water at 100°C?
10. Why does a pressure cooker cook foods faster than a regular pot?
11. Why is the air inside your home usually very dry in the winter?
12. On a hot, humid day, when you first turn on the air conditioner in your car, water sometimes condenses on the windshield. Can you explain why?
13. A metal plate with a round hole in the middle is heated. Will the hole become larger or smaller? Why?
14. An iron ring has an inner diameter equal to the diameter of an iron sphere. If you heat the ring, could you pass the sphere through the ring?

CHAPTER 11

1. What happens to the pressure in the tires when a car is driven some distance? Is there a change in the volume of each tire?
2. Why should you inflate the tires of your automobile to a slightly higher pressure than recommended in the owner's manual if the car has been driven for several miles?
3. You might have noticed that the bottom part of a bicycle pump gets hot as you pump air into a tire. Use the gas laws to explain why.
4. Which has more entropy, seven grams of water or one 7-g ice cube?
5. You enter your room and turn the lights on. Has entropy increased?
6. Does the entropy of the filament of a light bulb increase or decrease when the light is turned on?
7. Does entropy change if you drop a ball on the floor?
8. Could you cool your kitchen by leaving your refrigerator door open?
9. The two chambers of a container with a partition are filled with oxygen at 2 atm. Does the entropy of the oxygen gas increase if the partition is removed? Explain.
10. If you have ten molecules in a container, the entropy will increase by a measurable amount after a certain time. If you have one million molecules, the increase in entropy after the same length of time will be negligible. Explain why.

CHAPTER 12

1. Explain why a glass rod attracts a gold leaf before it touches it but repels it after the leaf touches the rod.
2. A glass rod that has been rubbed with silk is placed near a small, uncharged metallic sphere. Is the sphere attracted or repelled by the rod? Explain.
3. If a glass rod acquires charge after it is rubbed with silk, does the silk acquire any electricity?
4. You are told that a sewing needle is electrically charged. How can you determine whether the charge is positive or negative?
5. After you rub a balloon on your hair, it will stick to a wall. Explain why.
6. Two small metal spheres are charged by an amber rod that has been rubbed with wool. Do the spheres attract or repel each other when placed close together?
7. If two objects attract or repel each other electrically, do both need to be charged?
8. When you bring a charged glass rod near an uncharged gold leaf, the leaf is attracted to the rod. If you briefly let the rod touch the leaf, why is the leaf repelled by the rod now?
9. Is it possible for two electric field lines to cross? Can a single electric field line loop around and cross itself?
10. The electric field between two parallel and oppositely charged plates is constant. Does this mean that an electron would feel the same force when it is located halfway between the plates as when it is two thirds of the way, closer to the negative plate?
11. Is the electric field of a positive point charge the same at all points around the charge?
12. Is it possible to move an electron inside an electric field without doing any work?
13. If an electron is moved closer to a proton, does the potential energy of the proton decrease or increase?
14. If an electron loses energy as it moves in an electric field, is it moving in the same or in an opposite direction to that of the electric field lines?
15. Does the capacitance of a capacitor depend on whether the capacitor is charged?

CHAPTER 13

1. When you comb your dry hair, it usually becomes electrically charged. This does not happen when the hair is wet or when the air is humid. Explain why.
2. Old phonograph records (still in use today) attract a great deal of dust. Some record care products use water to keep the cleaning pads moist while the record is wiped. Why do they clean better when they are wet?

3. Wiping dust off a TV screen with a dry cloth sometimes proves quite difficult. The dust moves around but stays on the screen. Explain why this happens. How could this annoying problem be remedied?
4. Is a conductor needed for an electric current to exist?
5. If you replace a flashlight bulb with one of greater resistance, is the current drawn greater, smaller, or the same? Explain.
6. If your car's 12-V battery dies, could you start the car with eight 1.5-V flashlight batteries?
7. Why aren't birds electrocuted when they stand on bare high-voltage power lines?
8. There are heat losses associated with the flow of current through a conductor. To minimize the losses, power plants deliver electric power at very high voltages, at least 20,000 V and sometimes up to 600,000 V. Can you explain how this works?
9. What is the resistance inside the evacuated chamber of a linear particle accelerator?
10. Germanium is a semiconductor with valence 4. If you wanted to make a p-type semiconductor with germanium, what impurity atom should you use, one with valence 3 or one with valance 5?

CHAPTER 14

1. You are given two identical-looking metal rods and told that only one of them is a permanent magnet. Without the aid of any other piece of metal, compass, or instrument, could you identify the magnet?
2. How should you place two magnets if the force between them is to be zero?
3. If a magnet is cut in two, two smaller magnets are produced. What will happen if you bring together two magnets with their opposite poles touching? Is a single magnet formed?
4. Could there be a magnet in the form of a doughnut? Where would the north and south poles be in such a magnet?
5. Is it possible to use a magnetic field to accelerate a charged object?
6. Does the earth's magnetic field affect the direction of motion of the electrons in a CRT?
7. Would you be able to move a plastic comb that is electrostatically charged (perhaps by using it to comb your hair several times) by bringing a strong magnet close to the comb?
8. Would you expect to detect a small current in a metal ring aboard the space shuttle as it orbits the earth?
9. A paper clip can be picked up by a magnet, regardless of which end of the magnet is used. Explain why.
10. Combing your hair several times charges the comb with static electricity. Could you generate a magnetic field by simply moving the charged comb in circles?

11. Is an electric current generated in the class ring you are wearing due to the motion of your hand in the earth's magnetic field?
12. Could an electric current be generated in a metal ring in a constant magnetic field if the ring changed shape (from a circle to an oval, for example)? Discuss different possibilities.

CHAPTER 15

1. A row of upright dominoes will fall if the first one is pushed. Is this a wave motion?
2. Since waves carry energy and momentum, does this mean that the source of a wave motion loses energy and momentum to the wave?
3. Is energy transmitted in standing waves?
4. We learned that energy is transmitted in a wave motion. Is energy transmitted past a node in a standing wave?
5. When you walk with your morning coffee in your hand, circular standing waves are formed on the surface of the liquid. Explain why.
6. When you walk normally with a pan full of water, the water usually sloshes and spills; varying the length of the step stop the sloshing. Why?
7. When an automobile is out of alignment, it vibrates at certain speeds but not at others. Explain why.
8. You can start a basketball bouncing without picking it up by pounding on it with your hand rhythmically. Once small bounces start, they quickly develop into large bounces. Explain how this works.
9. Soldiers usually "break step" and walk when crossing a suspension footbridge. Could you explain why this is done?

CHAPTER 16

1. Does a child's telephone, made with two cans and a taut string, really work? Explain.
2. A bell enclosed in a waterproof jar is made to ring underwater. Would the frequency you hear be different if you are also underwater?
3. Does sound travel faster in cool air or in warm air?
4. Where is the speed of sound greater, in the lowlands or at the top of a mountain?
5. If you inhale helium from a balloon and attempt to speak, you'll sound like Donald Duck. Explain why.
6. If you count the seconds between the lighting flash and the thunder during a thunderstorm, you can estimate the distance to the flash in kilometers by dividing this time by three. Explain why.
7. Why is it that two different instruments, such as a piano and a violin, playing the same note with the same loudness, sound different?
8. Some people claim that they sound wonderful when they sing in the shower. Can this claim have any substance to it?

9. Using what you have learned about vibrating strings and resonating cavities for musical instruments, explain why women's voices generally have a higher pitch than men's.

10. Is it possible to determine the speed of rotation of a merry-go-round by analyzing the sound from a horn that a child on the merry-go-round is playing?

11. Could the Doppler effect be observed with water waves? Explain.

12. A tape recorder is placed on a bench, recording the sound of a passing fire truck. When the tape is played later, do you hear the drop in pitch?

13. If the obstacle on which a sound wave bounces off were moving away from you, would the frequency of the echo perceived by you change due to Doppler effect?

14. You are standing some distance away from a parked car blowing its horn. If there is a wind coming toward you from the direction of the parked car, would there be a change in frequency because of the Doppler effect?

15. You are driving along a stretch of a highway that is parallel to a railroad, with the same velocity and slightly behind the engine of a train. Is the sound you hear from the whistle Doppler-shifted?

16. Two airplanes are flying at the same altitude, one at Mach 1 and the other at Mach 2. Which one is farther away when you hear its sonic boom? If they are side by side when they fly overhead at your location, which sonic boom do you hear first?

17. You hear a sonic boom from an airplane flying nearby. Does that mean that the airplane just crossed the sound barrier?

CHAPTER 17

1. When you look at a mirror, do you see the mirror or the light reflected from it?

2. Why can you see the reflection of your face in a mirror but not in the wall?

3. Why does a plane mirror reverse left and right but not up and down?

4. Does a plane mirror magnify?

5. If you want to be able to see your entire body reflected in a mirror, do you need a mirror that is as long as your body?

6. You need to buy a mirror that would allow you to see you entire body but the store where you are shopping is cramped and you can't stand back from the mirrors to try them. Does it matter? Explain.

7. A person submerged in water appears shorter but not thinner than she really is. Why?

8. Why does light travel at a lower speed in water than in a vacuum?

9. Does it matter which side of a magnifier faces the object?

10. The two sides of a lens have different curvatures. Are the two focal lengths different? Explain.

11. Does the focal length of a lens change while it is immersed in water? Explain.

CHAPTER 18

1. In 1729 the English astronomer James Bradley, using a telescope mounted vertically, detected an apparent shift of 20.5 seconds of arc in the position of *every* star. This shift could not be attributed to stellar parallax because stars are all at different distances from the earth. Explain this statement.
2. Different colors of light refract at different angles when passing through a prism. This also happens when light passes through a lens. Would you expect the focal points for all colors to be different?
3. The sun is a hot, huge ball of gas; 74% of the sun is hydrogen, 25% is helium, and 1% other elements. Why is the solar spectrum continuous with dark lines?
4. Could you obtain interference with two flashlights?
5. Red light is used in Young's double-slit experiment. How would the interference fringes change if the red light were replaced with blue light?
6. Light incident on two Polaroids placed with their transmission directions perpendicular to each other will not be transmitted beyond the second Polaroid. What would happen if a third Polaroid was inserted in between the other two at an angle of 45° with either one of them?
7. How would you determine if two pairs of sunglasses were polarized?
8. How would you determine if one pair of sunglasses was polarized?
9. A Polaroid will transmit 50% of an incoming unpolarized beam of light. Since an unpolarized beam of light has all its electric field vectors oscillating in all directions, and the Polaroid transmits light oscillating in only one plane, why is the transmitted light 50% of the incoming light?
10. If the probability of emission is the same as that of absorption, how is population inversion achieved?

CHAPTER 19

1. A child riding a skateboard at a constant speed throws a ball straight up. Does the ball return to the child's hands or does it hit the ground behind? Describe the path of the ball as seen by the child and as seen by a person standing to the side of the child.
2. You are riding in a car which is passing a metal bridge at a constant velocity. If you want to drop a small stone so that it passes through a small hole in the bridge, when should you release it, before you actually get to it or the moment your hand is exactly above the hole?
3. If you are on an overnight train trip and you wake up in the middle of the night, can you come up with a way to figure out whether the train is moving or has stopped at a station? Assume you are in one of the last cars, away from any possible station lights, and that it is so dark that you can't see outside.
4. Even without looking outside, why can you always tell whether the vehicle you are in is slowing down?
5. Imagine that you are traveling in a spaceship with a constant velocity. A small meteorite hits the ship causing no appreciable damage. You notice that when it happened, the alarm on your watch coincidentally went off.

An engineer on a nearby space station is monitoring the flight of your ship and also registers both the meteorite impact and your alarm. Are these two events also simultaneous for the engineer?

6. An astronaut moving at a speed near that of light passes near the earth. Does he see a spherical planet?

7. According to Einstein, inertial mass and energy are equivalent. If you wind up a toy, does its mass increase?

8. Does a cool gas have a smaller mass than the same gas when heated?

9. Inertial mass increases with velocity. Does the density of an object in motion also increases? What about its volume? Explain.

10. A bright young student once proposed that you could accelerate the ends of a very long stick to a speed greater than that of light by rotating it from the middle. If the stick is long enough, the ends are going to move at a speed greater than that of light, even if the speed at which you rotate it is not too large. How would you respond to this student?

CHAPTER 20

1. If the inertial mass were different from the gravitational mass, how would objects of different masses fall to the ground?

2. If the inertial mass were different from the gravitational mass, would two objects of the same mass but different composition fall to the ground with the same acceleration?

3. If you had access to extremely accurate instrumentation, how would you go about determining the curvature of space near the surface of the earth?

4. If two friends agree to meet at a certain location exactly one month later and one of them—an astronaut—takes off on a short trip aboard a spaceship traveling at very high speed, would the two friends meet at the agreed time?

5. Would you expect the bending of a beam of light to be greater near Jupiter than near the earth? Explain.

6. If time runs more slowly in a gravitational field, would it not be possible to distinguish a gravitational field from an acceleration?

7. Would a clock run more slowly near Jupiter than on earth? Explain.

8. Would a clock aboard the space shuttle in orbit run faster or more slowly than a clock on the ground?

9. Is the escape velocity from earth greater for a more massive object than for a lighter object?

10. The space shuttle does not start its ascension to orbit at the escape velocity of 11.2 km/s (40,000 km/h), but rather at a much lower velocity. Why?

CHAPTER 21

1. Can we say that an ice cube in an otherwise empty glass radiates energy?

2. Are all quanta the same?

3. Based on the absorption and reflection of dark and light-colored objects, which color should you wear when it is very hot?

4. Red light has smaller frequencies than blue light. Which Christmas tree light bulb emits quanta with greater energy, a blue bulb or a red bulb?
5. When the energy of the light incident on a metal surface is greater than the work function, are electrons always emitted?
6. Would a very bright red light produce electrons?
7. Light of a certain frequency strikes a metal and no electrons are emitted from the metal. If the intensity of the light is increased, could electrons be released from the metal?
8. A metal surface is illuminated with a beam of light of a frequency that releases electrons from the metal. The light is then replaced by a second beam of light of the same intensity but lower frequency. Which light releases more electrons from the metal?

CHAPTER 22

1. In the attempt at watching the electrons in the double-slit experiment, why is it not possible to shine a light of low enough intensity that it practically does not disturb the electrons but that is still bright enough for us to see them?
2. Does the uncertainty principle have any appreciable effect on the motion of a baseball or the firing of a gun? Why? (Hint: consider the magnitude of Planck's constant.)

CHAPTER 23

1. Is it possible to have an isotope of hydrogen decay by alpha emission?
2. Is it possible for the same nuclide to decay by both alpha and beta emission?
3. Sodium-24 has a half-life of 15 h. How long do we have to wait for 1 kg of sodium-24 to be reduced to 0.5 kg?
4. The half-life of iodine-131 is 8.1 days. This means that if we have 100 grams of this nuclide, in 8.1 days we would only have 50 grams. How much would we have in 16.2 days?
5. The half-lives for some nuclei are given in hundreds of thousands of years and even in billions of years. Since nobody has been around long enough to measure these half-lives, how do you think they have been determined?
6. The half-life of lead-214 is 26.8 minutes. If you were able to isolate a single lead-214 atom, would you still have the same atom half an hour later? Explain.

CHAPTER 24

1. Can two photons with total energy equal to the rest energy of one electron collide and produce a positron–electron pair?
2. Does the electromagnetic force ever become zero?

3. How could the mass of the meson be calculated by knowing the range of the nuclear force?
4. In this chapter we calculated the maximum range of the virtual particle. What makes this range "maximum"?

CHAPTER 25

1. Would the rotation of a snowflake through an angle of 30° be a symmetric operation? Explain.
2. If you rotate a long cylinder through an angle of 23° around its long axis, would this be a symmetric operation? How about a rotation of 50°? Explain.
3. If you rotate a cube through an angle of 20° around one of its three axes you will not get the original orientation. What compensating change can you perform to restore the symmetry?
4. If every proton in the universe were changed into an electron, would the electric force remain invariant? Explain.
5. Rotating an ice cube through any arbitrary angle does not maintain its symmetry at the molecular level because the water molecules in ice form a hexagonal lattice (see Chapter 10). What spontaneous symmetry-breaking mechanism takes place in this case to restore their symmetry; that is, to allow for a rotation through any angle and still maintain the symmetry at the molecular level?

CHAPTER 26

1. Where does the energy for the creation of the virtual particles that surround electrons and quarks come from?
2. At a large distance from an electron, another electron senses the entire charge of the first electron. Why does the cloud of virtual particles surrounding each electron not shield the electronic charge in this case?

Glossary

Absolute zero: Temperature at which no thermal energy can be extracted from an object. It is the minimum temperature attainable. It is equal to −273.15°C.

Acceleration: The rate at which velocity changes. The SI units of acceleration are m/s².

Alpha decay: If a nucleus contains too many protons, it is unstable and emits an alpha particle, which is a nucleus of helium-4.

Alpha particle: See *alpha decay*. A stable nuclear particle that consists of two protons and two neutrons. An alpha particle is th e nucleus of helium.

Amorphous solid: See *crystal*.

Ampère's law: An electric current produces a magnetic field.

Amplitude: The amplitude of an oscillation is the maximum displacement of the medium from its equilibrium position.

Angular momentum: A measure of the rotation of an object. It is the tendency of a rotating object to keep rotating because of its inertia. Angular momentum can be expressed as $L = I\omega$. The angular momentum of a body is conserved if the net external torque acting on the object is zero. This is the law of conservation of angular momentum.

Angular velocity: The rate of change of angular displacement with time.

Antimatter: Antimatter particles, called antiparticles, are identical to ordinary matter except that if the particle has electric charge, its antiparticle would have the opposite charge. If a particle has no charge, like the photon, it is its own antiparticle.

Archimedes' principle: An object partially or completely submerged in a fluid is buoyed up by a force equal to the weight of the fluid displaced by the object.

Atomic mass unit, amu: A unit of mass used in the atomic realm. 1 amu = $1.6605402 \times 10^{-27}$ kg.

Atomic number: The total number of protons in a nucleus.

Average speed: Defined as the total distance traveled divided by the time taken to travel this distance. The units of speed are units of distance divided by units of time. The SI unit is the meter per second (m/s).

Baryon: A subatomic particle composed of three quarks held together by the color force.

Beta decay: If a nucleus contains too many neutrons it is unstable and decays by emitting a beta particle, which is an electron.

Beta particle: See *beta decay*. An electron emitted from the nucleus of an atom undergoing beta decay.

Binding energy: The total energy of the nucleus is less than the total energy of its separated nucleons. This energy difference is called binding energy.

Black body: An object that absorbs all radiation incident upon it. It is also a perfect emitter of radiation.

Black hole: If a star has a mass of more than 3 solar masses, gravitational compression will make the star so dense that the escape velocity from it becomes greater than the speed of light. The star contracts to a single point, called a *singularity* or black hole.

Boyle's law: See *ideal gas law.*

Brane: A membrane-like object in M-theory with spatial dimensions ranging from one to ten.

Buoyant force: Upward force exerted by a fluid on a floating or immersed object as a reaction to the force exerted by the object to displace the fluid.

Calorie: The amount of heat required to raise the temperature of 1 gram of water by 1°C. It is equal to 4.186 J.

Capacitor: A device for the storage of electrical energy. It consists of two oppositely charged metal plates separated by an insulator.

Centripetal acceleration: An object moving with a constant speed v in a circular path of radius r has an acceleration directed toward the center of the circle called centripetal acceleration. It has a magnitude v^2/r.

Chain reaction: A reaction in which some of the products initiate further reactions of the same kind allowing the reaction to become self-sustaining.

Charles's law: Another name for Gay-Lussac's law (q.v.).

Coherent radiation: Electromagnetic radiation such as is seen in radio waves and laser beams, where all the radiation is of a single frequency and all the photons are in phase (in step). The *coherence length* is a measure of the quality of the coherence; the *bandwidth* is another.

Color charge: A measure of the strength of the strong interaction.

Compound: If the atoms retain their identities while they attract each other owing to the mutual attraction of their respective ions (*ionic* bond), the atoms are said to form a compound.

Concave mirror: A curved mirror in which the interior surface is the reflecting surface.

Conservative and nonconservative forces: When the work done by an unbalanced force acting on a body depends only on the initial and final positions of the body, the force is said to be a conservative force. If the work done depends on the path taken by the body, the force producing this motion is said to be nonconservative.

Constructive interference: See *interference.*

Convex mirror: A curved mirror which has the exterior surface as the reflecting one.

Coulomb: SI unit of electric charge. One coulomb is the charge of 6.25×10^{18} electrons or an equal number of protons.

Coulomb's law: The force exerted by one charged object on another varies inversely with the square of the distance separating the objects and is proportional to the product of the magnitude of the charges. The force is along the line joining the charges and is attractive if the charges have opposite signs and repulsive if they have the same sign. If we call q_1 and q_2 the magnitudes of the two charges and r, the distance between their centers, we can state Coulomb's law as an equation: $F = k(q_1q_2/r^2)$. The value of Coulomb's constant k is 9×10^9 Nm^2/C^2.

Covalent bond: A type of chemical attraction that depends on the fact that the presence of two electrons in a certain region of space is energetically advantageous. In a covalent bond, atoms are bound together by sharing electrons.

Critical angle: When light is passing from a medium with a higher index of refraction to one with a lower index of refraction, the critical angle is the angle of incidence for which the angle of refraction is 90° (i.e., the emergent beam travels along the interface).

Crystal: The forces that bind the atoms together in a solid are strong enough for the solid to maintain its shape. If the atoms arrange themselves in a pattern that is repeated through the substance, the solid is called a crystal. Solids that do not form these patterns are said to be amorphous.

De Broglie wavelength: The wavelength associated with a particle, equal to the ratio of Planck's constant to the momentum of the particle.

Decibel: The unit of sound level, a measure of relative sound levels.

Density: The mass per unit volume of a substance. The SI unit is the kilogram per cubic meter (kg/m^3).

Destructive interference: See *interference*.

Dew point: See *humidity*.

Diffraction: The spreading out of waves on passing through a narrow aperture.

Diode: A device that acts like a switch in an electric circuit, permitting the flow of current in only one direction.

Doppler effect: The change in frequency perceived by a listener who is in motion relative to a source of sound.

Efficiency: The ratio of the useful work performed by a machine to the total amount of energy required to operate it.

Elastic collision: A collision in which the kinetic energy is conserved.

Elastic potential energy: In a spring of force constant k stretched a distance x, the elastic potential energy is $\frac{1}{2}kx^2$.

Electric charge: A property of subatomic particles that is responsible for electric and magnetic phenomena. The fundamental charge is the charge of one electron or one proton and has a magnitude of 1.602×10^{-19} C.

Electric current: The rate at which charge flows in a conductor. If during a time t an amount of charge q flows past a particular point in a conductor, the electric current is $i = q/t$. The unit of current is the ampere (A), which is a fundamental SI unit.

Electric field: Property of space around an electric charge. The electrostatic force per unit charge.

Electric potential difference: See *voltage*.

Electromagnetic wave: Propagation of oscillating electric and magnetic fields through space.

Electron: A fundamental particle, one of the main constituents of matter. The electron has an electric charge of -1.602×10^{-19} C and a mass of 9.1094×10^{-31} kg or 5.486×10^{-4} amu.

Electron-volt (eV): Unit of energy used when dealing with atoms or electrons. 1 eV = 1.602×10^{-19} J.

Electrostatic force: See *Coulomb's law*.

Electroweak force: Unification of electromagnetism and the weak nuclear force. The triplet of massive bosons W^+, W^-, and Z, along with the massless photons, are the mediators of this force.

Energy: The capacity to do work or the result of doing work.

Entropy: From a Greek word that means *transformation*, entropy is a measure of the disorder of a system.

Escape velocity: The escape velocity of an object on the surface of the earth is the minimum velocity that we must impart to the object so that it escapes the gravitational grasp of the earth.

Event horizon: See *Schwarzchild radius*. A sphere around a black hole with a radius equal to the Schwarzchild radius. No particle inside this sphere can escape the gravitational attraction of the black hole.

Faraday's law of induction: The induced voltage in a circuit is proportional to the rate of change of the magnetic field.

Field: The concept of field is used to specify a quantity for all points in a particular region of space. The electric field describes the property of the space around an electrically charged object. A charged body distorts the space around it in such a way that any other charged body placed in this space feels a force that is given by Coulomb's law. The electric field strength $\mathbf{E}$ is the Coulomb force felt by a test charge q_0 that is placed in the field divided by the magnitude of this test charge q_0, $\mathbf{E} = \mathbf{F}/q_0$. The direction of the electric field vector is the direction of the force on a positive test charge.

Fission: See *nuclear fission*.

Fluids: Substances in which the binding forces are weaker than in solids, so that the atoms or molecules do not occupy fixed positions, and move at random. Liquids and gases are fluids.

Focal point: The point at which all of the rays gathered by a lens, curved mirror, or optical instrument pass.

Frame of reference: See *reference frame*.

Frequency: The number of wave crests that pass a given point per second. The unit of frequency is the hertz, Hz.

F-stop: The ratio of the focal length of a lens to the diameter of its aperture.

Fundamental charge: The charge on one electron or one proton, $e = 1.602 \times 10^{-19}$ C. The electric charge on a charged object always occurs in integral multiples of the fundamental charge.

Fundamental forces: There are four fundamental forces in nature: gravitational, electromagnetic, strong, and weak. The strong force, a short-range force, holds the nucleus together. The weak force, also a short-range force, is responsible for radioactive beta-decay processes.

Fundamental units: The fundamental quantities in mechanics are *length*, *mass*, and *time*. The corresponding fundamental SI units are *meter*, *kilogram*, and *second*. The other fundamental SI units are *ampere*, *lumen*, *kelvin*, and *mole*.

Fusion: See *nuclear fusion*.

Galilean principle of relativity: The laws of mechanics are the same in all inertial frames of reference. This means that there is no special or absolute reference frame. Thus, there is no absolute standard of rest; uniform motion has to be referred to an inertial frame.

Gamma decay: If a nucleus is left in an excited state after an alpha or beta process, it will decay to the ground state by emitting one or more photons, called gamma rays.

Gamma rays: Electromagnetic radiation with frequencies greater than about 3×10^{19} Hz and wavelengths smaller than about 10^{-11} m.

Gauge symmetry: Physical theories that remain invariant under changes taking place everywhere in the universe are said to obey global gauge symmetry. When the changes are different at every point in space, the theory is said to obey local gauge symmetry. To maintain a local symmetry, in which different changes take place at different points or to different objects, a compensating change must take place.

Gay-Lussac's law: See *ideal gas law*.

Gravitational force: See *law of universal gravitation*.

Gravitational potential energy: The energy that a body possesses by virtue of its separation from the earth's surface. For an object of mass m situated at a height h above the ground, the gravitational potential energy is $PE_{grav} = mgh$.

Gravitational red shift: Einstein showed that, according to general relativity, time runs more slowly in a gravitational field and that celestial objects in a strong gravitational field would show a spectral shift toward longer wavelengths.

Gravity: See *law of universal gravitation*.

Hadrons: Particles that participate in the strong interaction. Hadrons are not fundamental particles; they have a definite extension. Hadrons that decay into a proton and another stable particle are baryons. The remaining hadrons are the mesons.

Half-life: The time required for half the nuclei in a given radioactive sample to decay.

Heat: The thermal energy transferred from a warmer object to a cooler object. The SI unit of heat is the joule; another unit of heat, defined during the times of the caloric theory, is the calorie, cal, which is the amount of heat required to raise the temperature of 1 gram of water by 1°C. The relation between calories and joules is 1 cal = 4.186 J.

Heat capacity: The heat required to increase the temperature of a mass m of the substance an amount ΔT. The specific heat capacity, or heat capacity per unit mass, is *the heat required to raise the temperature of a unit mass of a substance by one kelvin*. If Q is the amount of heat required; m, the mass of the substance, and ΔT the change in temperature, the specific heat capacity $c = Q/(m \, \Delta T)$. The SI unit of specific heat capacity is the joule per kilogram kelvin (J/kg K).

Heat of fusion: See *latent heat*.

Heat of vaporization: See *latent heat*.

Heisenberg's uncertainty principle: It is not possible to measure the exact position and the exact momentum of a particle simultaneously.

Hertz: The SI unit of frequency. One cycle per second.

Higgs field: See *Higgs mechanism*. A quantum field with very special properties that allow it to give mass to the elementary particles when they interact with this field.

Higgs mechanism: See *Higgs field*. The Higgs mechanism is a special set of circumstances such that the state in which the Higgs field has its lowest energy is one of broken symmetry.

Hologram; holography: The recording of an image by recording the standing-wave pattern caused by the interaction of two coherent beams of radiation, one having been modified by interaction with the object.

Humidity: Measure of the amount of water present in the air at any given time. *Absolute humidity* (*AH*) is the total mass of water vapor present in the air per unit volume, generally given in *g/m³*. *Humidity at saturation, HS*, is the mass per unit volume of water vapor required to saturate the air. *Relative humidity, RH*, is the ratio of the absolute humidity to the humidity at saturation: *RH = AH/HS*. The temperature at which the air saturates is called the *dew point*.

Ideal gas: Any gas in which the cohesive forces between molecules are negligible and the collisions between molecules are perfectly elastic. Collisions between the molecules of an ideal gas conserve both momentum and kinetic energy. If an ideal gas is kept at constant temperature, the pressure is inversely proportional to its volume; that is, the product of the pressure and the volume is a constant. This is Boyle's law. The constant is the same for all gases. If the pressure of an ideal gas is kept constant, a change in volume is proportional to the change in absolute temperature. This is Gay-Lussac's law, also known as Charles' law.

Ideal gas law: By combining Boyle's law and Gay-Lussac's law into one single expression, we obtain the ideal gas law, $PV = \text{constant} \times T$.

Image: Strictly, optical image. Formed where light rays intersect or where they appear to have originated from. A real image is one formed by light rays actually intersecting, whereas a virtual image is an image formed by light rays which appear to come from a point in space.

Index of refraction: For a transparent substance, the ratio of the speed of light in vacuum to that in the substance.

Inelastic collision: A collision in which kinetic energy is not conserved. Momentum is conserved.

Inertia: The tendency of an object to resist any change in its state of motion.

Inertial reference frame: A reference frame in which the law of inertia holds. Inertial reference frames move at constant velocities.

Infrared radiation: Electromagnetic radiation in the region of the spectrum between about 4×10^{14} and 10^{11} Hz and wavelengths between about 7.5×10^{-7}m and 3×10^{-3} m.

Instantaneous speed: The speed that an object has at any given instant.

Intensity: The rate at which a wave transports energy.

Interference: Two or more trains of the same frequency sharing the same space will interfere constructively or destructively, depending on whether the combined amplitude is greater or less than the component wave amplitudes.

Internal energy: The internal energy, U, of a system is the sum of all forms of energy, thermal energy and potential energy. *Thermal energy* is the sum of all the random kinetic energies of the atoms and molecules in a substance and *potential energy* is the energy stored in the molecules, atoms, and nuclei of a substance.

Invariance: If a system remains unchanged after some operation is performed on it, we say that the system is invariant under that operation.

Inverse square law: A mathematical expression in which a value for a quantity varies inversely with the square of the distance.

Ion: An atom or molecule with a net electric charge.

Ionic bond: Bonding due to the electrical attraction between oppositely charged ions.

Ionizing radiation: This is produced by a particle or a photon with enough energy to remove an electron from an atom.

Isobaric process: A process that occurs at constant pressure. In this case, the work done is $W = P\,\Delta V$.

Isotope: One of several forms of an element having the same number of protons but a different number of neutrons.

Joule (J): The SI unit of energy or work, equal to a newton meter (Nm).

Kepler's laws of planetary motion: In the early 1600s, Kepler discovered the three laws of planetary motion that bear his name:

> *Law of orbits:* Each planet moves around the sun in an elliptical orbit, with the sun at one focus.
>
> *Law of areas:* A planet moves around the sun at a rate such that the line from the sun to the planet sweeps out equal areas in equal intervals of time.
>
> *Harmonic law* or *law of periods:* The squares of the periods of any two planets are proportional to the cubes of their average distances from the sun.

Kinetic energy: The energy that an object has by virtue of its motion. It is equal to one-half the product of the mass (m) and the square of the speed (v): $KE = \frac{1}{2}mv^2$.

Laser radiation: Laser is an acronym for light amplification by stimulated emission of radiation. A laser is a device which produces a narrow beam of single-wavelength, coherent radiation (q.v.) by stimulated emission of photons.

Latent heat: The heat *absorbed* or *released* by one kilogram of a substance during a phase transition. If the transition is from solid to liquid or vice versa, it is called *latent heat of fusion*, L_f. If the transition involves the liquid and gas phases of a substance, it is called the *latent heat of vaporization*, L_v. The heat required to melt a solid of mass m is given by $Q = mL_f$; the heat required to vaporize a liquid of mass m is $Q = mL_v$.

Law of reflection: The angle of reflection is equal to the angle of incidence. The incident ray, the reflected ray and the normal are all in the same plane.

Law of refraction (Snell's law): The angle of refraction is in a constant relationship to the angle of incidence, and the incident and refracted rays are in the same plane as the normal.

Law of universal gravitation: Any two objects of mass M and m, separated by a distance r, will attract each other with a force proportional to the product of their masses and inversely proportional to the square of their distance apart. The constant of proportionality is the universal constant G, with a value of 6.67×10^{-11} Nm^2/kg^2.

Laws of thermodynamics: These are numbered zeroth, first, second, and third:

0   If two objects are each in thermal equilibrium with a third object, they are in thermal equilibrium with each other.

1   In an isolated system, the total internal energy of a system remains constant, although it can change from one kind to another.

2   Heat does not pass spontaneously from a cold to a hot object. Another way of stating this is: All natural changes take place in the direction of increasing entropy.

3   It is impossible to reach absolute zero temperature in a process with a finite number of steps.

Length contraction (relativistic): An observer in motion relative to an object measures the length of that object along the direction of motion to be contracted when compared with the length measured by an observer at rest relative to the object.

Leptons: Particles that interact via the weak force. All leptons are truly elementary particles, without internal structure. Leptons are classified in three generations, each containing a charged lepton and a neutrino.

Lever arm: The perpendicular distance from the center of rotation to the point of application of a force.

Light year: The distance traveled by light in one year. It is equal to about 9.5×10^{12} km.

M-theory: An extension of superstring theory. M-theory is formulated in an 11-dimensional space-time. A full set of equations to describe M-theory have not yet been discovered.

Magnetic field: The property of space around a magnet. Since magnetic poles exist only in pairs, magnetic field lines do not start or end anywhere, an essential difference from electric field lines, which start on positive charges and end on negative charges.

Magnetosphere: The volume around the earth that is influenced by the earth's magnetic field. It is believed to be caused by motions of the metallic core which produce electric currents in the hot, electrically conductive material; these currents flow upward and are in turn carried around by the earth's fast rotation. The magnetosphere traps some matter from the solar wind (q.v.).

Magnification: In a lens or mirror system, the ratio of the size of the optical image to the size of the object.

Mass number: The total number of protons and neutrons in a nucleus.

Maxwell's equations: The four equations by which James Clerk Maxwell described the relationship between electricity and magnetism and provided a model for the propagation of electromagnetic radiation. To state them precisely requires a use of vector calculus.

Meniscus: The curved surface of a liquid in a container produced by the cohesive forces between the liquid molecules and the adhesive forces between the liquid and the container.

Meson: Particle with a mass that falls between the mass of the electron and the mass of the proton. Mesons are combinations of a quark and an antiquark.

Metastable state: An excited energy state of an atom with a longer life time than that of regular excited states.

Meter: SI unit of length. The meter is defined as the distance traveled by light in 1/299792458 of a second.

Microwaves: Electromagnetic radiation in the region of the spectrum between about 3×10^8 and 3×10^{11} Hz and wavelengths between about 10^{-4} m and 1m.

Molecule: A structure formed when atoms combine in such a way as to share some of their electrons.

Moment of inertia: The moment of inertia of a body measures its resistance to change in its state of rotation about a given axis.

Muon: A fundamental particle with a mass about 200 times that of the electron. A muon is a lepton that decays into an electron and neutrinos.

Newton: The SI unit of force. One newton is one kilogram meter per second squared $(kg \times m/s^2)$.

Newton's laws of motion:

> *Newton's first law*: Every body continues in its state of rest or of uniform motion in a straight line unless it is compelled to change that state by forces impressed upon it.
>
> *Newton's second law*: The acceleration of an object is directly proportional to the net force acting on it and inversely proportional to its mass.
>
> *Newton's third law*: To every action there is always an equal reaction. The mutual actions of two bodies upon each other are always equal and directed to contrary parts.

Nöther's theorem: The theorem that establishes the connection between symmetry and the laws of physics. It can be stated as follows: For every continuous symmetry in the laws of physics there exists a corresponding conservation law. A continuous symmetry is one in which the corresponding transformation can be varied continuously, as in a rotation.

Nuclear fission: The event which occurs when a heavy nucleus splits into two smaller nuclei (called fission fragments).

Nuclear fusion: This takes place when two light nuclei are fused together to form a heavier nucleus.

Object and image distances (conjugate foci): The distance from the object and image, respectively from an optical mirror or lens. For a plane mirror, the object distance is always equal to the image distance.

Ohm's law: The current flowing through a conductor is directly proportional to the voltage V that exists between the two ends of the conductor, or $i = V/R$, where R is the resistance of the conductor.

Pair annihilation: See *pair production*. When a particle encounters its antiparticle, they annihilate each other, disappearing in a burst of photons.

Pair production: See *pair annihilation*. The inverse process, in which high energy photons create a positron–electron pair.

Pascal's principle: This states that the pressure applied to a liquid is transmitted undiminished to all points of the liquid and to the walls of the container.

Period: The time required to complete one cycle of a periodic motion. Period is the inverse of frequency (q.v.).

Photoelectric effect: Light of a certain frequency incident on a substance causes electrons to be emitted from the substance.

Photon: Quantum of light. The photon is the mediator of the electric and magnetic forces. The character of the force depends on the polarization of the exchanged photon. Longitudinal and time-like polarizations mediate the electrostatic force while the magnetic force is mediated by the other two polarizations. The photon is a boson with spin 1 and zero mass.

Pion: The particle that mediates the strong force. Also called the π-meson.

Polarization: Orientation of the oscillation vector of a wave or of the rotation axis of a spinning object.

Positron: An antielectron. A fundamental particle with the same mass as the electron and a positive electric charge of the same magnitude as that of the electron. A constituent of antimatter.

Potential energy: The energy that an object has by virtue of its position in a field.

Power: The rate at which work is done or energy is released.

Pressure: Force per unit area. The SI unit of pressure is the pascal (Pa). One pascal is one newton per square meter (N/m^2). Normal atmospheric pressure at sea level is 101.3 kPa or 1 atm.

Principle of equivalence: This states that it is impossible to distinguish the effects of accelerated motion from the effects of gravity. It extends the relativity principle to accelerated frames of reference. The principle of equivalence can be stated in the alternative form: Gravitational mass and inertial mass are equivalent and no experiment can distinguish one from the other.

Proper time: The time interval measured by an observer in his own reference frame.

Proton: One of the main constituents of matter. The proton has an electric charge of 1.602×10^{-19} C and a mass of 1.6726×10^{-27} kg or 1.007276 amu.

Quanta: Packets of energy, introduced in 1900 by Max Planck to explain the behavior of the radiation emitted by a hot body. Albert Einstein generalized this revolutionary concept and stated that light behaves both as a wave and as quanta of energy or photons (q.v.).

Quantum chromodynamics (QCD): The theory that explains the color interactions between quarks. This theory is modeled after quantum electrodynamics or QED.

Quarks: Quarks are believed to be truly fundamental particles. Hadrons are thought to be composed of quarks. There are six flavors or varieties of quarks: *Up* and *down*, *strange* and *charm*, and *bottom* and *top*. Like leptons, quarks are also grouped in generations. The strong force arises from the interaction between quarks. Quarks possess a kind of charge called color charge: red, green, and blue. All hadrons are color-neutral or "white."

Radian: The central angle in a circle subtented by an arc length equal to the radius of the circle. It is equal to 57.3°.

Radio waves: Electromagnetic radiation in the region of the spectrum smaller than about 3×10^8 Hz and wavelengths greater than about 1 m.

Radioactive decay: See *radioactivity*.

Radioactivity: Emission of several particles such as electrons, photons, neutrons, neutrinos, or positrons, due to the decay of unstable nuclei.

Resistance: In a conducting medium, the ratio of the voltage to the current is a constant, the resistance of the conductor. The units of resistance are ohms, Ω.

Schwarzchild radius: See *event horizon*. The radius to which a given object must be reduced so that its escape velocity equals the speed of light. It is the event horizon of a black hole.

Semiconductor: Material with electrical conductivity that is intermediate between that of conductors and insulators. The conductivity of a semiconductor is changed by the addition of small amounts of impurities to its crystal structure.

Solar wind: The stream of energetic particles emitted by the sun.

Sound: A mechanical longitudinal wave that propagates through a medium with frequencies that range from a fraction of a hertz to several megahertz. Audible waves are sound waves with frequencies between 20 and 20,000 Hz. Sound waves with frequencies below 20 Hz are called infrasonic waves and those with frequencies above 20,000 Hz are called ultrasonic waves.

Space-time: A four-dimensional geometry consisting of the three coordinates of space and one of time.

Specific heat: The heat required to increase the temperature of 1 gram of a substance by one degree Celsius.

Speed: See *velocity*. The magnitude of the velocity.

Speed of sound: This depends on the elastic and inertial properties, and temperature, of the transmitting medium. In air, the speed of sound at 20°C is 343 m/s.

Spin: A quantum property of elementary particles. The mathematics that describes the spin property resembles that of a spinning ball, except that it only takes certain quantized values.

Standard Model: A description of the behavior of the non-gravitational forces and particles that make up the universe.

Standing waves: In general, the resultant of two identical wave motions of equal amplitude and wavelength traveling in opposite directions. The points that do not move are called nodes. Standing waves can be seen on a stretched string fixed at both ends. The natural frequencies of vibration form a harmonic series. Any two coherent beams of radiation whose paths cross will generate a standing waveform, and this is the principle of holography (q.v.).

Strong force: See *nuclear force*. Holds the nucleons together in a nucleus. It is a short-range force, becoming negligible at distances greater that 10^{-15} m. The strong force acts on protons and neutrons but not on electrons, neutrinos, or photons. The strong force is 137 times stronger than the electromagnetic force.

Supergravity: Theory that attempts to unify gravity with the other three forces with the use of a powerful new gauge symmetry that unites quarks and leptons with messenger particles called supersymmetry. Supergravity theories are formulated in more than four space-time dimensions.

Superstring theory: A theory that promises to provide a unified description of all the forces of nature. According to this theory, all elementary particles are represented by strings, open or closed, no more than 10^{-35} m in dimensions.

Supersymmetry: A gauge symmetry that unites quarks and leptons with messenger particles. In supersymmetry, the laws of physics are symmetric with the exchange of fermions and bosons.

Surface tension: The intermolecular forces that act on the molecules on the surface of a liquid to make the surface of the liquid as small as possible. These forces are also responsible for the rising of the liquid in very thin tubes, a phenomenon known as capillarity.

Symmetry: If something remains unchanged after some operation is performed on it, it is said to be symmetric under that operation.

Temperature: A measure of the average random kinetic energy per molecule of a substance.

Thermal expansion: The proportional change in length or area or volume when a change in temperature has occurred.

Time dilation: Time in the moving reference frame always flows more slowly than in the stationary reference frame.

Torque: The product of the applied force and the lever arm length. The ability to rotate an object depends on the applied torque.

Total internal reflection: The reflection that occurs when light is incident from a medium with a high index of refraction to one with a low index of refraction at angles of incidence greater than the critical angle (q.v.). The light beam then obeys the laws of reflection (q.v.).

Transistor: A semiconductor device that can act as a current switch and as an electronic amplifier in a circuit.

Ultraviolet radiation: Electromagnetic radiation in the region of the spectrum between about 7.5×10^{14} and 3×10^{18} Hz and wavelengths between about 4×10^{-7} m and 10^{-10} m.

Uncertainty principle: See *Heisenberg's uncertainty principle.*

Uniform motion: Motion at a constant velocity.

Uniformly accelerated motion: Motion with constant acceleration. One important example of uniformly accelerated motion is the vertical motion of an object falling toward the ground due to the gravitational attraction of the earth. Another is circular motion at constant speed.

Universal law of gravitation: See *law of universal gravitation.*

Vector quantities: Quantities that require both magnitude and direction for their complete specification, e.g. velocity.

Velocity: The speed and the direction of motion of an object.

Virtual particles: Particles that exist only for the brief moment allowed by Heisenberg's uncertainty principle.

Visible light: Electromagnetic radiation in the region of the spectrum between about 4×10^{14} and 7.5×10^{14} Hz and wavelengths between about 7.5×10^{-7} and 4×10^{-7} m.

Voltage: The change in electric potential energy of a charge divided by the magnitude of that charge.

Wave: A mechanism for the transmission of energy in which the medium itself does not travel. In particular, electromagnetic radiation does not need a medium to propagate.

Wavelength: The distance between two identical points of a periodic wave.

Weak force: See *strong force*. Responsible for radioactive β-decay processes, such as the transformation within the nucleus of a neutron into a proton or a proton into a neutron. The weak force controls many of the reactions that produce energy in the sun and the stars. The weak force is some hundred thousand times weaker than the strong force.

Weight: The gravitational force with which the earth attracts an object toward its center when the object is on or near the earth's surface.

Work function: The minimum amount of energy required to release electrons from a particular metal.

Work: When a constant force acts on an object along the direction of motion of the object, the work done on the object is equal to the product of the force and the distance that the object moves. The SI unit of work is the joule (J), equal to 1 newton meter (Nm).

X-rays: Electromagnetic radiation in the region of the spectrum between about 3×10^{16} and 10^{21} Hz and wavelengths between about 10^{-8} m and 3×10^{-13} m.

Index